Movable Bridge Design

Movable Bridge Design

Charles Birnstiel, William Bowden
and George A Foerster

Published by ICE Publishing, One Great George Street, Westminster, London SW1P 3AA.

Full details of ICE Publishing sales representatives and distributors can be found at: www.icevirtuallibrary.com/info/printbooksales

Other titles by ICE Publishing:

Bridge Launching, 2nd edition. M. Rosignoli. ISBN 978-0-7277-5997-7
Bridge Design to Eurocodes: UK Implementation. S. Denton.
ISBN 978-0-7277-4150-9
ICE Manual of Bridge Engineering, 2nd edition. G. Parke and N. Hewson.
ISBN 978-0-7277-3452-5
Bridge Design, Construction and Maintenance. ICE. ISBN 978-0-7277-3593-5
Current and Future Trends in Bridge Design, Construction and Maintenance. ICE.
ISBN 978-0-7277-3475-4

www.icevirtuallibrary.com

A catalogue record for this book is available from the British Library

ISBN 978-0-7277-5804-0

ICE Publishing is a division of Thomas Telford Ltd, a wholly-owned subsidiary of the Institution of Civil Engineers (ICE).

Commissioning Editor: Rachel Gerlis
Development Editor: Amber Thomas
Production Editor: Rebecca Taylor
Market Development Executive: Elizabeth Hobson

FSC
www.fsc.org
MIX
Paper from responsible sources
FSC® C013056

Typeset by Academic + Technical, Bristol
Index created by Indexing Specialists (UK) Ltd
Printed and bound in Great Britain by TJ International Ltd, Padstow

Contents

Preface

This book is intended for those involved in design, engineering, construction and maintenance of movable bridges. It was brought together by authors with knowledge and experience gained from working on a broad range of movable bridge projects. Movable spans of many kinds are treated in the book, from short spans over narrow navigation channels to the 340 metre span of the two-armed swing bridge over the Suez Canal in Egypt. It deals with bridges supporting only a few pedestrians, those loaded by 32 tonne (72 kip) lorries, those with an average daily traffic (ADT) of 200 000 vehicles, and others with multiple railway tracks designed for a running load of 15 tonnes per metre (10 kips per foot) of track. Movable bridges are considered herein that are part of the city street fabric and railway bridges so isolated that the nearest roadway is 15 km (9 miles) away. Various forms of operating machinery and controls are discussed from the viewpoints of design, inspection, testing, and maintenance.

The primary purpose of this book is to serve as a reference on past and current mechanical, electrical, hydraulic, and structural design of movable bridges. The emphasis is on the machinery required to operate the movable spans and to stabilise them, in both the open and closed positions. Among the subjects included are a brief historical account of the development of movable bridges citing significant structures, typical superstructure forms, engineering design specifications and standards, construction materials including physical properties and kinds failure, span drive and stabilising machinery arrangements, and mechanical, hydraulic, electrical and structural design.

The reality is that thousands of movable bridges exist, many over 50 years old, and that they have deteriorated since they were built due to environmental effects, traffic overloads and inadequate maintenance. Funding for replacing deteriorated bridges is scarce in many jurisdictions and besides the permitting process for new construction on new alignments, with its required environmental impact statements, can take decades. If the condition of the bridge is poor, with respect to safety and operating reliability, the practical solution is often rehabilitation of the structure, machinery and controls. Widening of the carriageway may be a feature of the rehabilitation plans. The authors are cognisant of this situation and have endeavoured to present the development of the systems during the 20th century in minimal space. Current technology is an outgrowth of prior engineering.

It is presupposed that readers have knowledge of elementary physics. To appreciate the chapters on materials, and mechanical, hydraulic and electrical design, the reader having a background in statics and dynamics, strength of materials, structural analysis, fluid mechanics, and electrical circuits – as generally taught at a university undergraduate level – will have an advantage. Mechanical design is presented at an elementary level. Advanced solid mechanics topics are not stressed. The book considers a wide range of topics within an interdisciplinary subject and so, because of space limitations, some aspects are not treated comprehensively. However, in such cases reference is made to multiple publications that treat those topics in depth.

Acknowledgements

Thanks are due to the many persons who have contributed to this book in some way, especially William Bowden, the late Robert L Cragg, George A Foerster, James M Phillips III and Robert J Tosolt. Also thanks to the writers' families for forbearance during the manuscript preparation process.

The impetus for writing this book came two decades ago from the late Jerome SB Iffland, the President of Iffland Kavanagh Waterbury. Also recognised are the contributions of the late Alexander H McPhee who served as adviser on movable bridge machinery systems in the early 1980s when that technology was closely held by a few bridge engineering firms.

Thanks also to agencies that facilitated inspections of their movable bridge stock by assigning senior staff to arrange for and accompany Charles Birnstiel on bridge tours. Some of the agencies and personnel in the UK were:

- Kingston upon Hull City Council – Keith Stubs and Barrie Young.
- The Manchester Ship Canal Company – Ray Howells and David Ogilvie.
- Humber Bridge Board – Roger Evans.
- Port of Tyne Authority – Donald Graham and George Fenwick.
- Newport, Wales, City Council.
- Corporation of London – Charles Harrison in co-operation with Rexroth staff members Nigel Hart, Anneliese Herteux and Erich Wirzberger.

On the Continent, movable bridge tours were arranged by:

- Hamburg Strom und Hafenbau – Karl Hermann Jonetzki.
- Landesamt für Strassenbau and Verkehr Schleswig-Holstein – Volker Richter and Jürgen Bätcke.
- Danish Road Directorate – Jörn Lauridsen and E. Stoltzer.
- Port of Copenhagen – Ulla Bladt.

In the USA, arrangements for Dr Birnstiel's visits to special movable bridges and on-site guidance were provided by Michael P Freeman, Hans Hutton, Chris T Keckeisen, M Kevin Moran, Steven Rogers, and Jeffrey Routson. Their help is much appreciated.

Appreciation is expressed to those who read drafts of the manuscript chapters and commented on them, including James F Alison III, Dr Martin Bechtold, Roger Evans, the late Dr Sviatoslav Liapunov, Kenneth H Moyer, Thomas E Secules and Paul M Skelton.

Also recognised are Lisa M Weiser for word processing and Rachel Gerlis, and Rebecca Taylor, editors at ICE Publishing, who guided the book preparation.

Charles Birnstiel, May, 2015

About the authors

Charles Birnstiel

Charles Birnstiel studied civil engineering at New York University in New York City, taking his bachelor's and master's degrees in 1954 and 1957, and earning his doctorate in 1962. His engineering experience includes 20 years of teaching and structural mechanics research at New York and Polytechnic universities and 23 years leading a multidisciplinary engineering firm specialising in the design, inspection and testing of movable bridge machinery, followed by employment with Hardesty and Hanover, LLC, a prominent movable bridge engineering firm headquartered in New York City.

Besides his structural stability research at university, Dr Birnstiel has been active on technical committees of the American Society of Civil Engineers (ASCE). He was chairman of the ASCE Standards Committee responsible for the ASCE/SEI 19-10 *Structural Applications of Steel Cables for Buildings* Standard. He is an emeritus member of the American Railway Engineering and Maintenance-of-Way Association (AREMA) Committee 15 – Bridges, and participated for many years in updating the AREMA *Manual for Railway Engineering* – a recommended practice. He is a licensed professional engineer in several US states and a Fellow of the Institution of Civil Engineers (FICE) and of ASCE.

William Bowden

William Bowden is the founder and president of Link Control Systems, a firm that has been in the movable bridge industry for 35 years and is a recognised leader in the field. William Bowden has been involved with the design and supervision of bridge control systems during this time and has completed over 200 movable bridge control systems. He has also taught classes on control system design and maintenance for the Institute of Electrical and Electronic Engineers and has presented technical papers covering various aspects of bridge control system design. William holds a Master of Science degree from Fordham University and two Master of Arts degrees from the Seminary of the Immaculate Conception.

George A Foerster

George A Foerster is an associate senior mechanical engineer in the New York City office of engineering firm Hardesty and Hanover, LLC, which specialises in bridge design and inspection. George graduated from the State University of New York at Stony Brook, NY, in 1991, with a Bachelor of Engineering degree, majoring in mechanical engineering. Since college, George has accumulated over 22 years' experience in the design, condition inspection and construction inspection of machinery for movable bridges and heavy movable structures. As a senior mechanical engineer George is involved with technical concept development, design development, specifications, cost estimating, analysis, inspections, condition reports, construction inspection reports, emergency response, and proposed repairs. Project experience includes all major movable bridge types, ferry transfer bridges, movable inspection platforms and movable stadium roofs. In the USA, George is a licensed professional engineer in several states and in

Washington, DC. He is a member of Heavy Movable Structures (HMS) and the American Society of Highway Engineers (ASHE). He has presented several HMS papers on topics involving movable bridges and the mechanical components and systems associated with them.

About the contributors

Robert L Cragg

The late Robert L Cragg received a Bachelor of Science degree in mechanical engineering from Drexel Institute of Technology in 1947 after military service in World War II, including the Battle of the Bulge. He received the Bronze Star Medal for bravery and was named *Chevalier* of the *Légion d'honneur* for his service in the liberation of France. After graduation he associated himself with the rolling element bearing manufacturers in the Philadelphia, PA area and eventually joined Earle Gear and Machine Co., in Philadelphia. He was Vice President of Earle when that firm was acquired by Steward Machine Co. of Birmingham, AL, in 1985. Until December 2012 he remained an active consultant to Steward. He authored manuals on movable bridge inspections and related topics and participated in conducting movable bridge inspectors' maintenance training schools. The energy-absorbing span lock system now marketed by Steward was patented by Cragg. Cragg was a licensed professional engineer in several states. Until his passing on 16 January 2013 he was a respected member of the movable bridge engineering profession.

James M Phillips III

James M Phillips III is a bridge practice leader for Hardesty and Hanover, LLC, in Tampa, FL. His career includes more than 32 years of experience in movable bridge inspection, planning, design and construction engineering with leading consulting engineering firms such as Parsons Brinckeroff, EC Driver and Associates, and URS. James has participated in the planning and design of dozens of new movable bridge projects throughout the USA as well as internationally. In 1982 he earned a Bachelor of Science degree in civil engineering from the University of Florida, College of Engineering. He is a licensed professional engineer in many states. In addition to his formal education in civil engineering, he has more than 32 years of practical experience in design of movable bridge mechanical and hydraulic systems and is certified by the Fluid Power Society as both a fluid power specialist and a fluid power engineer. In addition to movable bridges, James's experience includes design of machinery and fluid power drives for navigation locks, flood control gates, and other large movable structures. He has been responsible for design of hydraulic drives of numerous configurations, including open loop hydraulic motor drives, open loop hydraulic cylinder drives, closed loop hydraulic motor drives, variable displacement pump control, flux-vector pump control, proportional valve flow control and others.

Robert J Tosolt

Robert J (Bob) Tosolt is a practising professional engineer with over 23 years of experience working with mechanical operating and support systems on all types and vintages of movable bridges. He received a Bachelor of Science degree in mechanical engineering from Lafayette College, Easton, PA in 1992. He gained his initial employment under Charles Birnstiel from 1992 to 1995 and has since been employed at Stafford Bandlow Engineering, Inc. in Doylestown, PA, where he now serves as senior engineer and associate. Robert has been responsible for new machinery design for bascule, swing, and vertical lift bridge projects, and has extensive practical experience that has been gained through field services (including inspection, construction support, installation acceptance, operational testing and troubleshooting, machinery start-up testing, emergency call-out, and failure analyses). He is a licensed professional engineer in 11 states, and is an active member of AREMA and HMS.

Notation

SYMBOLOGY

Latin letters

A_b	hydraulic cylinder bore area (in.^{2}); keyway bearing area
$A_{effective}$	effective metal cross-sectional area of wire rope
A_r	hydraulic cylinder rod end area
A_s	key shear area
B/L	gearset backlash (in.)
C	compressive force; gearset centre distance (in.); degree Centigrade
C_D	size factor
C_H	hardness ratio factor
C_M	miscellaneous factor
C_R	reliability factor
C_S	surface roughness factor
C_T	temperature factor
$C_{basic\ rating}$	bearing basic load rating
C_d	orifice coefficient
C_f	surface condition factor for pitting resistance
C_p	elastic coefficient for steel
c	bolt torque coefficient, also known as the bolt 'K factor'
D	diameter of curved surface; shaft diameter; gear pitch diameter (in.); sheave diameter
Ded	gear tooth dedendum (in.)
D_b	diameter of cylinder bore (in.)
D_o	diameter of an orifice (in.)
D_r	diameter of cylinder rod (in.)
D_{tread}	tread diameter of sheave grooves
d_w	diameter of outer wire
d	diameter, shaft diameter; diameter of roller; nominal bolt diameter; displacement of a pump or motor (in.^{3}/rev); pinion pitch diameter (in.); rope diameter
E	modulus of elasticity, psi (taken as 29×10^6 psi for steel)
E_s	secant modulus of elasticity
E_w	tensile modulus of elasticity of steel wire
Eff	power transmission efficiency
F	force (lb); specified dwell; tooth net face width (in.); degree Fahrenheit
FS	factor of safety
F_1	test force
F_E	Euler bucking load (lb)
F_a	applied axial load
F_b	design buckling load for a hydraulic cylinder (lb)
F_i	bolt initial preload
F_o	preliminary test force
F_r	applied radial load
F_y	minimum material yield strength
f	AREMA shaft extreme fibre stress, tension or compression; distance between P and assumed cantilevered beam point of fixity
f.c.c.	face-centred cubic crystal structure
F	tooth net face width (in.)

g	acceleration due to gravity; distance to shear lock from assumed cantilevered beam point of fixity
I	geometry factor for pitting resistance
J	geometry factor for bending strength
K	effective length factor for buckling (unitless); impact factor
K_B	rim thickness factor
K_F	fatigue stress concentration for normal stress
K_{FS}	fatigue stress concentration for shear stress
K_R	reliability factor
K_S	tooth size factor
K_T	temperature factor
K_V	dynamic factor
K_{bolt}	bolt 'K factor'
K_f	stress correction factor
K_m	load distribution factor
K_{my}	load distribution factor for overload condition
K_o	overload factor
K_t	theoretical stress concentration factor for normal stress
K_{ts}	theoretical stress concentration factor for shear stress
K_y	yield strength factor
L	length; bearing journal length; length of hub-bearing surface
$L_{allowable}$	allowable length of shaft
L_{10}	anti-friction bearing life in hours
ℓ	distance from mid-span to trunnion axis
M	bending moment
M_b	moment necessary to move the leaf solely to overcome imbalance
m	mass; distance from front live load support to trunnion access
m_G	gear ratio
N_G	number of teeth in the gear
N_P	number of teeth in the pinion
n	number of teeth; shaft speed (rpm); rotational speed (rpm)
n_c	critical shaft speed (rpm)
n_i	load modifier
P	concentrated live load; direct load on wire rope; pressure (psi)
PA	pressure angle (degree)
P_b	pressure acting on the blind end of a cylinder piston (psi)
P_c	circular pitch (in.); shrink-fit pressure
$P_{equivalent}$	equivalent load on bearing
P_{hp}	horsepower
P_{nd}	normal diametrical pitch (in.)
P_o	operating load
P_r	pressure acting on the rod end of a cylinder piston (psi)
p	circular pitch (in.)
p_d	pitch diameter
Q	flow (gpm)
Q_i	force effect
q	fatigue design notch sensitivity factor
R	nominal outside radius of inner member; radius of roll
R_n	nominal resistance
R_r	factored resistance
r	disk-projected bearing radius; radius of shaft at friction surface

Movable Bridge Design
ISBN 978-0-7277-5804-0

http://dx.doi.org/10.1680/mbd.58040.001

Chapter 1
Introduction

Charles Birnstiel

1.1. Need for movable bridges

A movable bridge (sometimes called an opening bridge) has one or more spans that can be moved by machinery that is permanently mounted on the bridge in order to temporarily increase vertical under-clearance, usually over a navigable waterway. The machinery may be mounted on the movable span and its foundation or on adjacent spans of the bridge. Movable bridges are built where an acceptable gradient for a fixed railway or roadway crossing is not feasible because of cost, environmental impact or aesthetics. For railway crossings, costs include that of approach viaducts or fills necessary to provide adequate waterway clearance over the legal channel and the cost of energy for hauling trains on steeper grades.

Construction and maintenance costs of a movable bridge are generally higher than those for a fixed bridge of like span. For that reason and because of the inconvenience of bridge openings to the travelling public, bridge owners prefer fixed to movable bridges for new crossings or replacements where vertical clearance requirements can be satisfied by long approaches. However, if few bridge openings per year are anticipated, economics may favour a movable bridge.

Removable, as opposed to movable, bridges are built where the need for only occasional openings to accommodate large vessels is anticipated. They are built with a span that may be temporarily removed and then replaced, usually by floating cranes. Removable bridges will not be covered in this book.

Thousands of movable bridges exist with movable spans ranging from a few metres to over a hundred metres in length. They have been built since Antiquity and replaced as necessary to accommodate increasing vehicle loads and traffic, because of deterioration and wartime destruction, and changes in transportation patterns. The actual quantity is unknown because few countries have, or make available publicly, an inventory of their movable bridges. The huge size and topography of the USA necessitated the construction of many movable bridges. There are about 3000 active movables in North America, of which some 2000 are highway bridges and about 1000 support railway tracks. A fairly comprehensive list of movable bridges in North America was compiled by Koglin (2003). The Canal and River Trust and Scottish Canals together have about 400 under their purview (mostly with movable spans of only a few metres in length) and there are about 600 across other waterways in the British Isles. The canals of Belgium are crossed by about 200 movables. Finland has about 40 movable bridges and some 25 operate in Sweden. There are about 100 movable bridges in Germany. In the city of St. Petersburg, Russia, there are some 20 major movable bridges across the Nova River and main canals that are operated nightly during the navigation season. Hundreds more exist elsewhere in Europe. A few large bascule and vertical lift bridges were recently built in the Netherlands, France and Spain.

New movable bridge construction is underway in Australia, China, Canada and the USA and in countries with developing economies. One reason for this is economic globalisation that has resulted in more ocean transport and a concomitant increase in vessel size. The growth in popularity of cycling and walking has created a need for small movable pedestrian bridges. They offer design opportunities for non-traditional forms of architectural expression. In summary, much movable bridge engineering will be needed in the future to maintain and replace the existing stock of ageing bridges in the developed world and for new bridges required in developing regions. While undertaking such work, the bridge designer should consider the conclusion of a bridge owner representative that ‘Although it may be intellectually more interesting to design new equipment or a new structure each time, it would be better to build bridges while conserving the equipment and the particular details that have proven their solidity through the years’ (Berthelot, 1991). Berthelot’s conclusion was based mainly on the economics of maintaining movable bridges. However, it also applies to the bridge design engineer’s risk.

1.2. Basic motions of movable spans

The motions of all movable bridge spans are rotation, translation or a combination of these; the differences between bridge types are due to the axes selected for these displacements (Birnstiel, 2008b). In terms of primary displacement and axes of displacement, movable spans may be categorised as follows

- rotation about a fixed transverse horizontal axis (simple trunnion bascule, balance beam bascule, Strauss heel trunnion bascule)
- rotation about a transverse horizontal axis that simultaneously translates longitudinally (rolling bascule, also referred to as rocking bascule, rolling lift and Scherzer)
- rotation about a fixed vertical axis (swing)
- translation along a fixed vertical axis (vertical lift)
- translation along a fixed horizontal axis (retractile or transversing, and transporter)
- rotation about a fixed longitudinal axis (gyratory)
- rotation about multiple transverse horizontal axes (folding).

The transverse horizontal axes mentioned above are usually oriented at 90° to the longitudinal axis of the bridge, but there are bridges for which the transverse axis of rotation is oblique to the longitudinal axes. The displacements mentioned are termed primary because there are other, secondary motions associated with some types of movables that are necessary for securing and releasing the movable span. Most of the movable bridge types listed above have subtypes, some of which are described subsequently. The most common categories of movable bridge are shown in Figure 1.1(a)–(g) (adapted from Birnstiel, 2008b).

Figure 1.1(a) shows a single-span, simple trunnion bascule bridge having a counterweight fixed to the bascule girders, which may fully or partially balance the leaf (flap). The trunnions are usually fastened to the girders and rotate in bearings mounted on the bascule pier. The bearings form a fixed horizontal axis about which the leaf rotates similar to a seesaw (*bascule* in French). For bascules with low over-water clearance, an open pit within the confines of the bascule pier enables the counterweight to remain dry during leaf rotation so that its effectiveness would not be reduced by buoyancy were it to descend into water. There are variations of simple trunnion bascules distinguished by location of counterweight and type of trunnions, as described in Chapter 2.

Figure 1.1(b) is a single-leaf rolling bascule (also called rocking bascule or rolling lift). The most common rolling bascules are those based on a series of patents granted to William and Albert Scherzer of Chicago, starting in 1893. Almost all rolling bascules are nominally balanced by counterweights fixed to the bascule girder or truss. The counterweights may be located above the deck, or below it, or outboard of the moving span. The deck type rolling bascule shown has a counterweight

Figure 1.1 (a) Simple trunnion bascule; (b) rolling bascule; (c) swing; (d) vertical lift

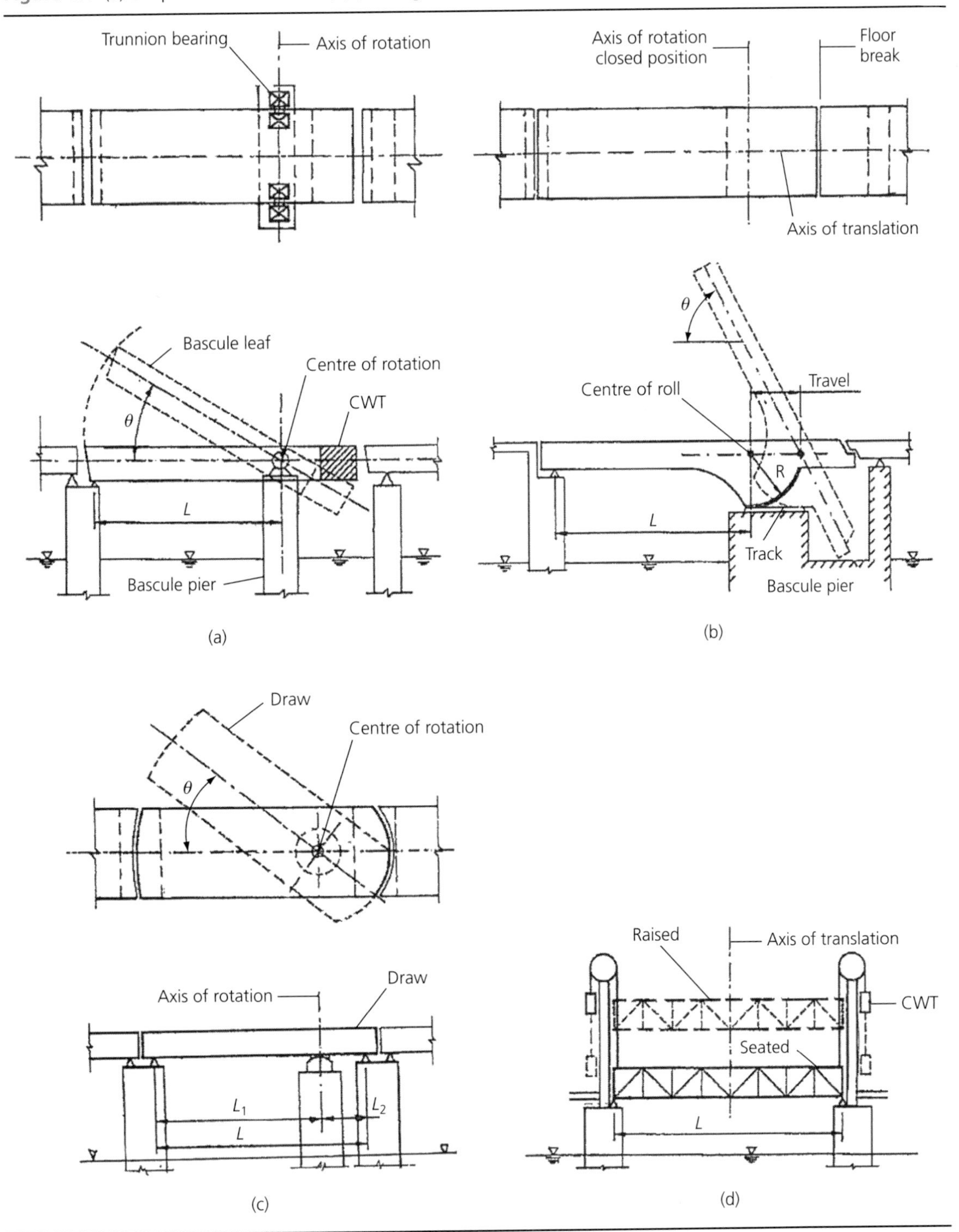

Figure 1.1 (e) Retractile bridge; (f) gyratory bridge; (g) folding bridge

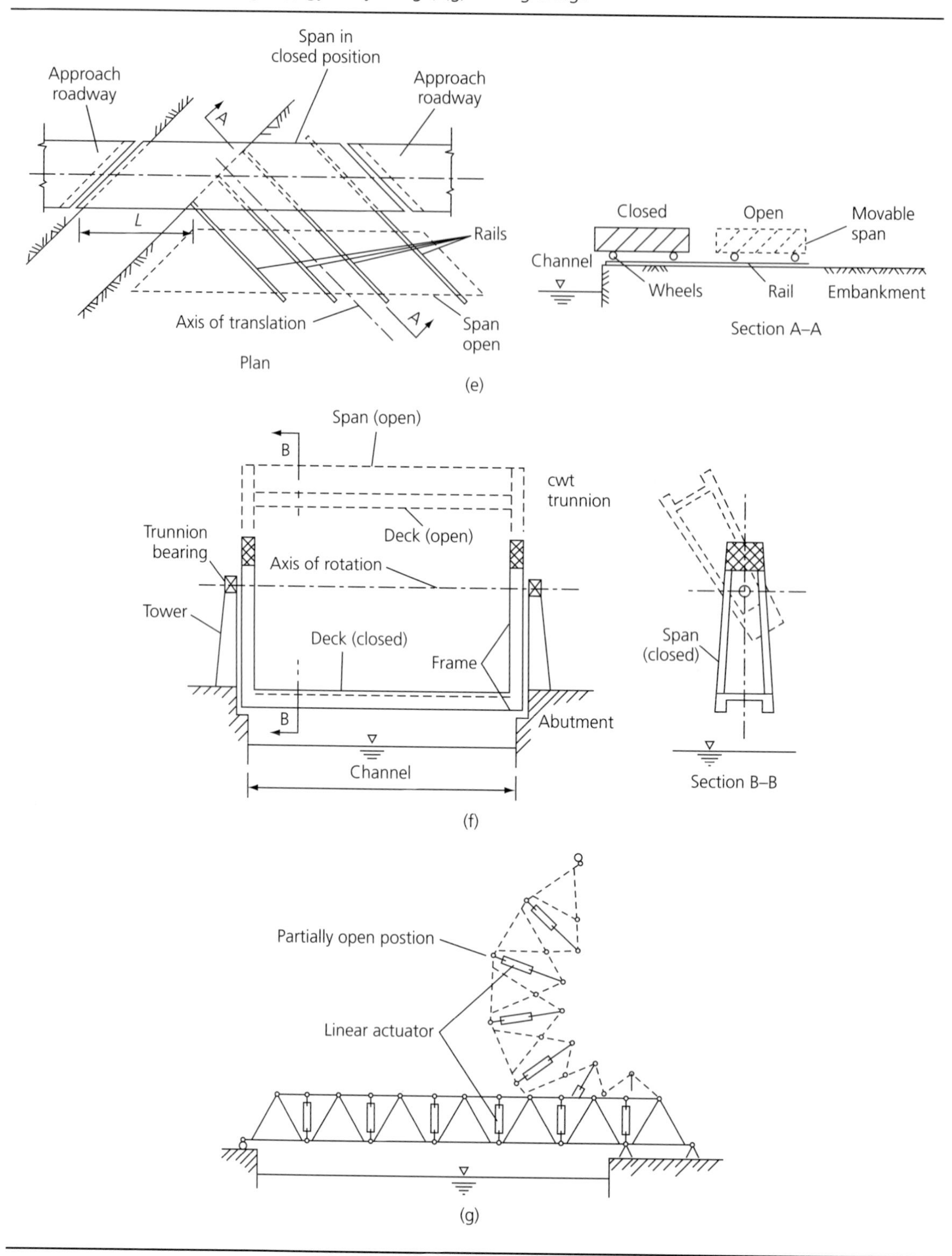

that descends into a pit. However, as will be shown in Chapter 2, most rolling bascules are configured so that pits in bascule piers are unnecessary.

Figure 1.1(c) depicts a swing bridge which rotates about a vertical axis after temporary supports at the ends of the draw are released. If L_1 equals L_2, the swing span (draw) is said to be symmetrical (usual for most swing bridges); otherwise it is known as a 'bobtail' (asymmetric) swing span. The pivot pier may be located in water or on land (the latter often being the case for swing bridges crossing canals or narrow rivers), but it is solidly founded. Swing bridges have subtypes based on the type of rotational bearing at the pivot pier.

A span that translates along a fixed vertical axis in order to obtain additional vertical clearance for navigation is shown in Figure 1.1(d). It is termed a vertical lift bridge. There are many subtypes, distinguished mainly by location of the span drive machinery. The modern vertical lift bridge is credited to JAL Waddell who designed the South Halstead Street Bridge across the Chicago River, which was completed in 1895 along the lines of Waddell's USA patent.

Figure 1.1(e) shows a retractile bridge. Retractile, or traversing, bridges translate horizontally, usually along a straight path which may be normal to, or at an angle to, the channel. If the translation is at an angle to the channel (approximately 45°), as in the figure, the movable span does not have to be lifted or depressed to clear a path for motion. This is a desirable feature for heavy spans. If the movable span does not roll on rails at ground level as depicted, but rather rolls on rails mounted on an overhead structure that completely spans the channel, it is called a transporter bridge. In the latter case the retractile platform (or cabin) is hung from a carriage that runs along the bottom of the overhead spanning structure.

Figure 1.1(f) illustrates a concept for a gyratory bridge, based on a US patent granted to E Swensson in 1909. It is a version of a simple trunnion bridge, but the span rotates about a longitudinal axis instead of a transverse axis. As of 2013, only one major bridge of the gyratory type exists, Gateshead Millennium Bridge, across the River Tyne in Newcastle upon Tyne. At Gateshead the trunnion bearings are located at a lower elevation, near the waterline.

Figure 1.1(g) shows one concept for a folding bridge that has been executed. Rigid deck panels fold about multiple transverse horizontal axes, only one of which is at a fixed location. The other folding axes translate, of course. At least one folding bridge has been built in which all of the transverse folding axes translate during motion.

The basic primary motions and types of movable bridges have been described in the preceding paragraphs. Subtypes and pertinent features for each will be discussed in Chapters 2, 3 and 4.

1.3. Early movable bridges

Movable bridges were built in Antiquity. A submersible vertical lift bridge in China was used for military purposes about 500 AD (Needham, 1971). It was a suspension bridge which could be lowered into the water to prevent enemy access to fortifications, much as balance beam bascules were built at the entrances to French forts in the Middle Ages. To obstruct the foe, the Chinese lowered the ribbon bridge deck into the water; the French raised the bascule leaf. The earliest accessible illustrations of movable bridges of which the writer is aware are the 1436 painting entitled 'Madonna with Chancellor Rolin' by the Flemish artist Jan van Eyck and a drawing showing plan, elevations and cross-sections of a bridge over the Loire River at Orleans, France, which was painted after the English siege of 1423 (Murray and Stevens, 1996). The earliest architectural drawings of movable

Figure 1.2 Balance beam bascule at fortification. (Adapted from Belidor, 1729)

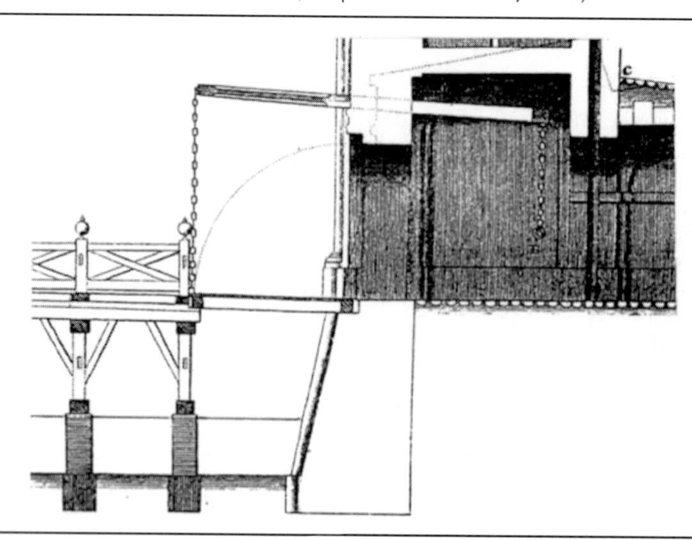

bridges that are readily available are those of de Belidor (1729) and those of Gauthey (1843(a)–(c)) that were published by his nephew CLMH Navier. Schultz (1994) also reviewed the historic development of movable bridges.

1.3.1 De Belidor's contribution

The fifth chapter of Book IV of de Belidor's work (1729) contains drawings of movable bridges at the entrance to military fortifications. Figure 1.2 is a cross-section through a tower showing a balance beam bascule bridge in the closed position. De Belidor devised another version of bascule that was more convenient for operation in fortifications, as shown in Figure 1.3. The counterweight rolls down a track having a profile such that the bascule leaf is nearly balanced at any opening angle. The type is known as a Belidor bascule and such bridges have been built during the past 150 years in the UK, the USA and Australia (see Chapter 2).

The distribution of movable bridge types in North America is approximately 50% swing bridges mounted on piers, 33% bascule bridges, 13% vertical lift bridges and 4% other types.

1.3.2 Gauthey's contribution

Emiland-Marie Gauthey (1732–1806) had the highest official post in French civil engineering during the last few years of the 18th century. He taught at the Ecole Nationale des Ponts et Chaussées and made important contributions to the study of building materials. Gauthey was involved in design and construction of French canals and river improvements, a navigation system that now has a length of about 9000 km. In his engineering positions he dealt with many movable bridges, as engineer and as a government official. He wrote about the design and construction of bridges and canals. His nephew CLMH Navier, assembled and published Gauthey's notes in three volumes (Gauthey, 1843a–c). Chapter IV in Volume 2, 'Des Ponts Mobiles', is accompanied by a Plate VII comprised of 13 figures showing various movable bridge types (in their closed positions). Figures 1.4(a)–(c) are adaptations of some of Gauthey's (1843b) diagrams of movable bridges. The action of each type of design was clearly described by Hovey (1926) and the following discussion is based on his text.

Figure 1.3 Belidor bascule. (Belidor, 1729, reproduced with permission from the Avery Architectural Library of Columbia University, New York)

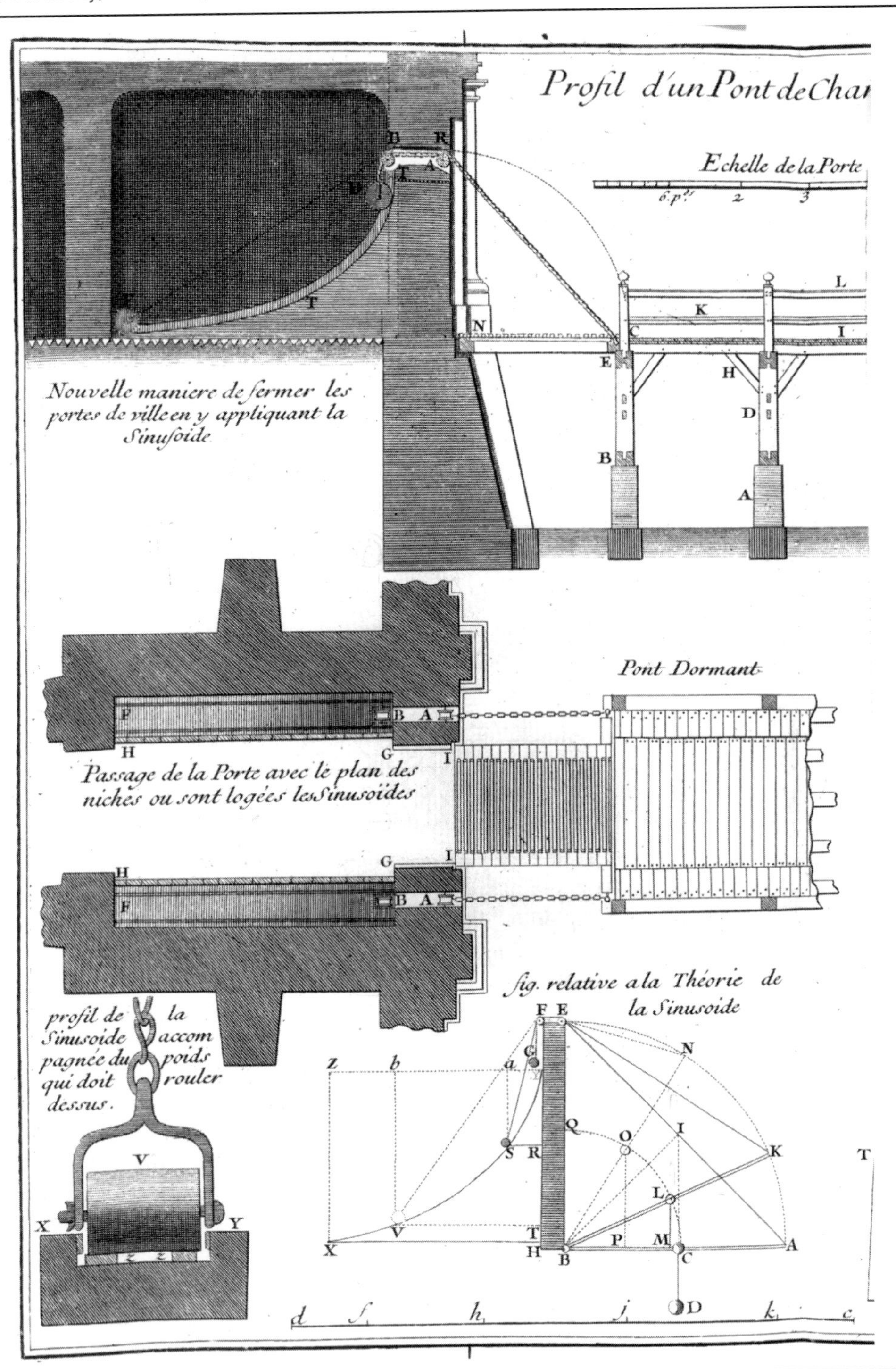

Figure 1.4 (a) Balance beam bascule; (b) rolling bascule; (c) centre bearing swing. (Adapted from Gauthey, 1843b)

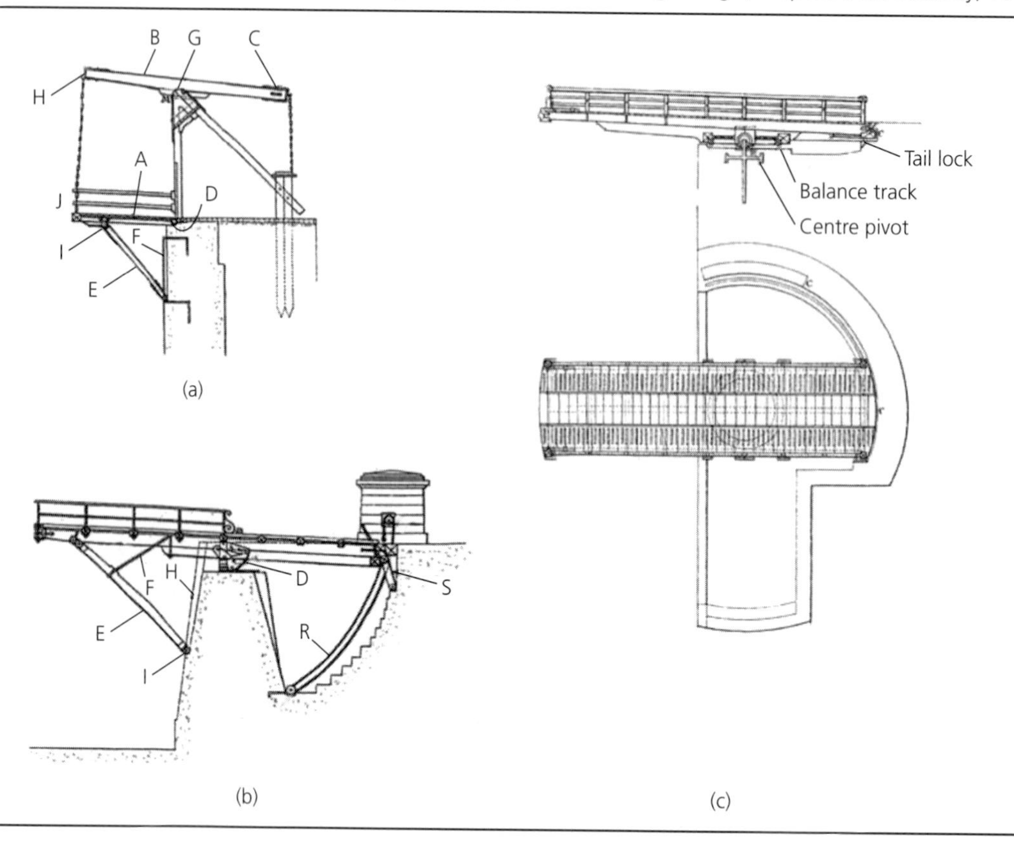

Figure 1.4(a) depicts one half of a double-leaf balance beam bascule. The closed leaf (platform) A is supported by the diagonal struts E and the trunnion D. To open the bridge, the balance beam B with its counterweight C is rotated about bearing G by pulling downward on the chain at C. Another chain, J–H, pulls the toe of the leaf upward taking strut E along for the ride because it is pinned to the platform at I and guided by the wrought-iron bar F. If points J–H–G–D form a parallelogram then the force at C necessary to continue opening the bridge is theoretically constant (assuming no wind or change in friction).

Figure 1.4(b) shows one half of a double-leaf rolling bascule. The heel end of the leaf rests on quadrant D which has teeth that engage the rack on which it rolls. The counterweighted end of the bascule lowers into the pit during opening, being pulled downward by the chain winch R. When the rear of the leaf is subjected to live load it is prevented from rotating at D by the tail struts S. During opening, the strut E (hinged at I) is pulled shoreward into the vertical slot H in the canal wall by the chain F.

Figure 1.4(c) comprises cross-section and plan diagrams of a double-draw swing bridge which was reportedly built at Cherbourg, France, in about 1625. A description of the bridge based on a de Belidor publication was published by Gauthey (1843b), and later translated by Hovey (1926). The clear waterway was 13 m wide and the roadway width 3.73 m. The bobtail draws were not counterweighted (Hovey, 1926, vol. 1, p. 13) but were furnished with wheel-like devices running in metal raceways embedded within the masonry of the rear walls. The raceways had a channel-shaped cross-section. The device was a tail lock that could provide an upward or downward reaction as needed by the live load distribution on the draw. The pivot housing was cast iron. The lower casting was embedded in the pier masonry. The upper casting was fastened to the timber centre girder of the arm. Rotation was accomplished on a bearing comprised of a metal disk sliding on a metal disk. Tipping of the arm on the centre pivot was controlled by balance wheels running on a track that was about 3 m in diameter. The ends of the longer arms had a tongue and groove interlock that could transfer vertical shear between the two draws when the bridge was closed.

1.4. Notable 19th-century movable bridges

Movable spans in the Renaissance were built of timber with mostly wrought-iron fittings. The British Industrial Revolution (approximately 1750 to 1850) introduced new materials into bridge construction. Expansion of trade following the voyages of discovery led to an expanding and wealthier population that demanded more and better goods. Advances in smelting iron, using coke as fuel instead of charcoal, made practical larger and heavier iron castings. They were used for fixed bridges (Troyano, 2003) and the columns of large warehouses at British docks. With the development of trade with the West Indies, it became necessary to build movable bridges to accommodate wider navigation channels at the docks. This necessitated larger bridges built of newer (stronger and more durable) materials. Cast iron became more available in the middle 1700s and improvements in producing wrought-iron shapes (rolling) made these materials practical for larger bridge construction in Britain and Europe. Cast iron was first used for swing bridges in Britain about 1810 (Taylor *et al.*, 2009), probably engineered by John Rennie (the elder) at the London Docks.

A sketch of a double-draw cast-iron swing bridge completed at St Katherine's Docks, London, in 1818 is shown in (Hovey, 1926). The clear span between walls was 44.75 ft (13.6 m). Each counterbalanced bobtailed draw was rim-bearing on 24 rollers and when the bridge was fully closed it formed an arch which supported live load positioned on the deck between springings. Similar bridges are extant at Hull and Liverpool. From this time onward many cast-iron swing bridges were installed in Britain.

1.4.1 Albert Dock Bridge, Liverpool and Wellington Street, Victoria Dock, Hull

Albert, the Prince Consort of Queen Victoria, had a strong interest in education and applying science to industry. As part of his public promotion of British manufacturing and trade he opened the Albert Dock in Liverpool in 1846, although construction was not completed until 1847. It was part of a large port development scheme initiated in 1837 by Jesse Hartley, a civil engineer, for a combined dock and warehouse system (Stammers, 2010). He and Philip Hartwick, also a civil engineer, drew the plans for Albert Dock, which included the double-draw swing bridge shown in the closed position in Figures 1.5 and 1.6. Each bobtailed draw is counterweighted and the entire self-weight is supported by a rim bearing comprised of a set of rollers running on a circular track. The span rotates about a centre post. In the closed position, the chords of the channel arms form a three-hinged arch that supports the live loads placed on the deck between the springing reactions. Live load placed rearward (land-side) of the arch reactions is supported by the rim bearing. The arch reaction blocks were cast integrally with the roller track as shown in Figure 1.6. Note that there are also reaction blocks for the arches when the draws are in the open position, although they are not theoretically subjected to thrust from live loads.

Referring to Figures 1.5 and 1.6, the live load thrust at the arch springing is equilibrated by the horizontal and vertical components of the springing reactions. At the centre pivot there is sufficient diametrical clearance so that when the draw is rotated the arch can ride past the detent on the reaction block. Operation is by a manual gear drive. Part of the cast-iron rack is visible in Figure 1.6.

Figure 1.5 Albert Dock Swing Bridge, Liverpool – overall view. (Photograph by Charles Birnstiel)

Figure 1.6 Albert Dock Swing Bridge, Liverpool – rim bearing. (Photograph by Charles Birnstiel)

Figure 1.7 Wellington Street Bridge, Hull. (Photograph by Roger Evans)

The Wellington Street Bridge at Victoria Dock in Hull is of similar design, but the arch reaction blocks have timber pads. Its operation has been electrified. Figure 1.7 shows one draw in the fully open position, and the modern operator's house. This bridge was part of the dock design by JB Hartley, the son of Jesse Hartley who was responsible for Albert Dock in Liverpool. During a recent rehabilitation some of the original cast-iron trusses were replaced by steel plate trusses (Taylor *et al.*, 2009).

1.4.2 Mississippi and Missouri River swing bridges

By 1850 westward expansion of the USA beyond the Mississippi River was well underway. That river was a barrier to the nascent railways' expansion in order to bring settlers to the vast western areas under the rubric Manifest Destiny (DeVoto, 1943). The stretch of river upstream from St Louis, MO, to St Paul, MN, the 'Upper Mississippi', some 900 km long, was an issue because it had to be kept navigable during the summer. Although thousands of horse teams crossed over the ice every winter, a more reliable support was necessary for railway track. But fixed bridges without sections that could not be moved quickly to enable river navigation were not acceptable to the War Department.

The first crossing of the Mississippi south of St Paul, MN, with a movable span was a railway swing bridge opened for travel in 1856 at Rock Island, IL. The authority for construction (westward to the centreline of the deepest navigable channel between Rock Island and the Iowa shore) was granted by the State of Illinois. It appears that the State of Iowa did not authorise the bridge within its jurisdiction. The original piers on the Iowa side were structurally inadequate and when it became necessary to replace them in 1858 the navigational interests objected. The Secretary of War appointed a Board of

Engineers to study the matter and they concluded 'that the bridge was not only a material obstruction, but one materially greater than was necessary' (Warren, 1878). In 1866 an act was approved for removing the bridge and replacing it with one adapted to highway as well as railway use, at a location which would not interfere with operation of the new US Arsenal on Rock Island. In 1872 the new bridge at Rock Island was finished and the first one was removed.

The second railway bridge across the Upper Mississippi was opened to traffic by the Galena and Chicago Union Railroad (G & CU) between Fulton, IL and Clinton, IA in 1865 (Birnstiel, 2008a). It had been built without federal authority but was legalised *post facto* by Congress in 1867. The overall length of the crossing was 4200 ft (1280 m) comprised of a 300 ft (91.5 m) symmetrical wrought-iron swing bridge over the West Channel, 1950 ft (595 m) of fixed wooden trusses, 1375 ft (419 m) of pile trestle, and 575 ft (175 m) of island embankment. The symmetrical draw comprised two Bollman patent truss bridges joined end to end over a pivot bearing assembly with the free ends stayed by chains radiating from the top of a central cast-iron tower. There were four sets of wrought-iron chains. At each free end of the trusses one set of chains was connected at the top chord and the other set at the bottom chord. Photographs of the draw in the open and closed positions were published (Birnstiel, 2008a). It was the first draw over the Mississippi operated by steam power. Because of ever-heavier railway loading the cable-stayed Bollman truss draw was replaced in 1887 by a Pratt truss draw of the same length.

The Civil War demonstrated the need for transportation improvements, especially along the Mississippi River. The Rivers and Harbors Act of 1866 required the Army Engineers to investigate 'the question of bridging the Mississippi in such a manner as to preserve navigation' (Warren, 1878). The task was assigned to Major GK Warren who submitted his report in 1878. During the 12-year interval between assignment and report delivery, at least a dozen railway and combined railway and highway crossings were constructed over the Upper Mississippi, all of them of the swing type. Although successive Rivers and Harbors Acts gave increased power over crossings to the Secretary of War, an official standard for navigation-opening widths seems not to have been set. However, with the exception of one bridge completed in 1871 all the newer drawbridges had draw lengths of 360 ft (110 m) or more, each providing two openings of 160 ft (49 m) for navigation. The wrought-iron railway swing bridge completed at Louisiana, MO, in 1873 had a symmetrical draw 444 ft (135 m) long giving two waterway openings, each 200 ft (61 m) wide.

The first bridge across the Missouri River at Kansas City was opened to the public on 3 July 1869 (Chanute and Morrison, 1870). It had a symmetrical swing draw 361 ft (111 m) long and was built for one railway track which was planked over for the use of horse and wagon during the intervals between train operations. Built by the Keystone Bridge Company, it was similar to the drawbridge that the company had erected over the Mississippi River at Dubuque a year earlier. A Seller's pivot (a tapered roller-thrust bearing) supported the self-weight of the draw in the bridge-open position and self-weight plus live load when closed. A balance-wheel assembly 30 ft (9 m) in diameter stabilised the span against tilting under wind loading and may have also supported live load. Figure 1.8 is a partial elevation view of the draw.

Construction of the draw required 496 000 lb (225 tonnes) of wrought iron and 122 000 lb (55 tonnes) of cast iron and the timber plank flooring. The bridge may be considered typical of single-track railway drawbridges constructed over the Mississippi and Missouri Rivers before 1880. Lithographs showing the railway swing bridge at Dubuque, dated 1870 and 1872, were printed in Reps (1994). Although not engineering drawings, they clearly show the swing bridge form in the closed and open positions. Note that the navigation channel is usually not at mid-stream, but hugs a shore.

Figure 1.8 Kansas City Swing Bridge. (Adapted from Chanute (1870))

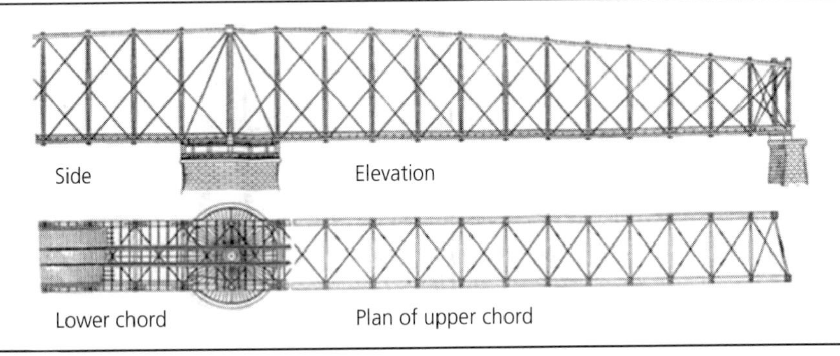

1.4.3 Newcastle upon Tyne swing bridge

The River Tyne has been used continuously as a port for over 2000 years (Humber Oak Consultants, 1995). By 123 AD, there was a bridge across the Tyne at Newcastle, then under Roman rule. This bridge was probably a timber deck on stone piers which was reportedly destroyed by fire in 1248. It was replaced by a stone arch bridge, known as the 'Medieval Bridge', before 1320, and this remained until 1771 when it succumbed to a Tyne flood. In 1781 the Medieval Bridge was replaced by a new stone bridge with nine arches on the alignment of the original Roman bridge. This, the third bridge, called the Town Bridge, was a low-level bridge that obstructed shipping headed upstream to the rapidly expanding industries along the banks of the River Tyne.

In 1850 the conservancy of the river was transferred to a Tyne Improvement Commission (TIC). In 1860 Mr John F Ure, the Chief Engineer of TIC, proposed that a new low-level bridge with a swing span be constructed to replace Town Bridge. This was approved and foundation construction commenced according to plans prepared under the supervision of Ure. However, when bids for the iron superstructure were sought the tenders were 'unsatisfactory'. According to Homfray (1878) '... the matter was placed in the hands of Sir W.C. Armstrong & Co. of Elswick who designed and constructed the present bridge, the foundations having been previously prepared by the River Tyne Commissioners'. According to an anonymous source,

> It is hardly surprising that Armstrong should obtain the contract to build the bridge, for not only did they have a vested interest in upriver navigation, but also, they were the most important builders of hydraulic machinery in general and hydraulic swing bridges in particular.
>
> Anon. (1999)

Based on Homfray (1878) the swing span is 85.65 m long (see the foreground of Figure 1.9). The two bowstring trusses are about 7.32 m deep at the centre and 2.52 m deep at the ends. The swing span is of the combined centre and rim bearing type with centre press relief (see Chapters 2 and 4). The weight of the draw was about 1300 tons (1321 kg). The draw rests on a drum girder of box-shaped cross-section, 1.22 m deep and 0.76 m wide, which transfers the superstructure load onto 42 cast-iron (with steel tires) tapered live rollers that are of 0.92 m diameter with a 0.36 m face width. In the draw-closed position the rollers support self-weight and live load. In the open position the rollers support about 450 tons (457 kg) self-weight and the hydraulic press pushes upward with a force of about 850 tons (864 kg). The purpose of the 1.6 m dia. press is to relieve load on the roller nest so as to minimise the torque required to rotate the draw. Two gravity hydraulic accumulators were installed, each of 0.43 m

Figure 1.9 Newcastle Swing Bridge. (Courtesy of the Port of Tyne Authority)

diameter with a 5.2 m stroke. They were located below the machinery room floor on the centre pier and are loaded by cast-iron weights. The operating pressure is about 48 bar.

The turning machinery design was based on the typical Armstrong system. Coal-fired boilers produce steam, and steam pumps force water into a gravity accumulator. To swing the draw, pressurised water from the accumulators drives reciprocating cylinder engines which rotate gearing that engages a cast-iron rack bolted to the drum girder. See Chapter 4 for additional details about British swing bridges.

The steam boilers and pumps were replaced by electrical equipment in about 1960. Apart from that, the present operating machinery is mostly original.

1.4.4 Tower Bridge, London

Tower Bridge was officially opened to traffic on 9 July 1894. It was the largest and most technically advanced simple trunnion bascule at the time (Birnstiel, 2000). Features of this bridge were later adopted by engineers in the USA, notably the City Engineer of Chicago, and elsewhere (Hardesty *et al.*, 1975; Hess, 2001; Holth, 2012). Figure 1.10 is an image of the bridge in the closed position taken from a vessel moving away from the bridge. The mid-span of the bridge in the partially open position is shown in Figure 1.11. Note that the mid-span lock bars have been completely withdrawn. Figure 1.12 is a cross-section through the original leaf of this double-leaf simple trunnion bascule (Cruttwell, 1896). The bascule trunnions are 226.5 ft (69 m) on centre horizontally giving a clear waterway opening of 200 ft (61 m). Each 49 ft (15 m)-wide leaf is comprised of four bascule girders that originally supported a buckle

Figure 1.10 Tower Bridge, London – overall view. (Photograph by Charles Birnstiel)

Figure 1.11 Tower Bridge, London – leaves partially open. (Photograph by Charles Birnstiel)

Figure 1.12 Tower Bridge, London – half cross-section. (Reprinted from Ryall, Park and Harding, 2000)

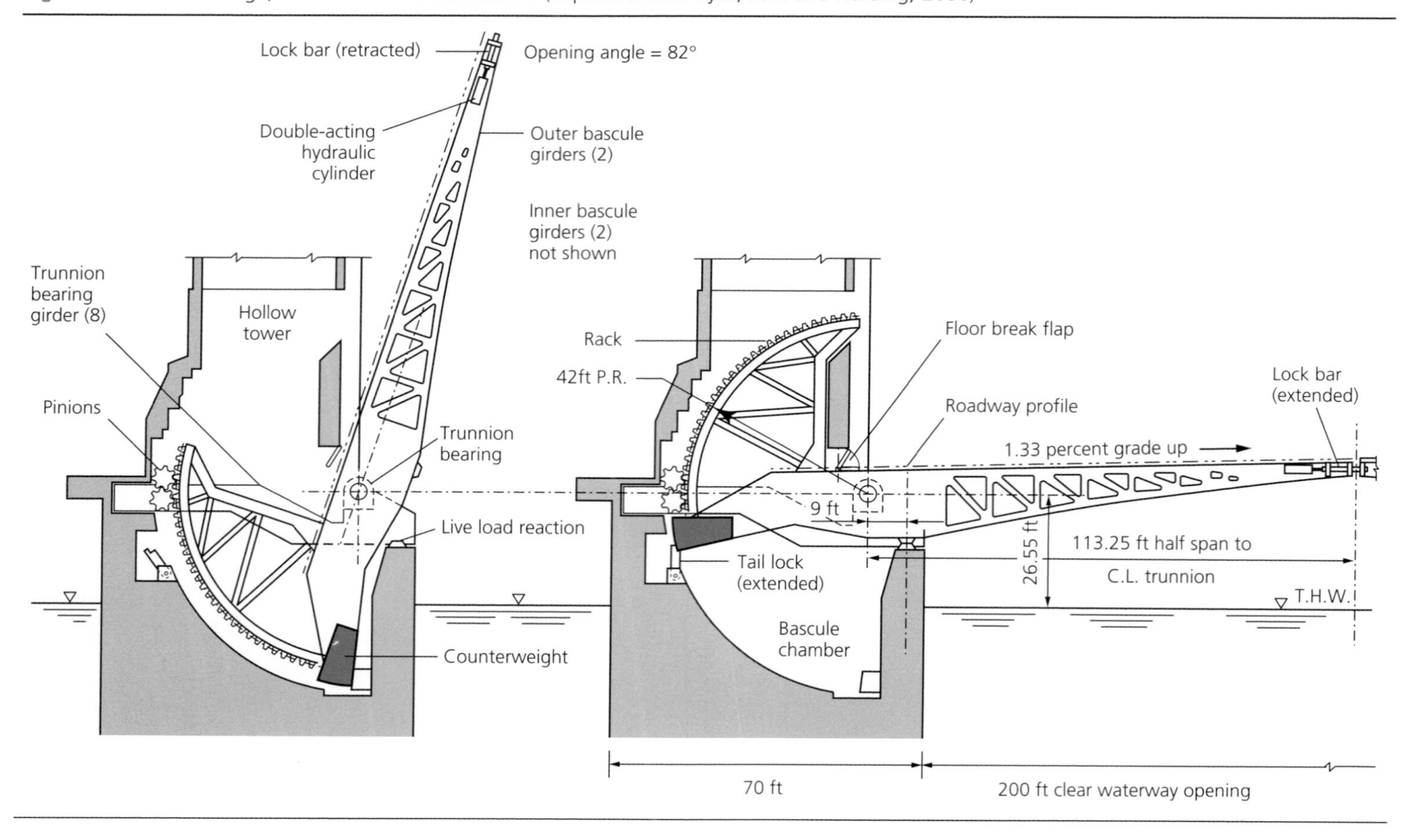

plate floor system which extended a considerable distance rearward of the trunnions. The leaf was counterweighted below deck level by cast-iron and lead blocks at the rear end of the bascule girders.

The stabilising machinery comprises the trunnion bearings, live load reactions, mid-span shear locks, and tail locks. Each trunnion is 21 in. (533 mm) in diameter and 48 ft (16.8 m) long, passing through the webs of the four bascule girders and keyed to each. Eight bearings supported each trunnion shaft. They were bespoke cylindrical roller bearings with live rollers $4\frac{7}{16}$ in. (113 mm) in diameter and $22\frac{1}{2}$ in. (577 mm) long, rolling in cast-steel housings. The trunnion bearings were supported by individual girders spanning between the front and rear walls of the piers. The live load supports were placed on the front wall of the bascule pier a distance of 9 ft (3 m) forward of the trunnions, which is only 8% of the bascule girder cantilever. Because the deck extended well rearward of the trunnion, rotating strut tail locks were provided. These were engaged or disengaged by double-acting hydraulic cylinders. They are under compression when live load on the deck rearward of the trunnion creates a moment about the trunnion exceeding that produced by live loads on the forward part of the leaf.

Span drive machinery was mechanical, powered by water hydraulic engines. Reciprocating steam pumps pressurised water to 700 psi (48 bar) which was stored in gravity-loaded accumulators (Homfray, 1896). A total of six accumulators were installed, two at the engine house and two in each bascule pier. The pressurised water operated the single-acting reciprocating engines which served as prime movers. There were two engines of unequal power at each side of the leaf, the smaller one for normal conditions. The intention was that under severe wind the larger engine should supplement the smaller. Through reduction gearing the engines on one side of the bridge turned one of the pinion shafts engaging racks mounted on the exterior bascule girders. The pitch radius of the rack was 42 ft (12.8 m), about 37% of the leaf span. The machinery on the other side of the house rotated the second pinion shaft. The span drive machinery was rehabilitated in 1976. Rotary oil hydraulic motors now power the bridge through geared speed reducers.

1.4.5 Vertical lift bridges including South Halstead Street, Chicago, IL

Although successful vertical lift bridges were built as early as 1849 across the Grand Surrey Canal (Hood, 1850), over the Erie Canal at Utica, New York, in 1873 (Tyrrell, 1912) and elsewhere, the origin of the modern vertical lift bridge is attributed to JAL Waddell (Nyman, 2002).

The vertical lift bridge across the Grand Surrey Canal comprised a wooden platform 23.5 ft (7.3 m) wide suspended by wire lifting ropes passing over sheaves on the tops of four independent towers. The towers were separated 34 ft (10.7 m) clear across the channel. At their lower ends the lifting ropes were attached to drums mounted on transverse shafts located below running rails (one shaft on each side of the channel). Mounted on the same shaft were six other drums of the same diameter about which the counterweight ropes were coiled. The other ends of the counterweight ropes were attached to circular disc counterweights which moved up and down within cast-iron cylinders that were sunk into the ground and that also served as part of the tower foundations. The transverse shafts were connected to gearing rotated by hand. Two operators were required to raise the platform, one on each side of the channel. The bridge was opened to rail traffic in June 1849 and was used for at least 60 years.

In 1872 Squire Whipple was granted a US patent for a vertical lift bridge of a subtype called a lifting deck (see Chapter 2). His first bridge was constructed across the Eire Canal at Syracuse, NY, in 1874. The lifting deck comprised a roadway platform 60 ft (18.3 m) long and 18 ft (5.5 m) wide which was suspended by rods spaced 10 ft (3 m) apart along the two sides of the deck and passing through the

Figure 1.13 South Halstead Street Bridge, Chicago. (Courtesy of Hardesty and Hanover)

hollow verticals of the fixed spanning truss above. The deck was counterbalanced by cast-iron boxes so that it could be lifted manually utilising reduction gearing.

Waddell's South Halstead Street Bridge over the East Fork of the South Branch of the Chicago River came about because a vessel demolished the then-existing swing bridge at that location (Waddell, 1895). South Halstead Street was one of the busiest streets in Chicago. Citizens clamoured for a replacement bridge but because 'there was no money in the City Treasury to pay for the bridge' (Waddell, 1895, p. 2), political problems, and refusal of the War Department to permit construction of a replacement swing bridge, the Commissioner of Public Works decided on a vertical lift bridge similar to that designed previously by Waddell for a proposed crossing of the ship canal at Duluth, but which was never built. The War Department required a vertical waterway clearance of 155 ft (47.2 m) above mean low water. Waddell's design, as built, is shown in Figure 1.13. It was a combination of connected-tower and tower-base drive vertical lifts, according to the classification system of Chapter 2.

As is evident in the figure, the lift span was a conventional Pratt through-truss span 130 ft (39.7 m) long and about 24 ft (7.3 m) deep which could be raised to a height of 155 ft (47.3 m) above mean low water.

The Pratt trusses were 40 ft (12.2 m) apart on centres supporting a 34 ft (10.4 m)-wide roadway and a 7 ft (2.1 m) clear-width sidewalk, cantilevered outboard from each truss. The roadway floor comprised wood block pavement laid on sand bedding overlying a 4 in. (102 mm) thick pine floor that rested upon I-beam stringers spaced 3.25 ft (0.5 m) on centres. There were a total of 32 counterweight ropes of 1 in. (25.4 mm) diameter, eight at each corner of the lift span. The main counterweights were cast-iron blocks moving up and down between vertical angle guides. They were intended to just balance the lift span. Beneath the roadway deck were four water ballast tanks with a total capacity of 19 000 lb (8607 kg). Their purpose was to enable balancing of the bridge due to variations in self-weight and secondly 'to provide a quick and efficient means of raising and lowering the span in case of total breakdown of the machinery'. In summary, the lift span and the counterweight ropes were balanced by

(*a*) cast-iron counterweights
(*b*) water ballast
(*c*) balance chains.

The original operating machinery was steam-driven, located in a machinery house with its floor level midway between high and low water. Electric motors were substituted for steam engines in 1907.

This bridge had many of the features that were later incorporated in lift bridge designs prepared by engineering firms headed by Waddell and his former associates. It had an enormous impact on American movable bridge engineering, much of it described in Waddell (1916).

1.4.6 Scherzer double-leaf rolling bascule at Van Buren Street, Chicago, IL

While employed by the Charles L Strobel organisation in Chicago, William Scherzer applied for a US patent for a rolling bascule bridge. It was awarded to him posthumously on 26 December 1893. Probably for commercial reasons (or possibly to conform to contemporaneous British terminology), he termed the bridge a lift bridge. It was also referred to as a rocking bascule in American engineering publications of the time. As mentioned previously, the rolling bascule type had been used in Europe for centuries before Scherzer filed for a patent.

The first Scherzer rolling bascule was completed in February 1895 across the south branch of the Chicago River at Van Buren Street (Anon., 1895). It was the second rolling lift bridge designed by William Scherzer, who died before the engineering design drawings were completed. It was a double-leaf bridge with a span of 114 ft (34.7 m) centre to centre of bearings (closed position) that supported two 18 ft (5.49 m)-wide roadways, two electrically powered streetcar tracks, and two sidewalks on three bascule

Figure 1.14 Rolling Bascule at Van Buren Street, Chicago. (Adapted from Anon. (1895))

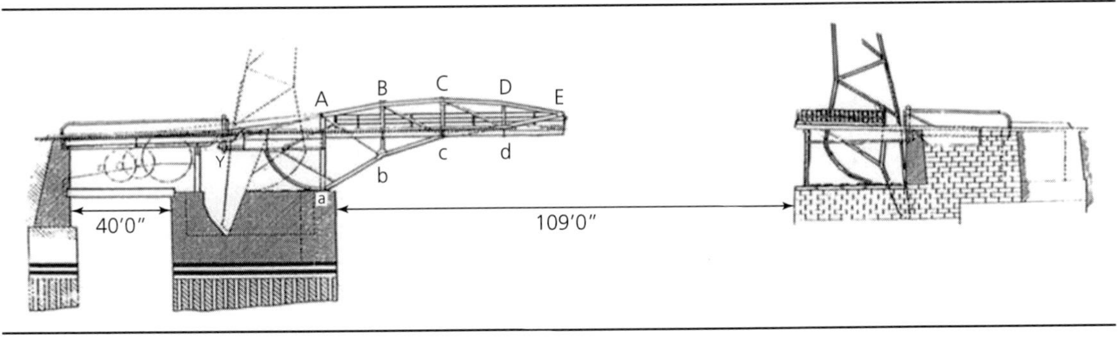

trusses of box-shaped cross-section. Figure 1.14 shows the leaves in the open and closed positions. The leaves were individually rotated by operating struts that engaged machinery mounted in the abutments. Each set of reduction gearing was driven by two 50 hp (horsepower) electric streetcar motors.

Cast-iron counterweight blocks were placed within the box-shaped bascule girders (then also known as tilt girders), and on platforms between them, such that the centre of gravity of the rolling mass was located somewhat rearward of centre of roll when the leaf was closed. Because of this 'counterweight-heavy' condition it was necessary to force the leaves into the fully closed position: an acceptable balance condition.

The radius of roll of the curved tread plates (also known as sole plates) was 16.0 ft (4.88 m). The tread plates were 3 in. (76 mm) thick. They were bent into a circular shape, and then pierced with 6 in. by 12 in. (152 m by 305 mm) holes in order to receive teeth protruding above the track castings on which the bridge rolled. The treads and the tracks formed large gearsets that defined the location of the movable leaf.

After applying for the rolling lift bridge patent, William Scherzer left the employ of Stroebel and opened his own consulting engineering office in Chicago. He proceeded with the design of the West Side Metropolitan Elevated Railroad Company (WSMERC) Bridge over the Chicago River, his first rolling bascule design, but did not finish it because he died on 21 July 1893. While construction was in progress a vessel crashed into the then-existing roadway swing bridge at Van Buren Street, only one block away from the WSMERC site. William Scherzer quickly developed a design for a roadway bridge at Van Buren Street based on his work for the WSMERC but he died before the tender documents were completed. The Van Buren design was his second design, but its construction was finished before that of his first design. Construction of both bridges was completed in January 1895 (Scherzer, 1901). Figure 1.15

Figure 1.15 Rolling bascules at Van Buren Street and Metropolitan West Side Elevated Railroad. (Courtesy of Columbia University Engineering Library)

shows the completed Van Buren Street Bridge in the foreground. It was replaced in 1956. The WSMERC Bridge with both leaves opened is depicted immediately behind it.

Because the rolling bascule bridge became an important movable bridge type early in the 20th century it is pertinent to define the Scherzer rolling bascule bridge by noting some of the principal claims in US Patent No. 511.713 granted to William Scherzer (deceased) on 26 December 1893. Claim 4 was 'A lift bridge having a movable span provided at one end with a curved part adapted to rest and roll upon a supporting surface, and teeth or projections on said curved part adapted to interlock with projections on the supporting surface to hold said curved part from moving or slipping on said surface, substantially as described'. Claim 7 reads in part '...said span being counterpoised...'. Basically, the patent was for a bridge with a counterweight mounted on one end of the bascule girder (truss) which was curved so as to be capable of rolling on a supporting surface with interlocking teeth to prevent the contacting parts from slipping.

Albert Scherzer and associates of the Scherzer company later received many patents for improvements to the 1893 design.

Among the advantages of the Scherzer rolling bascule type over the simple trunnion bascule type (such as Tower Bridge) are the following:

- The reaction for dead and live loads (bridge closed) can be brought closer to the channel for a Scherzer than for the simple trunnion bascule resulting in a decrease of cantilever span length of 20 to 30%, favouring the Scherzer.
- The dead and live loads are transmitted from the bascule girder directly to the foundation without passing through trunnions, trunnion bearings and trunnion towers.
- As the leaf of a Scherzer is opened it immediately starts to move away from the channel thereby increasing the channel 'horizontal air clearance'. This action was important at the beginning of the 20th century, when tall sailing vessels with wide mast-arms negotiated Chicago waterways.

The foregoing is also an indication of the difference in movable bridge design between British and American practice in the 19th century. British emphasis was on substantially constructed bridges. The geographical challenges and the economic realities of railway expansion caused American practice to concentrate on designing for lowest first cost. It was generally assumed that continuous maintenance of weak structures would tide the owners over for a few decades by which time higher vehicle loads would require bridge replacement anyhow, as happened. This approach was used by many railways during construction in the West and Midwest because federal government grants were contingent on miles of track laid on which vehicles could roll (Masterson, 1978).

The second difference between British and American movable bridge design practice was in mechanism design. Early British powered swing bridges were predominantly powered by hydraulic machinery due, in part, to the influence of the Armstrong Companies. Other contributing factors were the availability of public high pressure water utility service and the mild environmental climate. The preference for hydraulic operation in the UK and central Europe has been maintained to this day (now using oil as the fluid). In contrast, comparatively few hydraulically operated movable bridges exist in the USA and they are mostly located in the southern part of the country.

REFERENCES

Anon. (1895) A rocking bridge. *The Railroad Gazette*, **27**: 99–101.
Anon. (1999) *The Tyne Swing Bridge*. Port of Tyne Authority, North Shields, UK.

de Belidor BF (1729) *La science des ingénieurs dès la condite des travaux de fortification et d'architecture civile*. Chez Claude Jombert, Paris, France.

Berthelot P (1991) Concerning the effect of conception on the maintenance. In *Movable Bridges* (Mehue P (ed.)). Symposium of IABSE, 11–14 September. Association Française pour la Construction, Leningrad, USSR.

Birnstiel C (2000) Movable bridges. In *The Manual of Bridge Engineering* (Ryall MJ, Parke GAR and Harding JE (eds)). Thomas Telford, London, pp. 662–698.

Birnstiel C (2008a) The Mississippi River railway crossing at Clinton, Iowa. In *Historic Bridges: Evaluation, Preservation, and Management* (Adeli H (ed.)). CRC Press, Boca Raton, FL, USA, pp. 9–34, http://dx.doi.org/10.1201/9781420079968.pt1.

Birnstiel C (2008b) Movable bridges. In *ICE Manual of Bridge Engineering*, 2nd edn (Parke G and Hewson N (eds)). Institution of Civil Engineers, London, pp. 421–458, http://dx.doi.org/10.1680/mobe.34525.

Chanute O and Morrison G (1870) *The Kansas City Bridge*. Van Nostrand, New York, NY, USA.

Cruttwell GEW (1896) The Tower Bridge: Superstructure. *Minutes of Proceedings of the Institution of Civil Engineers*, **127**: 35–53, http://dx.doi.org/10.1680/imotp.1897.19436.

DeVoto B. (1943) *The Year of Decision, 1846*, Little, Brown and Company, Boston, MA, USA.

Gauthey E-M (1843a) *La Construction des Ponts. uvres de M. Gauthey*. Navier, Liège, Belgium (Nabu Public Domain Reprint).

Gauthey E-M (1843b) *La Construction des Ponts, uvres de M. Gauthey*, vol. 2. Navier, Liège, Belgium (Nabu Public Domain Reprint).

Gauthey E-M (1843c) *Les Canaux de Navigation, uvres de M. Gauthey*, vol. 3. Navier, Liège, Belgium (Nabu Public Domain Reprint).

Hardesty ER, Fischer HW and Christie RW (1975) Fifty-year history of movable bridge construction – Part 1. *Journal of the Construction Division*, **C03**: 551–557.

Hess JA (2001) Development of movable bridge technology in Chicago, 1890–1910. *Proceedings of the 7th Historic Bridge Conference*, pp. 37–47, Cleveland, OH, USA, 19–22 Sept.

Holth N (2012) *Chicago's Bridges*. Shire Publications, Oxford, UK.

Homfray SG (1878) *The Tyne Bridge*. Paper presented before the Tyne Engineering Society on 14 February, 1878. Port of Tyne Authority, South Shields, Tyne and Wear.

Homfray SG (1896) The machinery of the Tower Bridge. *Minutes of the Proceedings of the Institution of Civil Engineers*, **127**: 54–59, http://dx.doi.org/10.1680/imotp.1897.19437.

Hood RJ (1850) Description of a vertical lift bridge, erected over the Grand Surrey Canal, on the line of the Thames Junction Branch of the London, Brighton, and South Coast Railway. *Proceedings of the Institution of Civil Engineers (London)*, **IX**: 1849–1850: 303–309, http://dx.doi.org/10.1680/imotp.1850.24166.

Hovey OE (1926) *Movable Bridges*, vols I and II. Wiley, New York, NY, USA.

Humber Oak Consultants (1995) *Port of Tyne: The First Two Thousand Years*. British Publishing Company, Gloucester, UK.

Koglin TL (2003)*Movable Bridge Engineering*. Wiley, London, http://dx.doi.org/10.1002/9780470172902.

Masterson VV (1978) *The KATY Railroad and the Last Frontier*. University of Missouri Press, Columbia, MO, USA.

Murray P and Stevens MA (eds) (1996) *Living Bridges: The Inhabited Bridge: Past, Present and Future*. Prestel-Verlag and Royal Academy of Arts, New York, NY, USA, and Munich, Germany and London.

Needham J (1971) *Science and Civilization in China*. Cambridge University Press, Cambridge, UK, vol. 4, part 3.

Nyman WE (2002) Dr JAL Waddells' contribution to vertical lift bridge design. *Proceedings of the 9th Biennial Symposium of Heavy Movable Structures*. Daytona Beach, FL, USA, 22–25 Oct.

Reps JW (1994) *Cities of the Mississippi*. University of Missouri Press, Columbia, MO, USA.

Scherzer Rolling Lift Bridge Company (1901) *Catalog*, 2nd edn. Scherzer Rolling Lift Bridge Company, Chicago, IL, USA.

Schultz Jr JA (1994) Remember the past to inspire the future. *Proceedings of the Heavy Movable Structures, Inc., 5th Biennial Symposium*. Clearwater Beach, FL, USA, 2–4 Nov.

Stammers M (2010) *Liverpool Docks*. The History Press, Stroud, Gloucester, UK.

Taylor A, Plant C, and Dickerson J (2009) Wellington Street Swing Bridge, Hull, UK. *Proceedings of the ICE – Engineering History and Heritage*, **162(2)**: 67–79, http://dx.doi.org/10.1680/ehh.2009.162.2.67.

Troyano LF (2003) *Bridge Engineering: A Global Perspective*. Thomas Telford, London, http://dx.doi.org/10.1680/beagp.32156.

Tyrell HG (1912) *The Evolution of the Vertical Lift Bridge*. University of Toronto Engineering Society, Toronto, Canada.

Waddell JAL (1895) The Halstead Street Lift Bridge. *Transactions*, ASCE. **33(742)**: 1–60.

Waddell JAL (1916) *Bridge Engineering*, vols 1 and 2. Wiley, New York, NY, USA.

Warren GK (1878) Bridging the Mississippi River between St Paul, MN and St Louis, MO, USA. In Appendix X, *Report of the Chief of Engineers*. Government Printing Office, Washington, DC, USA.

Movable Bridge Design
ISBN 978-0-7277-5804-0

http://dx.doi.org/10.1680/mbd.58040.025

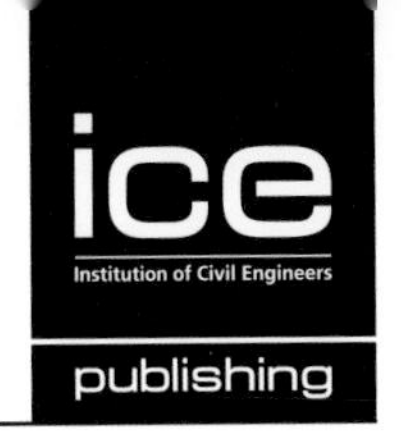

Chapter 2
Individual movable span forms

Charles Birnstiel

2.0. Introduction

The universe of movable bridges is vast and probably no two bridges are exactly alike. Space limitations preclude discussing more than the most important types. In Chapters 1–3, the common forms of movable bridges constructed since 1850 are described with comments on their origins and their advantages and disadvantages. This chapter deals with individual movable spans, either as single units or in parallel. Their characteristics usually also apply to bridges with two movable spans connected end-to-end in the closed position. Movable spans in series are coupled by stabilising machinery which forces the leaves to act as continuous or semi-continuous flexural members, or as arches, to support live load. Multiple movable spans in series are described in Chapter 3.

2.1. Simple trunnion bascules

The simplest movable bridge is the simple trunnion bascule bridge shown in Figure 1.1(a). Essentially, it is a beam rotating in the vertical plane about a horizontal axis – a pivot. One end (sometimes called the heel end) is usually shorter than the portion of the beam over the channel (the forward or toe end). A counterweight may be fastened to the shorter end of the beam of such weight as to balance the beam about the pivot, or nearly so, in order to minimise the energy required to rotate the leaf. The design imbalance varies, from only slightly span-heavy to very span-heavy because of the owners' operational preference and for reasons of safety. For example, in the USA most simple trunnion bascules have only a small span-heavy imbalance at the closed position, resulting in a dead load girder reaction at the live load supports of 1–3 tonnes for a 10–30 m leaf span. This minimises the power required to rotate the leaf open from the leaf-closed position. In Europe it is usual for the leaves to be much more span-heavy. One reason cited for selecting greater span imbalance is the supposed greater stiffness of a leaf that is firmly seated in the closed position, which may minimise vibrations due to vehicular traffic. Trunnion bascule balance considerations are discussed in Chapter 11.

Simple trunnion bascules with small, or without counterweights, are receiving increased attention from designers because of new seismic design criteria for movable bridges. In some geographical areas, bridges are being designed for 1000-year-return-period events, with horizontal accelerations approaching 0.5 g. Unbalanced leaves have an advantage in such situations because their mass may be less than one-half of that of a nominally balanced bascule. Lack of a counterweight is sometimes considered an architectural advantage. Bascule bridges that are severely span-heavy are often operated by hydraulic equipment, usually hydraulic cylinders.

Figure 2.1 is a diagram of a single-leaf simple trunnion bascule bridge with an under-deck counterweight that is operated by mechanical machinery, the usual situation in the USA. The superstructure comprises two steel girders each supported by two trunnion bearings straddling the girder. If the

Figure 2.1 Single-leaf trunnion tower bascule bridge

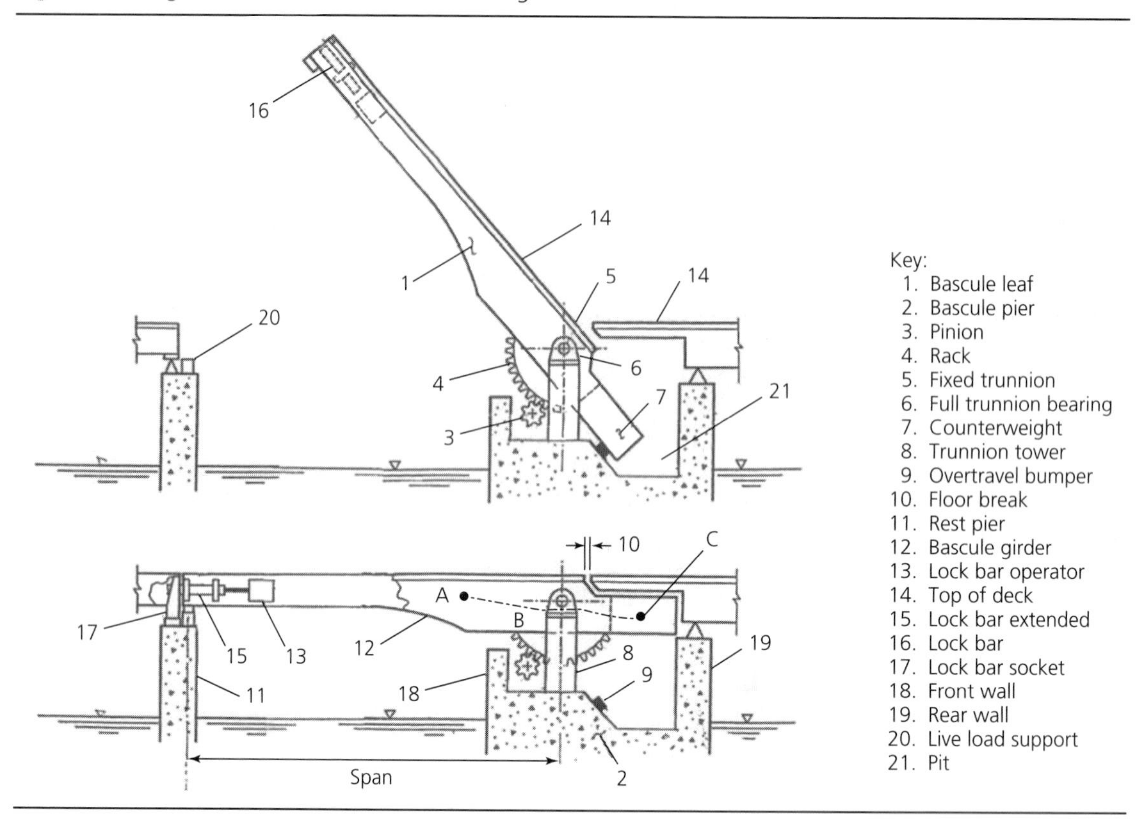

bearings are mounted on steel columns the bridge is sometimes known as a simple trunnion tower bascule. This is only one possible arrangement of trunnion bearings. Machinery mounted on the bascule pier engages curved racks (large gears) mounted on the bascule girders so as to simultaneously rotate them about the pivot. In Figure 2.1 the pinion (3) engages rack (4). The pivot is a trunnion (a large-diameter shaft) fixed to the steel plate web of the bascule girder. The trunnion rotates inside a trunnion bearing, which for this type of single-leaf bascule is a 'full' trunnion bearing; that is, the sliding surface of the trunnion can contract the sliding surface of the bearing sleeve for an angle of 360°. The trunnion bearing (6) is subjected to dead, live, wind and seismic loads. The centre of gravity of the leaf (except counterweight) is located at A; the centre of rotation at B; and the centre of gravity of the counterweight at C. The front (or toe) live load support (20) is loaded by some dead load (from span-heavy imbalance), live load, wind, and seismic. Mechanical span locks are often provided at the toe to resist possible uplift of the bascule girders in the closed position due to heavy concentrated live load applied upon the deck rearward of the trunnion bearings (between the trunnion bearings and the floor break), which would tend to lift the toe of the leaf. Figure 2.2 depicts a typical single-leaf simple trunnion tower bascule bridge.

The trunnion and counterweight in Figure 2.1 are located below the bridge deck. Sometimes, for longer spans, it is advantageous to locate the trunnion and counterweight above the deck, as for the railway bridge at Niantic, CT (Reid, 2010). Single-leaf simple trunnion tower bascules rarely have a span exceeding 40 m.

Figure 2.2 Rehoboth Avenue Bridge over Lewes-Rehoboth Canal – single-leaf trunnion tower bascule bridge (Courtesy of Hardesty and Hanover)

2.2. Balance beam bascules

In Chapter 1 it was shown that balance beam bascules were an important early movable bridge type. A major advantage of the type is that no pit is required at the bascule pier for bridges with low waterway clearance in the closed position because the counterweight can be located remote from the bascule leaf, often overhead. This was an advantage for canal crossings in the Low Countries, where balance beam bascules are ubiquitous.

In the Dutch language, balance beam bascules are usually termed *ophaalbrug* (drawbridge) (de Joode and Bernard, 1989), in French *pont-levis* (lever bridge) (Brignon and Bois, 1986), in German *Waagebalkenbrücke* (Saul and Humpf, 2007) or *Zugbrücke* (drawbridge) (Hawranek, 1936). For a recent US project, the design engineer termed a balance beam bascule a 'European/Dutch Style Heel Trunnion Bascule Bridge'. It has been noted by Koglin (2003) that 'bascules' having remote articulated counterweights are not, strictly speaking, bascules, because the word *bascule* in French denotes a beam with an implied directly attached weight.

Figure 2.3 illustrates the components of a modern balance beam bascule. It comprises a balance beam (5) with a counterweight (7) which rotates about the hinge (10) located atop the tower (11) during opening. A leaf (12) spans between abutments in the closed position supporting its self-weight and live loads. Connected to it on each side of the leaf, at the toe and at the heel, is a transfer beam (or, a spring beam) (13) which is not intended to support live load. The transfer beam is part of the parallelogram linkage A–B–C–D in which A and B are hinges of the counterweight link (6), C is the counterweight trunnion

Figure 2.3 Single-leaf balance beam bascule bridge

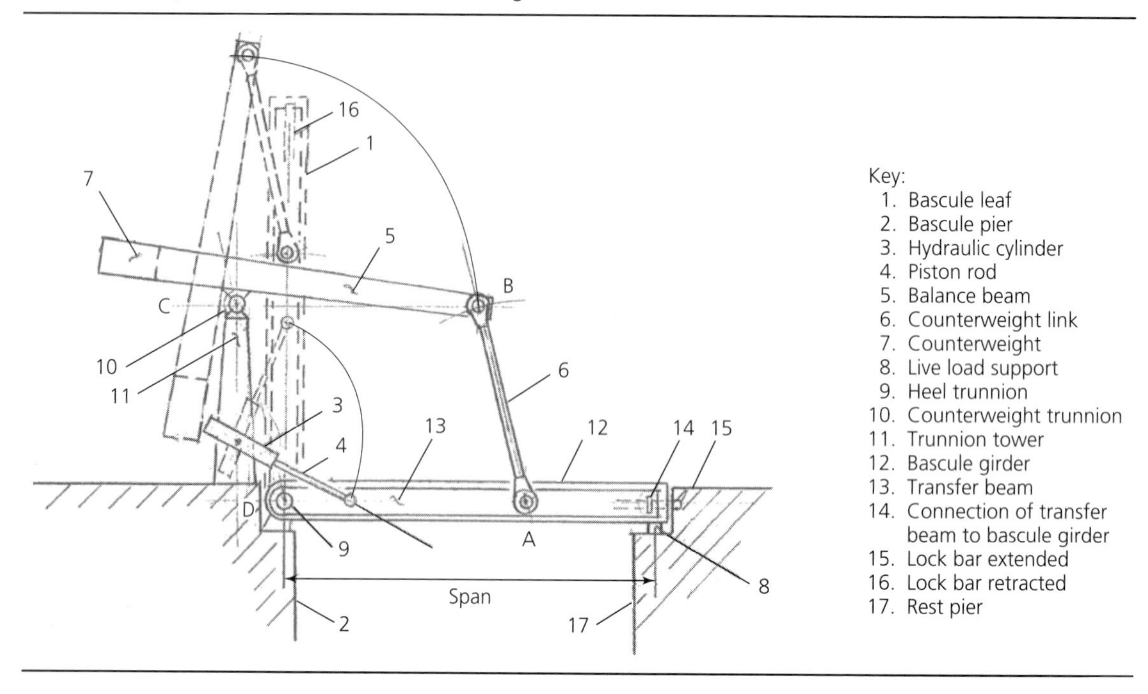

and D is the heel trunnion of the leaf, all forming one plane of the lifting mechanism. Many balance beam bascules do not have transfer beams, the counterweight links are connected directly to the leaf girders. For the latter case live load on the bascule girders causes them to deflect and the balance beam to tilt if not restrained by the operating machinery, resulting in rotation of the counterweight trunnion in its bearing with each passage of a live load on the leaf. These repeated minor rotations seem not to be problematic for sliding bearings, for example friction bearings with (copper alloy) sleeves, but they are sometimes considered a problem for anti-friction rolling element bearings. The transfer beam isolates most of the live load effects from the lifting linkage.

Transfer beams are sometimes used as leaf springs (hence the term spring beam) to force the leaf onto its structural bearings and keep it there. The linkage mechanism is over-driven during closing in order to create a downward force on the transfer beam and hence on the ends of the leaf through special end connections. The objective is to improve the dynamic response of the leaf under roadway traffic. To accomplish the same effect in situations without transfer (spring) beams the joint at the lower end of the counterweight link has a short sliding link containing a compression spring. At over-driving during closing it is compressed and forces the toe of the leaf downward.

Operation of the bridge may be by manual, mechanical, or hydraulic means. Figure 2.3 shows hydraulic cylinder operation with the piston rod (4) serving as the operating strut attached to the transfer (spring) beam. (Sometimes the operating strut is attached directly to the leaf, even though a transfer beam is present.) The linkage A–B–C–D should only be stressed by dead load, wind, and seismic because the piston rod is connected to the transfer beam. The resistance to wind forces may have an important influence on the required cylinder size so that the cylinder mounting geometry becomes

Figure 2.4 Estesperrwerk Bridge, Hamburg, Germany – balance beam bascule. (Courtesy of DSD Noell, Würzburg, Germany)

critical, especially for bridges with a large opening angle. An example of a hydraulically driven balance beam bascule is the road bridge at Geversdorf, near Bremen, Germany. This bridge has a torsionally rigid leaf of 11.5 m width that spans 30 m and which is operated by one double-acting hydraulic cylinder mounted outboard of one side of the leaf (Ortmann, 1990). A larger balance beam bascule operated by two double-acting hydraulic cylinders is the Estesperrwerk Bridge near Hamburg, Germany, shown in Figure 2.4, The span of the leaf is 50.5 m and the overall width is 12.8 m. The design opening angle is 86°, which provides a clear waterway channel of 40 m. The trunnion tower height is 22 m (Gische, 2013).

Alternatively, the cylinders can be mounted horizontally at a level slightly below and to the rear of the pylons with the piston rods acting on brackets fastened to the leaf girders, as at the Ennerdale Bridge across the River Hull in East Yorkshire shown in Figure 2.5 (Birnstiel, 2008).

Figure 2.6 depicts the Blue Bridge on Preston's Road on the Isle of Dogs, London. This bridge has hydraulic span drive machinery located inside the pylons with steel H-profile operating struts extending from the side of the deck girders up through the forward face (channel side) of the pylons. Sprockets on the output train of the machinery engage pins between the flanges of the strut (called a pin rack): see Figure 2.7. Rotation of the sprockets moves the pin rack and hence the leaf.

Another span drive scheme has the machinery mounted atop the counterweight with the operating struts hinged to the pylon base and extending upward to pinions located at the rear end of the counterweight (Sluszka and Kendall, 1991). Some mechanical balance beam bascules have machinery mounted at the base or the top of the pylon driving circular racks fastened to the transfer beams (or leaf girders) or the balance beams (Hawranek, 1936). The advantage of the circular rack arrangement is that operating struts are unnecessary.

Figure 2.5 Ennerdale Bridge, Hull – balance beam bascule (Ryall *et al.*, 2000)

Figure 2.6 Blue Bridge, Preston's Road, Isle of Dogs, London – balance beam bascule. (Photograph by Charles Birnstiel)

Figure 2.7 Operating strut at Blue Bridge. Note the pin rack staves made from circular rods. (Photograph by Charles Birnstiel)

2.3. Strauss articulated bascule bridges

The rapid growth of the City of Chicago around the Chicago River after 1850 led to the need to bridge it at many places. By 1871, the city had 27 movable bridges, mostly double-arm swing bridges (Grossman *et al.*, 2004). Their pivot piers were considered obstructions to the navigation of the ever larger vessels in Great Lakes trade and lead to the search for other movable bridge types that did not require structure within the channel. As described in Chapter 1, William Scherzer designed a rolling lift at Van Buren Street to replace a symmetric swing span destroyed by a ship.

Movable bridge design in Chicago was significantly affected by a massive municipal improvement, the Sanitary and Ship Canal, for which construction started in 1892. The project channel width was 160 ft (48.8 m) and the depth 21 ft (6.4 m) (Solzman, 2006). The waterway was put into operation on 2 January 1900.

Construction of the canal necessitated modifying the existing bridges or replacing them with longer spans to accommodate the wider channel. The Chicago City engineering department engaged various consulting engineers to assist with the movable bridge reconstruction programme, including George S Morison, Ralph Modjeski and Joseph B Strauss. The professional relationships between the eminent bridge engineer Morison and the younger Modjeski and Strauss are obscure. Regardless, Strauss established the Strauss Bascule Bridge Company (SBBC) in Chicago in 1904. Strauss had disagreed with Modjeski and others about the materials used for bascule counterweights. He advocated using stone concrete for counterweights claiming major cost savings over the cast iron used theretofore. Strauss was awarded his first USA movable bridge patent in 1893 (Strauss, 1893) for a trunnion bascule with a counterweight pivoted to the main span rearward of the main trunnion as shown in Figure 2.8. A link *D* is pinned to the counterweight *C* at one end and pinned to a fixed part of the structure at the other end. The link pin locations were to be such that the counterweight would remain horizontal during opening of the bascule thus keeping the leaf and counterweight in balance, regardless of friction

Figure 2.8 Strauss US Patent No. 738,954. (Strauss, 1893) (US Patent Office)

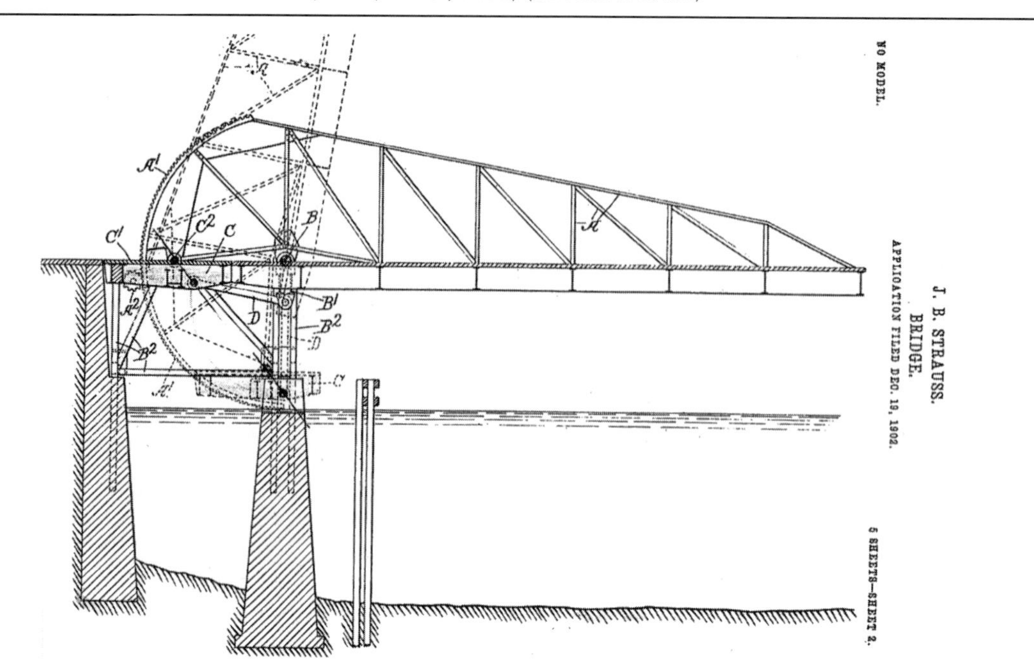

in the counterweight trunnion bearings. The author doubts that, practically, such pin locations could be found. Strauss claimed that with this concept the depth of the pit in the pier could be minimised. Note the absence of the iconic Strauss parallelogram linkage. The author could not find evidence of such a bridge having been constructed.

The first bridge designed by SBBC, based to some extent on the Strauss patent, was completed in 1905 for a single track railway at Cleveland, Ohio. It was a Warren truss with verticals having a span of 150 ft (46 m) with an overhead concrete counterweight and the parallelogram linkage (Hovey, 1926). A photograph of the Wheeling and Lake Erie Railroad Bridge appears in the Strauss *Catalog* (Strauss Bascule Bridge Company, 1920).

In 1908 Strauss received US Patent No. 894,239 for a bascule bridge (see Figure 2.9). This strange patent was for improving the rolling lift (Scherzer) bridge. He proposed to separate the counterweight from the bascule span of the Scherzer, locating it overhead and linking it to the end of the girder and a fixed tower in a manner such that the counterweight moved horizontally as the bascule opened, and the line of action of the counterweight remained vertical. His patent shows a tread (sole) comprised of cast-steel sections that could be replaced when worn, and the use of electromagnets as track teeth. The primary function of the electromagnets was to rotate the bascule girder, as a substitute for the then conventional mechanical span drive machinery. Although the patent had ridiculous components, Strauss did make one claim that was to be of value to his business. Claim 15 reads

> A bridge comprising a movable span adapted to rest and roll upon a support, said span provided with a pivotal counterweight, a link associated with said counterweight, a fixed part to which said link is connected, the parts arranged so that the counterweight remains vertical during all the various movements of the bridge.

Figure 2.9 Strauss US Patent No. 894,239. (Strauss, 1908) (US Patent Office)

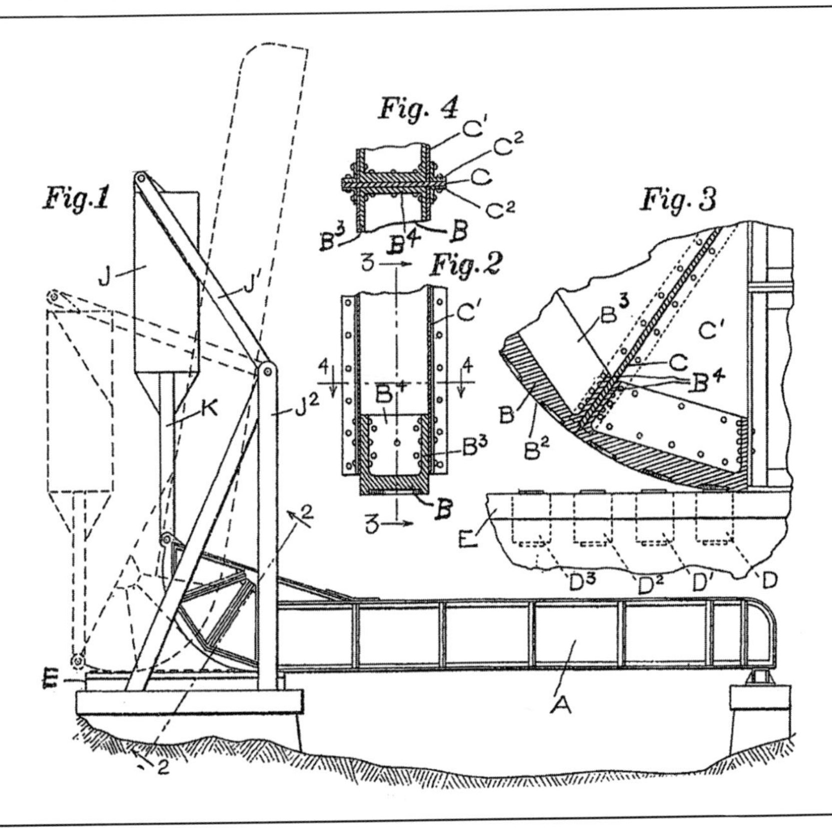

Strauss did not define the geometric requirements of the linkage in the patent, but it became the parallelogram linkage that he included in designs for most, if not all, of the Strauss bascule bridges he built (Hool and Kinne, 1943).

2.3.1 Vertical overhead counterweight (Strauss type 1)

The Strauss firm designed many single-leaf bascule bridges with vertical overhead counterweights along the line of Strauss's 1908 patent, except that they did not roll. Figure 2.10 shows the main elements of the type. The bridge balance is maintained during rotation if the lines through hinges B–C and D–E are always parallel, similarly for lines B–D and C–E. Many Strauss vertical overhead counterweight bridges were built, in the USA and in other countries. Early examples were in Copenhagen, Denmark (1908); Camden, New Jersey (1908); Stockholm, Sweden (1922); and the Fourth Street Bridge, San Francisco, California (1918). The Copenhagen and Camden bridges had architectural treatment (Strauss Bascule Bridge Company, 1920). The Fourth Street Bridge is still operational (in 2014), having just been refurbished. The Strauss vertical overhead bridge type was considered suitable for narrow channels in remote areas, and a few of these bridges were built across smaller rivers in New Jersey. An example is the bridge that spans the Mullica River at Green Bank. First built in 1926, with a roadway 20 ft (6.1 m) wide on a span of 40.67 ft (12.4 m), it was reconstructed with a wider roadway in 2003, at a slightly higher elevation as depicted in Figure 2.11.

Figure 2.10 Single-leaf Strauss vertical overhead counterweight bascule bridge

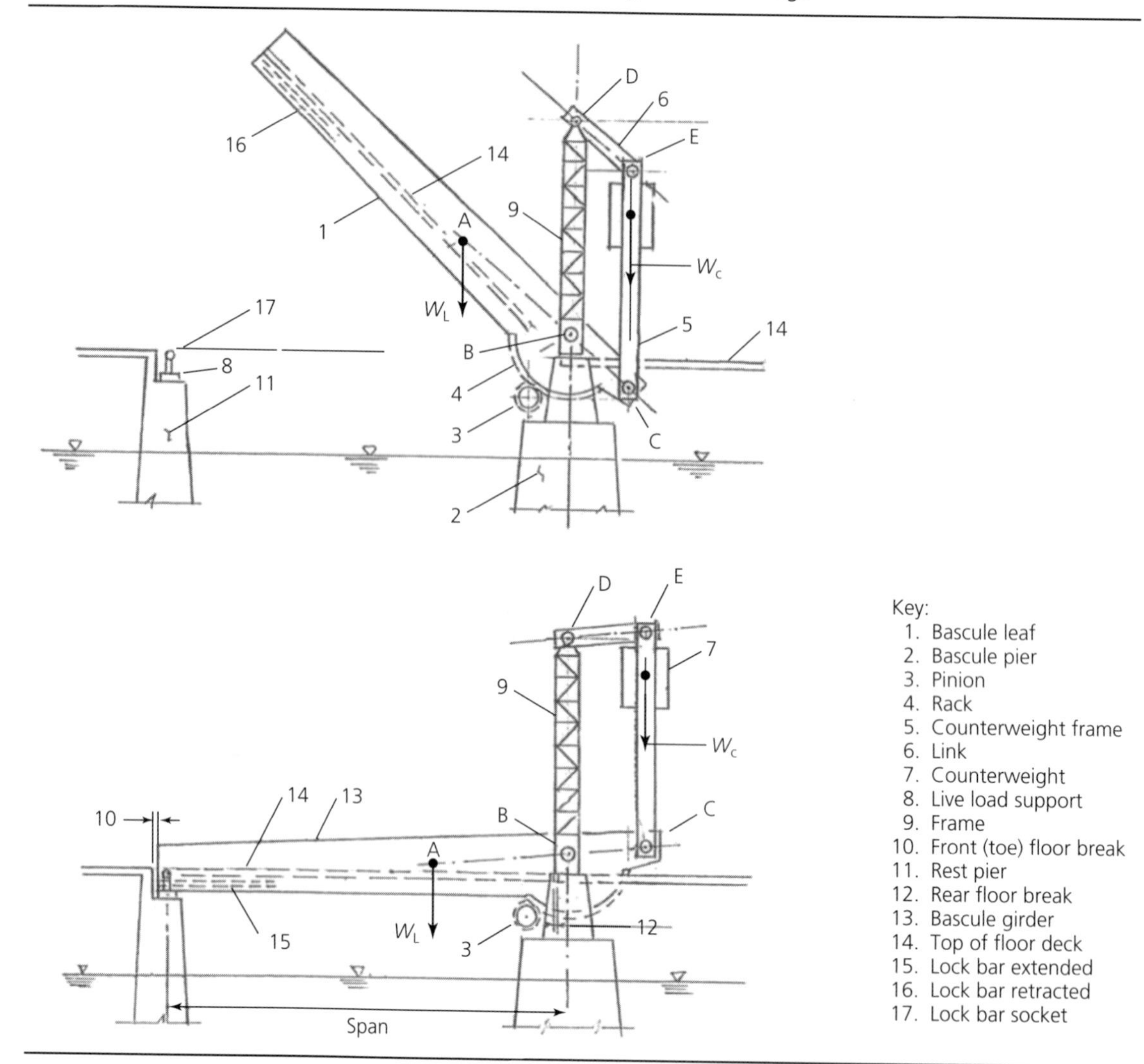

2.3.2 Underdeck counterweight (Strauss type 2)

The second in the series of Strauss patent bascules with the parallelogram linkage is the underdeck counterweight type shown in Figure 2.12. Balance is maintained during rotation of the leaf if line B–C remains parallel to D–E and plane B–D is always parallel to C–E. The configuration shown has the floor break in front (channel-ward) of the trunnions (the usual situation) and hence longitudinal slots are needed in the fixed deck in order to permit the bascule girder to rotate. The initially attractive feature of the bridge type was the suspended counterweight, having the counterweight free of the bascule girder. The result was that the length of the bascule girder rearward of the main trunnions could be made shorter than for bascules with counterweights fixed to the girder, and hence the pit piers could be shorter, and cheaper. To be sure, heavier counterweights were required, but that extra cost was offset by the smaller foundation. This advantage, and the conditions leading to the dropping of some Strauss hanging counterweights during operation were discussed by Altebrando (2000).

Figure 2.11 Green Bank Road Bridge, Green Bank, NJ. (Photograph by Charles Birnstiel)

On 24 June 1989 a counterweight of the Tayco Street Bridge over the US Canal at Menaska, Wisconsin, fell off a bascule girder while the bridge was opening, causing the span to drop into the channel. The cause of the failure was determined as inadequate lubrication (Schultz, 1990). Investigation disclosed that a feature of the original designer's lubrication channels in the machinery had been owner-modified years after the bridge was put into service in a manner such that lubricant could not reach a rubbing surface in the bearing during routine maintenance. This had caused the counterweight hanger (22 in Figure 2.12) literally to walk along the counterweight trunnion (10 in Figure 2.12), until it fell off its end because of the lack of a physical stop at the pin end. Strauss underdeck bascule bridges are still in use and are safe, if the machinery is properly lubricated.

2.3.3 Heel trunnion (Strauss type 3)

The third type of Strauss articulated bascule, known as a heel trunnion, is shown in Figure 2.13. An advantage of this bascule type is that the self-weight of the span and counterweight are divided between two piers, in contrast to types 1 and 2 in which one pier supports nearly all the dead load. In the figure, A represents the centre of gravity of the movable leaf and a portion of the operating struts (which varies with the opening angle), and C is the centroid of the counterweight and the counterwieght trusses. The balance condition is that plane 'A' – 'B1' be parallel to plane 'B2' – 'C', which is maintained during operation by the parallelogram linkage 'B1' – 'B2' – 'E' – 'D'. The vertical component of the reaction at the counterweight truss trunnion 'B2' is always directed upward (compression). At the heel trunnion 'B1' the dead load reaction may be negative (downward) at some angles of opening. This condition must be recognised during construction and while making inspections when loosening or removing the bearing cap bolts of the heel trunnion (Hovey, 1926).

Figure 2.12 Single-leaf Strauss underdeck counterweight bascule bridge

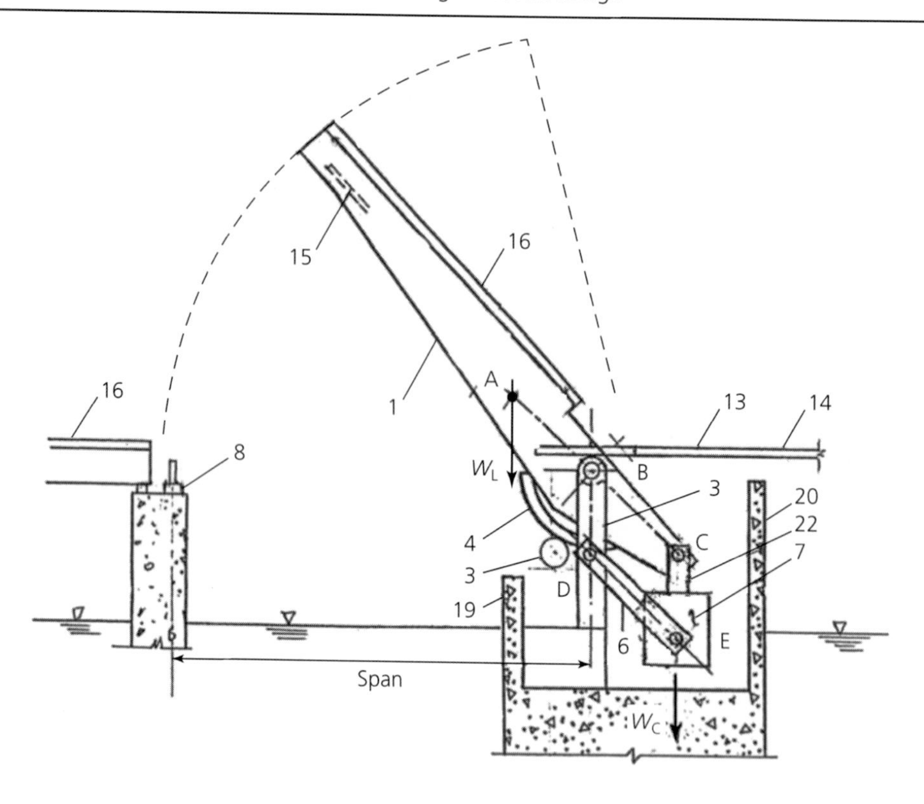

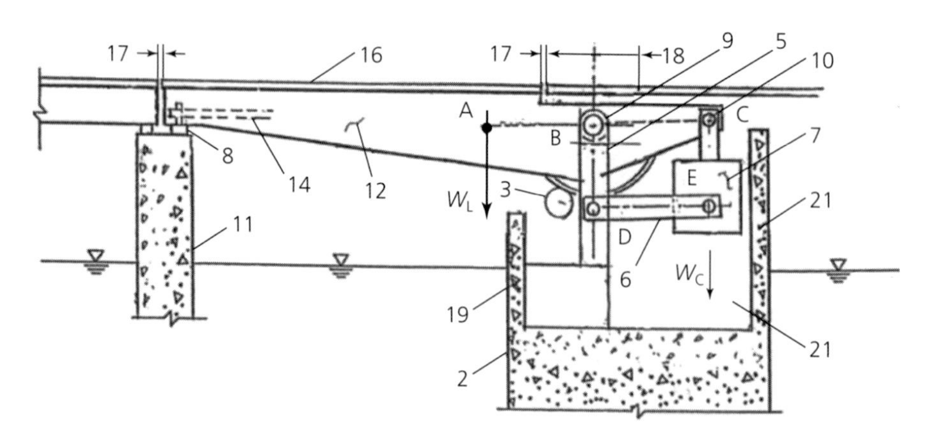

Key:
1. Bascule leaf
2. Bascule pier
3. Pinion
4. Rack
5. Trunnion tower
6. Counterweight link
7. Counterweight
8. Live load support
9. Trunnion
10. Counterweight trunnion
11. Rest pier
12. Bascule girder
13. Fixed deck
14. Lock bar extended
15. Lock bar retracted
16. Top of deck
17. Floor break
18. Slot in fixed deck for bascule girder clearance
19. Front wall
20. Back wall
21. Pit
22. Counterweight hanger

Figure 2.13 Single-leaf Strauss heel trunnion bascule bridge

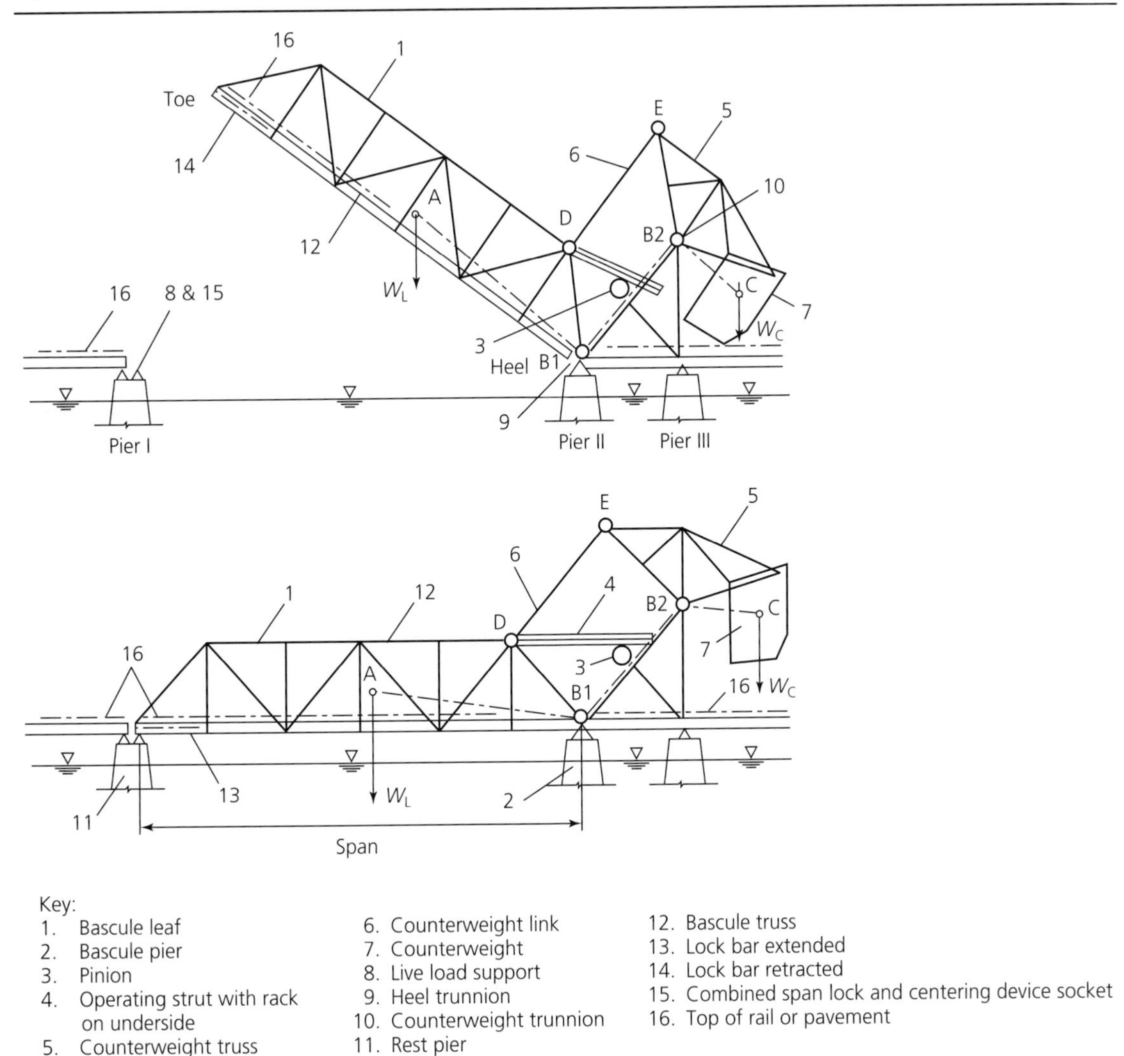

Heel trunnion bascules designed by the Strauss organisation, or others after expiration of the Strauss patents, were built by many railways in the USA because they favoured single-leaf bascules over double-leaf bascules for water channels up to 200 ft (61 m) wide. Many type 3 bridges were built in European and Scandinavian ports around the North and Baltic Seas.

A large Strauss heel trunnion bascule was erected in Antwerp Harbour. An unusual feature of the bridge is that it has three main trusses with the typical articulation but it has only two operating struts (de Joode and Bernard, 1989).

One of the longest Strauss heel trunnion bascule bridges is known as the St. Charles Air Line Bridge which was completed in 1919 across the Chicago River at 16th Street in Chicago, IL. It is a double track

Figure 2.14 South Front Street Bridge, Elizabeth, NJ – Strauss heel trunnion bascule. (Photograph by Charles Birnstiel)

railway bridge that was designed for Cooper's E60 loading. The original span was 260 ft (79.3 m). An unusual feature is the two concrete counterweights that flank the bridge. The main trunnions are 25 in. (635 mm) in diameter and 27 in. (686 mm) long. The counterweight trunnions are 46 in. (1168 mm) diameter and 44 in. (1118 mm) long. In 1930 the bridge was moved to its present location because the channel was relocated, and during that process the span was shortened to 220 ft (67.1 m). Provision for this eventual move was made in the original bridge design. Railway trains still cross the bridge, but because of a changed water traffic pattern it has not been operated as a bascule bridge for many years.

Figure 2.14 shows the South Front Street Bridge across the Elizabeth River at Elizabeth, NJ. A Strauss heel bascule type that was built for road and light rail service. The bridge is skewed in plan.

The Johnson Street Bridge in Victoria, British Columbia is shown in a partially open position in Figure 2.15. Two separate Strauss heel trunnion bridges on a common bascule pier are depicted: a road bridge in the foreground and a railway bridge behind it. The floor between the farthest two trusses supported a railway track. Since the photograph was taken, the farthest bridge has been removed.

2.4. Rolling bascules (Scherzer)

As noted in Chapter 1, rolling bascule bridges are an old form of movable bridge in Europe. In the USA the type was popularised by the Scherzer brothers. After the Van Buren Street and WSMERC rolling bridges in Chicago were completed, Albert Scherzer realised the value of his brother William's patent and founded the Scherzer Rolling Lift Bridge Company (SRLBC) in July 1897. He made many improvements to the rolling bascule design (Schultz, 1990). Many Scherzers were built in the USA, Europe, and Russia.

Figure 2.15 Johnson Street Bridge, Victoria, BC, Canada. (Courtesy of Dennis Robinson)

The first contract for SRLBC was design and supervision for a six-track rolling bridge across Fort Point Channel in Boston, Massachusetts. The superstructure steel and machinery were let to the Pennsylvania Steel Company of Steelton, PA and the contract for the substructure to a foundation contractor. Construction was completed in January, 1900. The three double-track bridges were on a 42 degree skew with the channel. The span length centre-to-centre of bearings was 114 ft (34.8 m) (see Figure 2.16). Each unit was operated individually. The bridge was made fixed in position in 1948 and most of the operating machinery was removed. After the bridges were removed, part of a Scherzer bascule truss and a track girder were erected as a monument in a park near South Station. Figure 2.17(a) depicts the end of the bascule girder sitting in its nearly closed position on the track girder. Lugs on the track plate and corresponding socket holes in the tread plate are shown in Figure 2.17(b).

The motion of rolling bascule leaves is due to rotation about a transverse horizontal axis which simultaneously translates in the longitudinal direction. The movement is akin to that of a rocking chair. Three common types of rolling bascules are the deck type, the half-through (pony) plate girder or truss and the through truss. They are illustrated in Figure 2.18.

Rolling bascule bridges are characterised by cylindrically curved parts of the bascule girders (or trusses) at the ends over bascule piers. These parts of the girders or trusses of some early Scherzer bridges were assembled from steel castings and the girders were called 'segmental girders', a term still in use. Each segmental girder may be viewed as a segment of a wheel. However, instead of rotating about an axle, these 'wheels' are part of the bascule leaf. As the 'wheels' roll along the tracks, the bascule leaf rotates. Slippage between the segmental girder treads and the tracks on which they roll is restricted by friction and lugs or teeth on the treads that mechanically engage sockets in the track, or vice versa.

Figure 2.16 Tower Bridges, Fort Point Channel, Boston, MA. (Encore Editions)

The rolling lift bridges shown in Figure 2.18 are all operated by a pinion located at the centre of roll which engages a fixed horizontal rack. Rotation of the pinion moves the leaf forward or backward. However, some rolling lift bascules built for railways have pinions located above the centre of roll because of the vertical clearance required above the running rails to accommodate trains and overhead catenary wires. An example is the Pelham Bay Amtrak Bridge across Eastchester Creek in the Bronx, New York. The racks of those bridges are not straight but are S-shaped in the elevation view.

Early Scherzer bridges such as Van Buren and Fort Point were operated differently, by movable rack struts which pushed or pulled the leaves to open the bridge. Similar action, but with hydraulic cylinders, is also used to operate rolling lift bridges. On some bridges they are located outboard of the segmental girder in a horizontal position with the piston rod end connected at the centre of roll. In such cases the cylinder usually needs to be supported at the head end to minimise bending of the piston rod. However, the cylinder need neither be horizontal nor connected at the centre of roll.

2.4.1 Hunters Point Avenue Bridge

Hunters Point Avenue Bridge crossing Dutch Kills in Long Island City, Queens, New York, is a single-leaf rolling bascule of the type shown in Figure 2.18(a). This roadway bridge spans 69.7 ft (21.3 m) with a 26 ft (11 m) wide roadway and two sidewalks, each 7 ft (2.1 m) wide. The present bridge was built in 1983 replacing a double-leaf Scherzer bridge built in 1910. The span stabilising machinery differs from Figure 2.18(a) in that the toe lock machinery is mounted on the rest pier and there are tailocks mounted on the rear of the front wall that engage sockets mounted on the bascule girder tails when the leaf is in the fully open position.

Figure 2.17 Tower Bridges Monument (Photograph by Robert Silman): (a) segmental girder and track girder; (b) Tower Bridges Monument, lugs and sockets

(a)

(b)

2.4.2 Halsskov Bridge

The movable span of Halsskov Bridge, located at Korsor in the county of West Zealand, Denmark, depicted in Figure 2.19, is a half-through rolling bascule bridge. The bascule span is 30 m and the overall width is 21 m. The orthotropic plate deck supports a 9 m-wide roadway with a railway track

Figure 2.18 Single-leaf rolling bascule bridges – Scherzer types

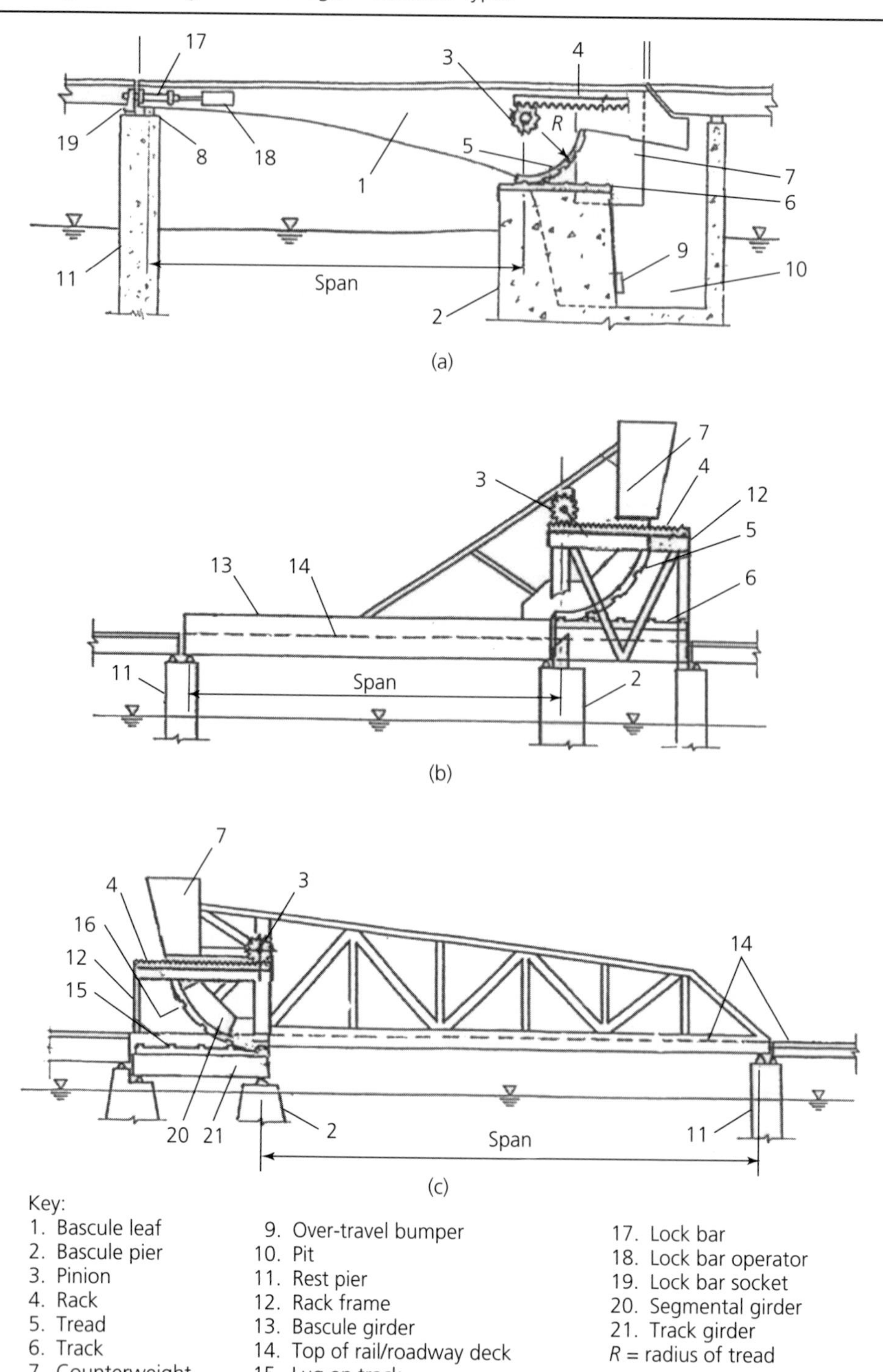

Key:

1. Bascule leaf
2. Bascule pier
3. Pinion
4. Rack
5. Tread
6. Track
7. Counterweight
8. Live load support
9. Over-travel bumper
10. Pit
11. Rest pier
12. Rack frame
13. Bascule girder
14. Top of rail/roadway deck
15. Lug on track
16. Socket to receive track lug
17. Lock bar
18. Lock bar operator
19. Lock bar socket
20. Segmental girder
21. Track girder

R = radius of tread

Figure 2.19 Halsskov Bridge, Korsor, Denmark – a half-through rolling bascule bridge. (Ryall *et al.*, 2000)

and pavements. The truss members are welded steel boxes. There is no horizontal bracing in the plane of the top chords of the pony trusses and the bridge has a clean, open appearance. COWI Consulting were the designers of the bridge, which was completed in 1985.

The Halsskov leaf is operated by hydraulic cylinders, one on each side of the bridge, alongside and below the track. This is an unusual, counterintuitive, configuration which was driven by architectural considerations. The pistons act on brackets that project downward alongside the trusses near the front teeth. To open the bridge piston rods push channel-ward, opposite to the direction of roll. The radius of roll is 6.40 m.

2.4.3 Pine Island Bridge

Pine Island Bridge shown in Figure 2.20 in the open position for water traffic is a half-through (pony) rolling bascule bridge of the type shown in Figure 2.18(b). It spanned the Intracoastal Waterway just west of Myrtle Beach, South Carolina, USA. It was designed and built *c.* 1936 by the American Bridge Company (ABCo) to support a single track railway and a 22 ft (6.7 m)-wide roadway. After the Scherzer patents expired, ABCo built many bridges of this type, mostly for railways. The photo shows the bridge in the normally open position prior to the 1995–1997 rehabilitation. The two bascule girders span 108 ft (32.9 m) when the bridge is closed.

Figure 2.20 Pine Island Bridge, South Carolina – half-through rolling bascule bridge. (Photograph by Charles Birnstiel)

During operation the bridge rolls on two box girders that span between the two concrete piers visible in Figure 2.20, one track girder on each side of the bridge. The roll radius is 20.25 ft (6.18 m). A segmental steel plate 28 in. (711 mm) wide and $9\frac{1}{2}$ in. (241 mm) thick (the track plate) is fastened to the top of each track girder. A single line of conical teeth (pintles) having a minimum diameter of $7\frac{3}{16}$ in. (178 mm) projects upward from the track plate. The teeth are spaced $23\frac{1}{2}$ in. (597 mm) on centres. As the leaf rotates, the teeth mate with socket holes bored into the bottom of a curved segmental steel tread plate fastened to the bottom of the rocker girder.

The bridge span drive is mechanical open gearing mounted on the movable span which is now powered by an electrical motor and a primary speed reducer that is located in the operator's house at roadway level. Articulated shafting transfers the torque from the house machinery to the secondary reduction gearing located on the movable span. The total self-weight of the leaf is about 2500 kips (1134 tonnes).

Figure 2.24 Madison Avenue Bridge over the Harlem River, New York City – rim bearing swing bridge: (a) the pivot pier from the fender; (b) close-up view of the tread plate. (Photographs by Charles Birnstiel)

(a)

(b)

2.5.4 Centre press bearing

British bridges are extant that were built by Armstrong, or for which the Armstrong system for the span drive and the stabilising machinery was installed. Examples are the Newcastle upon Tyne Swing Bridge and the Barton Swing Aqueduct (Birnstiel, 2000). In the Armstrong system, a central press serves as the

Figure 2.25 Swing bridge over Mississippi at Fort Madison, IA. (Photograph by Charles Birnstiel)

pivot axis and also is used to relieve dead load on the rim bearing during turning. Hydraulic motors turn the swing span. Figure 2.26 depicts the pivot press and the rim bearing of Barton Aqueduct over the Manchester Ship Canal. Originally the hydraulic pressure exerted an upward force approximately equal to the weight of the water in the trough when the bridge was rotated. Hence, the tapered rollers had to be designed to support only the dead load of the structure while rolling. An example of a later swing bridge with a hydraulic pivot bearing is the Harbor Island Swing Bridge across the Duwanish River near Seattle, Washington, USA (Mahoney, 1985) and (Green, 1991).

2.5.5 Slewing bearing

Slewing bearing swing bridges are rim bearing but without the central pivot. Rolling element thrust bearings of large diameter, first developed for military applications, are used instead of tapered rollers. Some bridge designers use the slewing bearings to support both live and dead loads of the bridge, as they are used in construction machinery applications. Other bridge designers use them only to support the draw while it is rotating. The latter design concept requires means for transferring the weight of the draw to a separate support system before turning the draw. This concept has been denoted as the lower-turn slewing bearing. Operation of this type will be described later in connection with the Selby Swing Bridge.

The A63 Selby Bypass crosses the River Ouse in England on a cable-stayed bobtailed draw with overall dimensions of 95.2 m by 17.96 m in plan. The main span from the pivot pier to the north landing pier is 55 m and the back-span is 40 m. It comprises two welded girders with an orthotropic deck, assisted by stays. Figure 2.27(a) is a view of Selby Swing Bridge from the north bank of the River Ouse.

This bridge is of the lower-then-turn type with a rolling element thrust pivot bearing that is only loaded with self-weight of the draw while turning. When the draw is in the fully closed position, ready for

Figure 2.26 Barton Aqueduct over Manchester Ship Canal – swing bridge. (a) Rim bearing; (b) centre press. (Ryall *et al.*, 2000)

(a)

(b)

highway traffic, the slewing bearing is unloaded. Describing the means for accomplishing this would require too much space here (see Birnstiel (2008) for further explanation). The slewing motors at the pivot pier are shown in Figure 2.27(b).

Rolling element slewing bearings have been used for swing bridges in situations where they support both dead and live load thereby avoiding the need to lift or lower the span prior to turning it. An example is the Naestved Swing Bridge described in Chapter 3. See also the pedestrian bridges over Bellmouth Passage at Canary Wharf.

2.6. Vertical lift bridges

The movable span of a vertical lift bridge is raised in order to provide additional vertical clearance for the passage of vessels, but at least one bridge was built with a lift span that lowers into the channel. The raising lift span is usually balanced by counterweights connected to wire ropes that pass over sheaves located atop the rest piers of the lift span. The relative amount of imbalance varies: in USA practice it is

Figure 2.27 Selby Swing Bridge, Selby, North Yorkshire – slewing bearing type. (a) View from north bank of River Ouse; (b) drive motors. (Parke and Hewson, 2008)

(a)

(b)

between 0 and 5% of the lift span weight; in European practice the design imbalance is greater. Five to 8% of lift span weight was mentioned by Hawranek (1936) although he suggested that a smaller percentage might be acceptable. A recently constructed French lift bridge has imbalance that is approximately 2% of the lift span weight – a decided departure from past French practice.

The counterweight sheaves are normally situated externally atop the towers, but they may be located internally. The primary purpose of the counterweights is to minimise the energy required to move the span but there is also a safety benefit. Less braking capacity is needed to hold a slightly imbalanced lift span stationary at an intermediate lift position in the event of a problem in the span drive. This type of bridge is known as a balanced vertical lift. Vertical lift bridges are categorised by the arrangement of the drive machinery. Simplified diagrams of some types of balanced vertical lift bridges, namely the span drive, the connected tower drive, the tower drive, and the tower base drive are shown in Figures 2.28 and 2.31. For heavy lift bridges the weight of the counterweight ropes may be so great that they themselves are counterweighted by an auxiliary counterweight system (see Chapter 10).

Since Waddell's design and patent of the first practical vertical lift bridge (Waddell, 1916), many such bridges have been erected. Due to Waddell's patents and the relationship between his and successor firms, and firms that were founded by a former partner of Waddell, most large vertical lift bridges built in the USA prior to 1950 were designed by two consulting engineering firms (Nyman, 2002).

2.6.1 Span drive vertical lift

Figure 2.28(a) depicts a span drive vertical lift bridge powered by wire rope hoist machinery located on the span, usually at mid-span above the roadway or railway tracks. The wire rope system hauls the lift span up or down. Because the four rope drums (one for each corner of the lift span) are usually geared to a common drive shaft, the span cannot skew appreciably longitudinally unless there is a problem with a haul rope. This type of span drive vertical lift bridge was perfected by the Waddell firm and many were built by railways. Auxiliary counterweights and many stabilising components are not shown in the figure. The Path Hackensack River Bridge shown in Figure 2.29 is a double track vertical lift bridge of the span drive type on which the Port Authority Trans-Hudson Corporation (PATH) commuter railway line crosses the Hackensack River at Kearny, New Jersey. The lift span is 322.5 ft (98.4 m) long between lift truss bearings and the maximum height of lift is 90 ft (27.4 m). A 'classic Waddell', it was designed by Waddell and Hardesty and placed into service in November 1930.

Wire rope span drive vertical lift bridges have also been built with the primary drive at mid-span, transmitting power by way of long line shafts to secondary speed reductions and hoist drums located at the ends of the lift span. An example of such a bridge is the double-track Burlington Northern Santa Fe Railroad Bridge at Portland, Oregon. The lift span has a length of 516 ft (157 m) and the normal lift is 146.25 ft (45 m), giving a vertical clearance of 200.7 ft (61 m) above low water

Span drive vertical lifts that are raised by a rack-and-pinion drive at the towers have been built for shorter spans, for example, those over the Illinois River, USA.

2.6.2 Connected tower drive vertical lift

A diagram of a connected tower drive vertical lift bridge is shown in Figure 2.28(b). The lift span is nominally balanced by counterweights. However, because this type of lift bridge is most suitable for short spans of low to moderate lift, the counterweight ropes are usually not balanced by auxiliary counterweight systems. The span drive machinery is mounted on the structure that connects the towers. Primary machinery is located in a machine room at mid-span, from which line shafts extend to secondary bevel speed reducers at the towers, midway between the sheaves. From the secondary reducers, shafts extend to pinions that engage the curved racks fastened to the counterweight rope sheaves. The force necessary to move or hold the lift span is transmitted between the sheaves and the counterweight ropes by friction.

Figure 2.30 depicts the New Jersey Route 13 Bridge which supports Bridge Avenue over the Intracoastal Waterway just northwest of Bay Head, New Jersey. It was completed in 1970. The span drive machinery is mounted on a platform spanning between the towers and drives all four counterweight

Figure 2.28 Span drive and connected tower drive vertical lift bridges

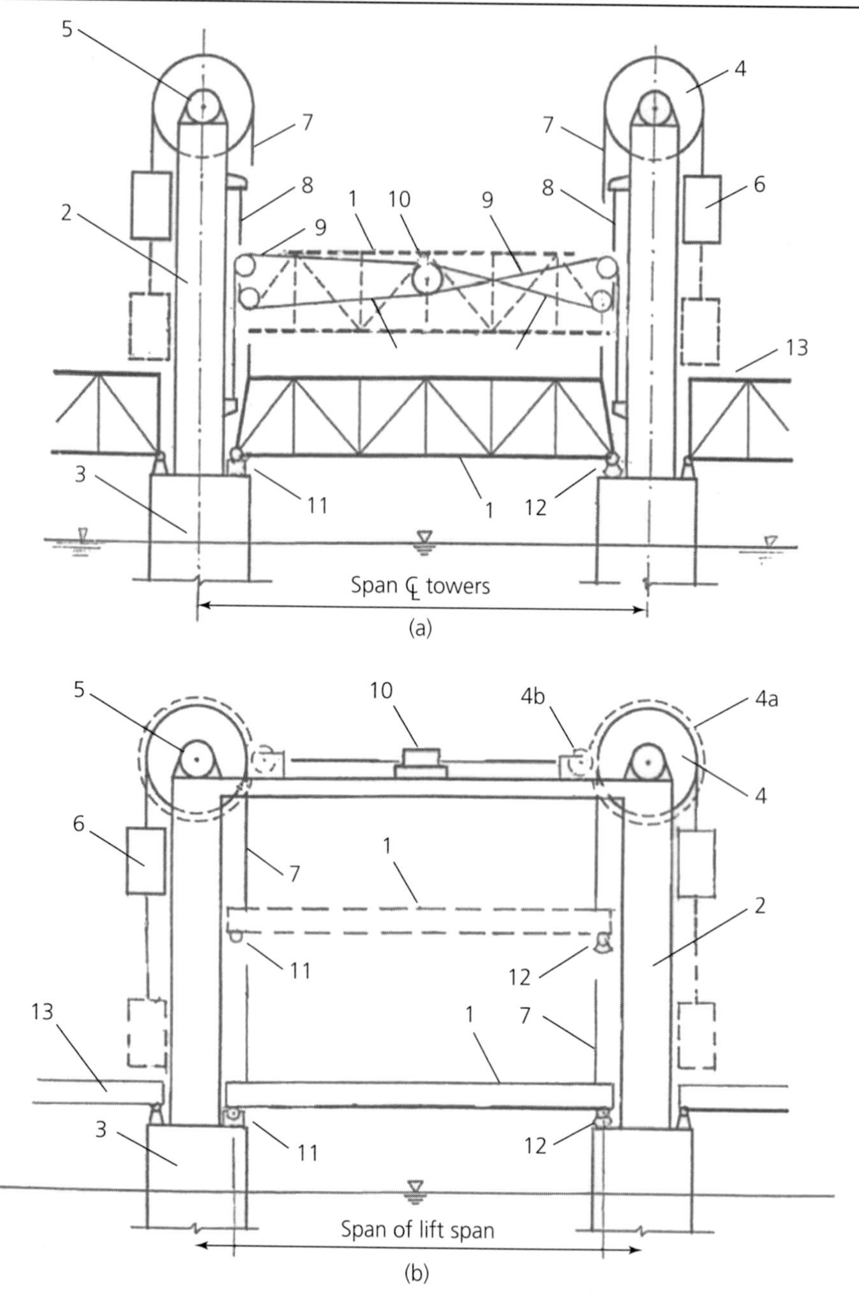

Key:

1. Lift span
2. Tower
3. Tower base
4. Counterweight sheave
4a. Rack
4b. Pinion
5. Trunnion bearing
6. Counterweight
7. Counterweight ropes
8. Uphaul ropes
9. Downhaul ropes
10. Operating machinery
11. Split hinge bearing 'fixed'
12. Rocker bearing
13. Flanking span

Figure 2.29 Path Hackensack River Bridge, Kearny, NJ – span drive vertical lift bridge. (Photograph by Charles Birnstiel)

Figure 2.30 New Jersey Route 13 Bridge over Point Pleasant Canal – connected tower vertical lift bridge. (Photograph by Charles Birnstiel)

sheaves simultaneously to raise or lower the lift span. The longitudinal distance centre to centre of towers is 92 ft (28.1 m) and the vertical distance the span can be moved is 35 ft (10.7m). The overall width is 62 ft (18.9 m), supporting four lanes of vehicular traffic and two sidewalks.

2.6.3 Tower drive vertical lift

The essentials of tower and tower base drive vertical lift bridges are shown in Figure 2.31. In the tower drive vertical lift bridge depicted in Figure 2.31(a), span drive machinery is mounted atop each tower and it rotates the counterweight sheaves. The forces necessary to raise the span are transmitted to the counterweight ropes by friction. The action is similar to that of a traction drive passenger elevator in a building. Both ends of the lift span should raise and lower at the same rate so that the lift span remains horizontal and does not wedge itself between the towers during motion. Various electrical and electronic means of controlling the drives in the two towers so that longitudinal skew is kept within permissible limits now exist. The synchrous tie system of electric motors was first installed on an American tower drive lift bridge, the Katy Bridge across the Missouri River at Boonville, Missouri, in 1931. It was designed to maintain the lift span reasonably level in the longitudinal direction during its vertical motion (Anon., 1932). With a lift span of 408.33 ft (124.46 m) it was the longest tower drive then constructed (see Figure 2.32).

At present, the longest vertical lift span, 558 ft (170 m), is a tower drive vertical lift that supports a single-track railway across Arthur Kill between Staten Island, New York and Elizabeth, New Jersey (Hedefine and Kuesel, 1959). The longest-span tower drive vertical lift highway bridge is the Marine Parkway Bridge over Rockaway Inlet, in the Borough of Queens, New York, with a lift span of 540 ft (165 m) (Birnstiel, 2000). It was erected in 1937.

Figure 2.33 depicts Broadway Bridge in New York City, viewed from the east. It is a very heavy tower drive vertical lift bridge with a 302 ft (92.7 m) long lift span between truss bearings. It supports two decks. The lower deck comprises two 34 ft (10.4 m)-wide roadways (one between each of the two outer trusses and the central truss) for vehicular traffic and two 8 ft wide sidewalks outboard of the exterior trusses. The upper deck has three railway tracks which are used by the IRT Division of the Metropolitan Transit Authority. The lift height is 111 ft (33.9 m). The bridge was opened to traffic in 1962.

The structural steel lift span of Broadway Bridge is suspended at each corner by two sets of wire ropes. Each set contains 12 wire ropes which are draped over a main counterweight sheave located atop the corresponding corner of a tower. There are a total of 96 ropes. The other ends of the wire ropes are connected to span counterweights. There is one span counterweight at each of the two towers. Raising and lowering the movable span is accomplished by rotating the main counterweight sheaves using a tower drive system. As the lift span is raised, the two counterweights descend. Because the counterweight ropes are so heavy, two sets of double auxiliary counterweight ropes were attached to each end of the lift span. Each set is run over an auxiliary counterweight sheave mounted on a tower and connected to an auxiliary counterweight. The lift span is raised and lowered by two identical tower drives, one located in a machinery room atop each of the two towers.

The New Jersey Route 7 Bridge shown in Figure 2.34 supports Belleville Turnpike across the Passaic River at Bellville in Essex County. This lift bridge may be considered a modern lift bridge of moderate span (Capers *et al.*, 2005). The longitudinal lift members are steel box girders 38.1 m long between bearing centres. The overall width of the lift span is 22.1 m and it weighs about 500 tonnes. Design lift height is 12.8 m. Because the weight of the counterweight wire ropes is small compared to the weight of the span, no auxiliary counterweights were installed.

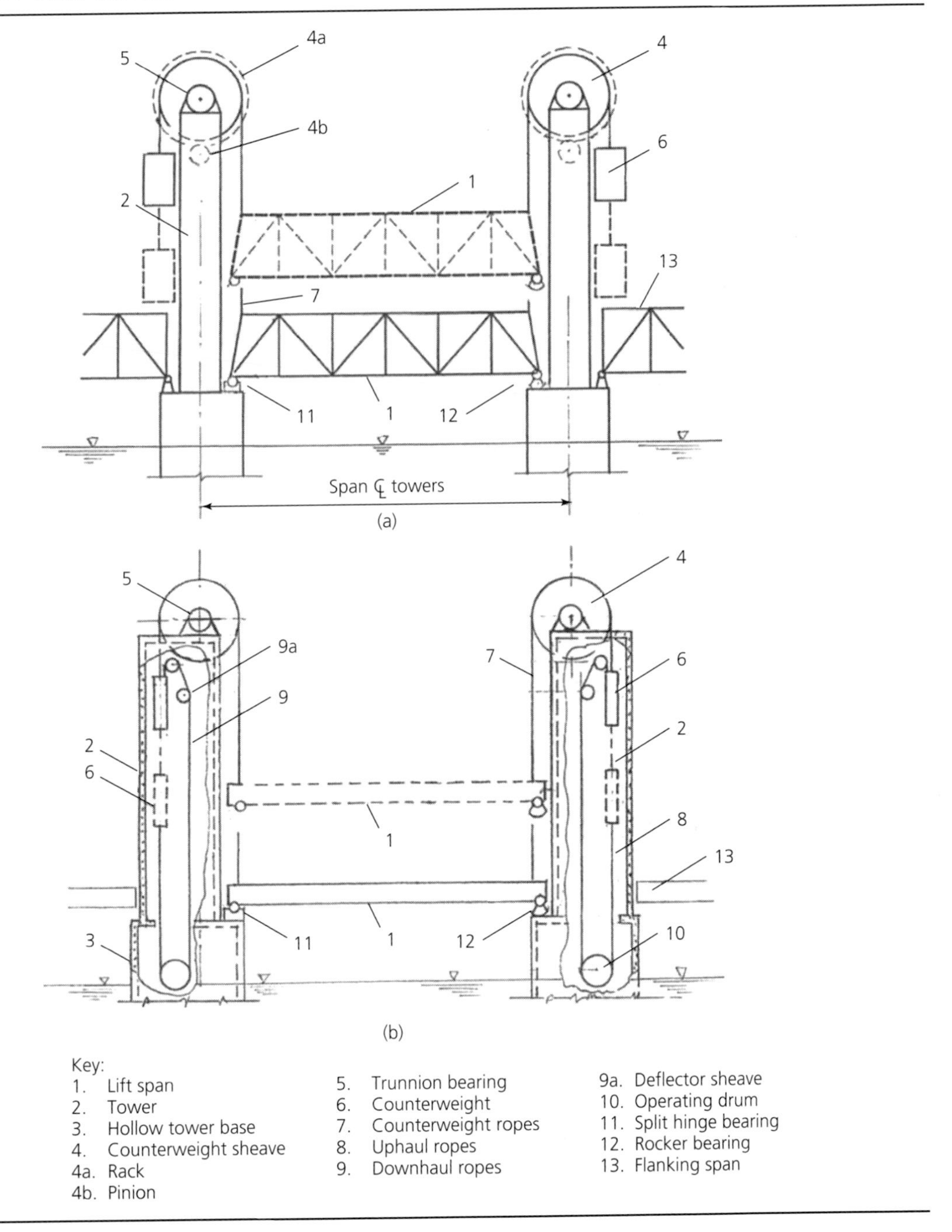

Figure 2.31 Tower drive and tower base drive vertical lift bridges

2.6.4 Tower base drive vertical lift

Figure 2.31(b) shows the arrangement of span drives of tower base drive vertical lift bridges. The type seems to have originated in Europe when Werkspoor applied a patented system in constructing the small De Ruijterkade lift bridge in Amsterdam (Hawranek, 1936).

Figure 2.32 Katy Bridge over the Missouri River at Boonville, MO – tower drive vertical lift. (Photograph by Charles Birnstiel)

Figure 2.33 Broadway Bridge over the Harlem River, New York City – tower drive vertical lift. (Photograph by Charles Birnstiel)

Figure 2.34 New Jersey Route 7 Bridge over Passaic River, Belleville, NJ – tower drive vertical lift. (Courtesy of Hardesty and Hanover)

The lift span deck was approximately 35 m long and 12 m wide and it was suspended at each corner by counterweight ropes draped over sheaves mounted on the tower tops. There were four counterweights but only two hoist drums. The span drive machinery, including the hoist drums, was located in a machinery room below the lift span on one pier (the 'powered' side), with the shafting running in the transverse direction. A cable equalising system similar to that installed on the New York State Barge Canal table lift bridges maintained the 'levelness' of the lift span during operation. To raise the span the two counterweights on one end of the lift span were pulled down by haul ropes attached to the bottoms of the two counterweights on the 'powered' side of the bridge. To lower the span, the haul cables were relaxed and the lift span lowered by the span-heavy gravity imbalance of the system.

Figure 2.35 depicts the Centenary Bridge in Manchester: a tower base drive vertical lift bridge with four towers. There are two span drives below the lift span, one on each side of the channel. The levelness of the lift span is maintained within a prescribed tolerance by electrical controls.

The Jacques Chaban-Delmas lift bridge at Bordeaux, France, is depicted in Figure 2.36. The lift span is 116.7 m long and 43 m wide and weighs 2889 tonnes. It is hung between four towers. There are span drives below the lift span level inside both tower piers. The haul ropes pull down on the counterweights to lift the span, as much as 50 m. In order to reduce the cost of machinery, the span-heavy imbalance (as a percentage of lift span weight) was less than the French norm, being reduced to 2%. However, the span drive has ropes to haul down the spans in the event that wind produces an uplift force on the span greater than self-weight. The rational for using a two-way hoist system and a lower imbalance despite a code requirement that the imbalance be sufficiently heavy to maintain a downward force through all conditions including environmental loads is described in (Cardin and Griesing, 2010).

Figure 2.35 Centenary Bridge, Manchester – tower base drive vertical lift bridge. (Parke and Hewson, 2008)

Figure 2.36 Jacques Chaban-Delmas lift bridge, Bordeaux, France – tower base drive vertical lift bridge. (Courtesy of Hardesty and Hanover)

or girder simple trunnion bascule design with a fixed centre of rotation for both single and double leaves. The story of the development of the Chicago-type bascule has been related often and will not be recounted here (see Wengenroth and Mix, 1975; Phillips, 2008; Holth, 2012; McBriarly, 2013). A description of a first-generation Chicago-type bascule appears in Anon. (1905). In the double-leaf classification of Figure 3.2, the first-generation Chicago-type bascule corresponds to Figure 3.2(a) and the Tower Bridge to Figure 3.2(c).

3.2.1 Motions of single-leaf trunnion and rolling bascule bridges

Figure 3.3 shows a single-leaf trunnion bascule and a rolling bascule in the closed and in the open positions with the leaves having been rotated through an angle, θ, from the horizontal. The trajectories of point A on the two spans differ because the trunnion bascule rotates about the fixed point O while the rolling bascule rotates the same amount while travelling rearward the distance, t. The recession, Δh_{T}, of the trunnion bascule is given by

$$\Delta h_{\mathrm{T}} = s \cdot \text{versine } \theta \tag{3.1}$$

in which

s = horizontal distance between O and A
θ = opening angle of bascule measured from the horizontal in degrees

During the rotation, θ, the centre of rotation of the rolling bascule moves rearward from O to C an amount

$$t = \frac{2\pi R\theta}{360} \tag{3.2}$$

The recession of A for the rolling bascule leaf is then

$$\Delta h_{\mathrm{R}} = s \cdot \text{versine } \theta + \frac{2\pi R\theta}{360} \tag{3.3}$$

In Equations 3.1 and 3.3 a positive number indicates a shoreward recession. A point on the girder of a trunnion bascule below a horizontal line through O, for example Point B, has negative recession at the start of opening. The increased recession of the rolling bascule toe over the fixed trunnion bascule toe was a selling point for Scherzer at channels with heavy sailing ship traffic. The motion, t, makes possible the ‘automatic’ mid-span shear lock for rolling lift bridges described later.

3.2.2 Supports for double-leaf trunnion bascule bridges

Some of the support arrangements for double-leaf trunnion bascules are shown in Figure 3.2. As shown in that figure the trunnions of these bridges are fixed to the girder webs and rotate in straddle bearings. However, the figure applies to other forms of trunnion bearings as shown in Figure 3.4. The straddle bearing arrangement is shown in Figure 3.4(a). The span drives are not shown because various drive types may be used to power these bridges.

The common arrangements of trunnion bascule bridge bearings shown in Figure 3.4 are used for single- and double-leaf bascules. The straddle bearing arrangement (Figure 3.4(a)) in which the trunnions are securely fixed in the web of the bascule girders is an old arrangement that is still used for heavily loaded bascule girders. Straddle bearings were also used for bascule leaves with multiple girders as late as Tower Bridge (London), in which the trunnion shafts are continuous over eight roller bearing pillow

Figure 3.2 Some support arrangements for double-leaf trunnion bascule bridges. (Adapted from WisDOT, 2002)

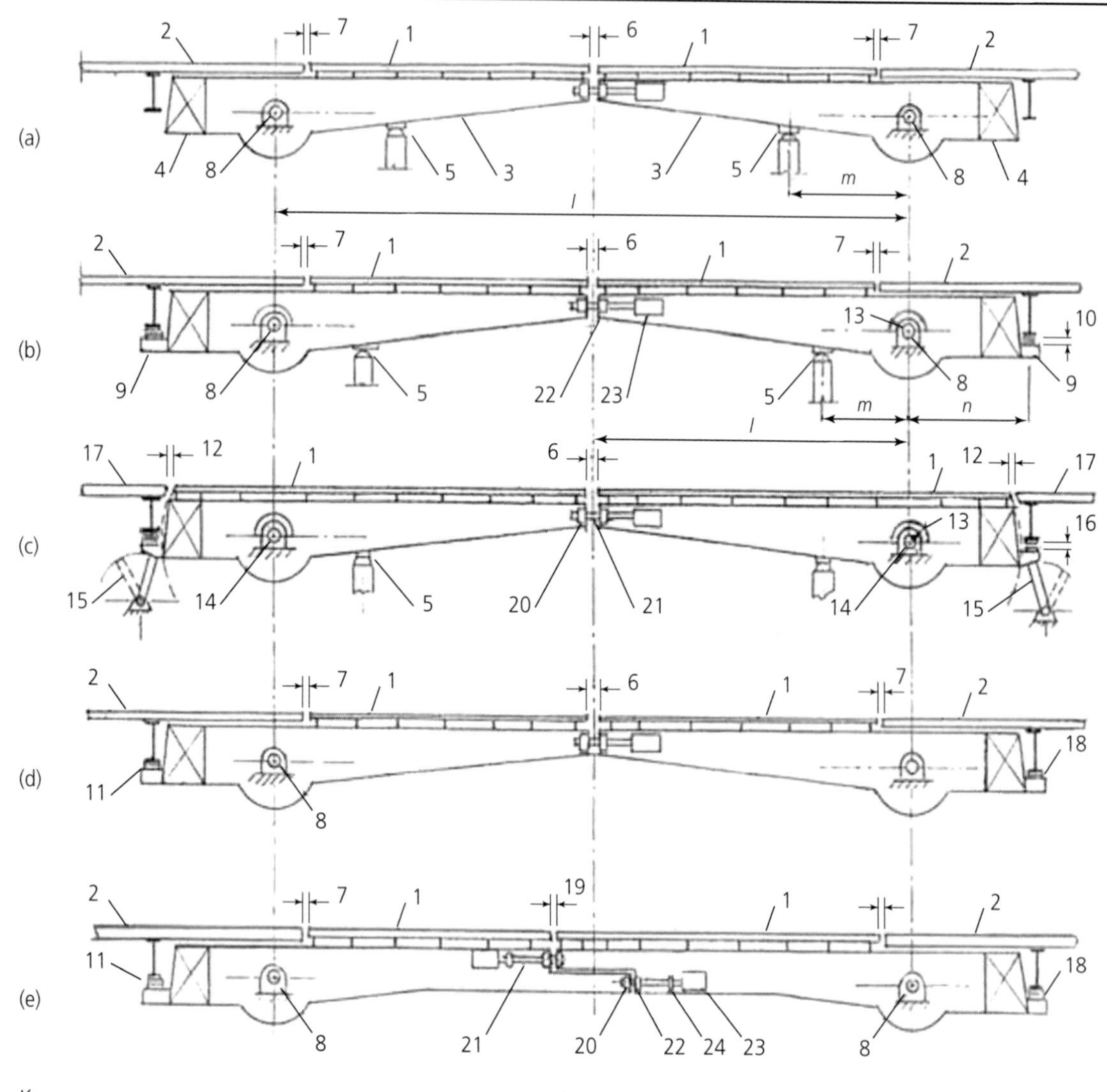

Key:

1. Deck supported on steel framing which transfers loads to order
2. Deck spanning transversely over bascule order
3. Bascule order
4. Counterweight
5. Live load support with shims
6. Central floor break
7. Rear floor break
8. Trunnion bearing
9. Bracket
10. Gap
11. Bearing plates and shims
12. Rear floor break
13. Gap
14. Bearing with loose cap
15. End lift (rotates in plane of girder)
16. Gap
17. Deck of flanking span
18. Hard contact
19. Off-centre floor break
20. Lock bar socket
21. Lock bar
22. Front lock bar guide
23. Lock bar operator
24. Rear lock bar guide

Figure 3.3 Comparative opening motions of fixed trunnion and rolling bascules

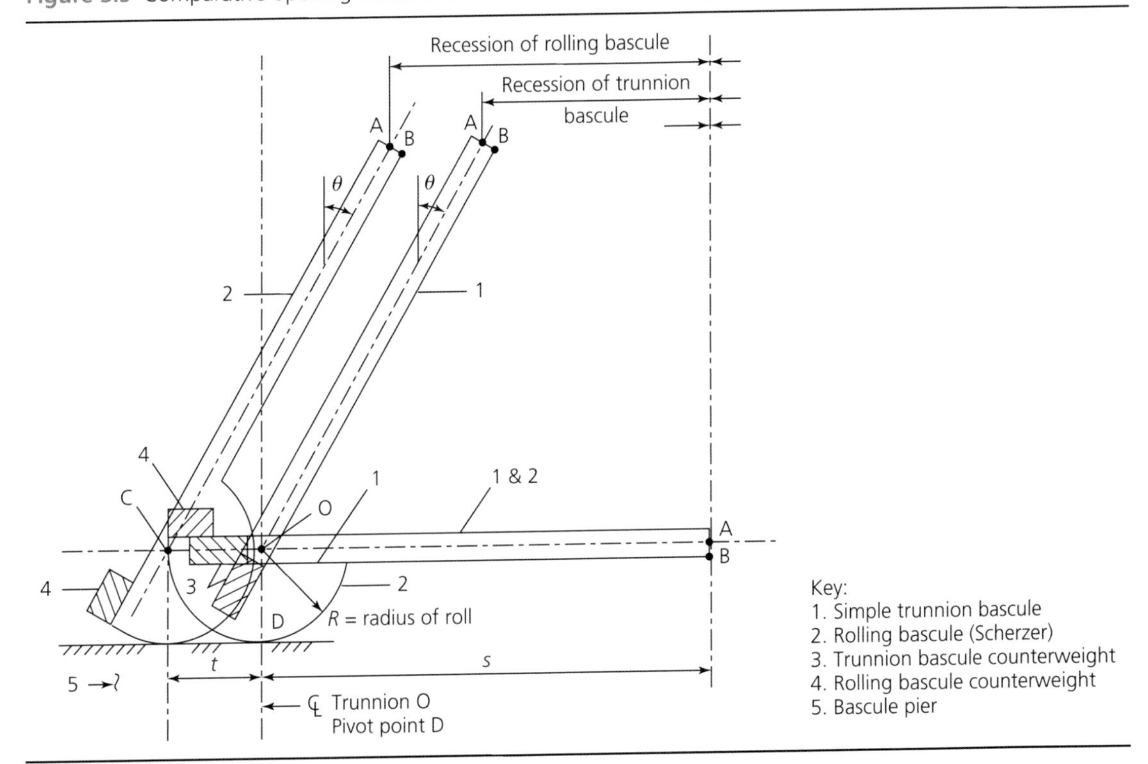

blocks to support the four bascule girders of each leaf (Birnstiel, 2000). Scott Street Bridge, completed in Hull in 1902, also has continuous trunnion shafts over multiple bearings.

Because aligning straddle bearings is often considered to be too difficult nowadays, short trunnions are sometimes cantilevered to bearings outboard of the exterior bascule girders, as shown in Figure 3.4(b). The advantage is that two fewer bearings are required at each leaf, but the trunnions still have to be aligned, which is usually accomplished with the aid of an adjustable feature at the interior end of the trunnion where it is fixed to the interior structural framing. This bearing arrangement is often referred to as the 'Hopkins' type, although the basic design was used long before the first Hopkins bridge was designed. However, because of the flexibility of the interior framing, wobble sometimes occurs resulting in wear of the trunnion sleeve bearings. In order to overcome this problem, spherical roller bearings or spherical plain bearings are now routinely specified, even for very large bascules. There are many small Hopkins bascules in the USA but the scheme has also been used for large bascules such as the Erasmus Bridge in Rotterdam, Holland (Reusink, 1996).

During the 1950s bascules of the trunnion girder type (Figure 3.4(c)) were built in the USA. The feature of this design is that the trunnions do not rotate; rather the bascule girders rotate about the trunnions. The plain bearings mounted within the bascule girder webs are inaccessible. Inadequate maintenance has led to excessive friction in the trunnion bearings on some bridges of this type, which is difficult to correct.

The tube trunnion shaft design (Figure 3.4(d)) was developed to overcome difficulties with the other types. The tube is mounted to the bascule girders and rotates with them in self-aligning trunnion

Figure 3.4 Trunnion bearing arrangements: (a) straddle bearings; (b) cantilivered trunnions; (c) trunnion girder; (d) tube trunnion shaft. (Parke and Hewson, 2008)

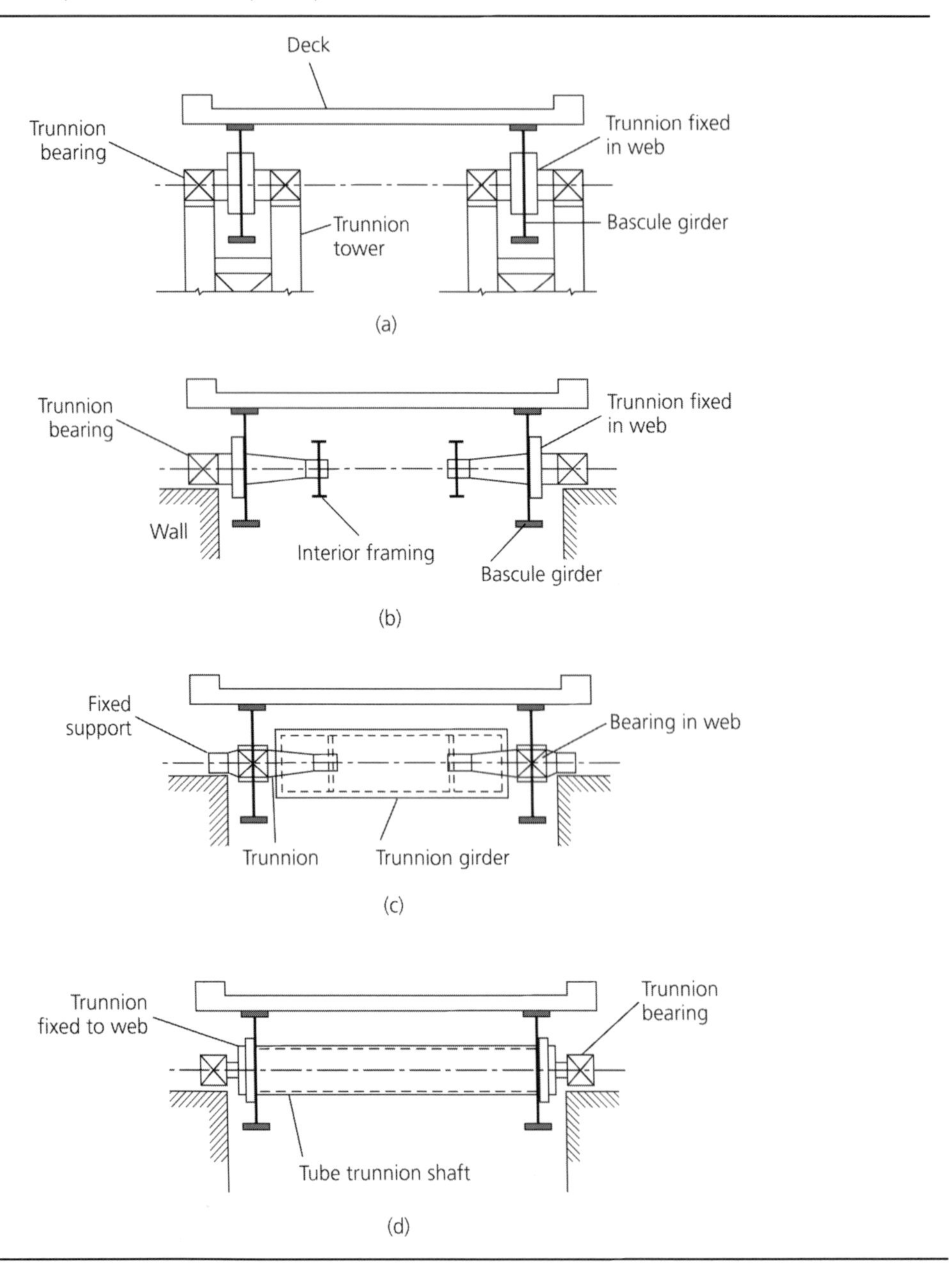

bearings. In essence, the trunnion shaft is full length as at Tower Bridge and Scott Street Bridge, but supported only by two outboard bearings.

Returning now to the double-leaf bascules of Figure 3.2, the most common type of double-leaf trunnion tower bascule is shown in Figure 3.2(a). The live load supports are shown forward (channelward)

Figure 3.16 Automatic mid-span lock for rolling bascules

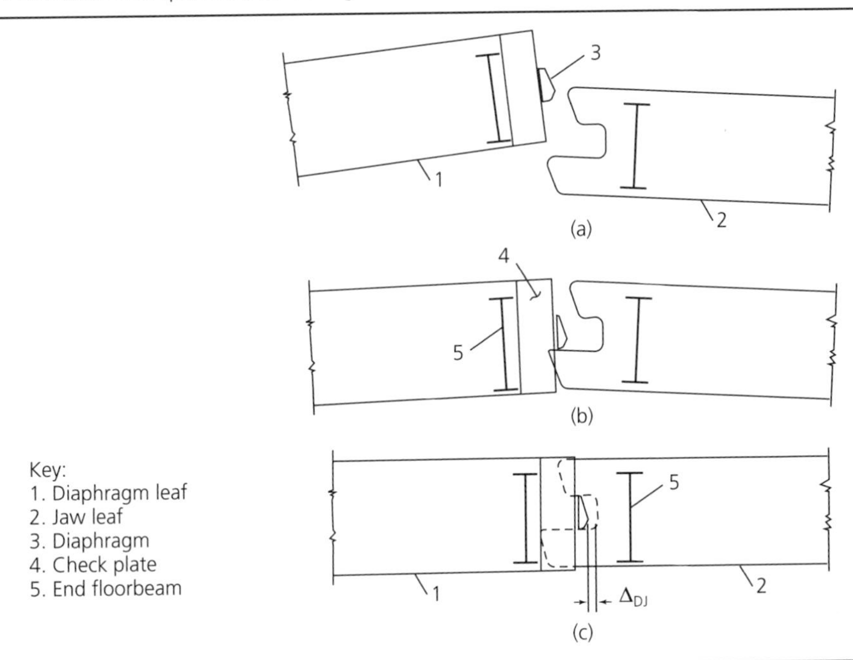

the span with time, due to foundation movement. The ability to lock the leaves without mechanical machinery is an advantage with regard to maintenance of double-leaf bascule bridges.

3.7. Tandem swing bridges

As discussed in Chapter 1, tandem swing bridges were an early form of movable bridge. Early bridges were of wood and later ones, such as those shown in Figures 1.5 and 1.7, were of cast iron. A large tandem iron lattice girder swing bridge was erected in 1861 spanning the Penfeld River in Brest, France (Price, 1879). The clear channel was 350 ft (107 m) – the widest passage crossed at that time. Each bobtailed swing span rested on a rim bearing turntable of 29.5 ft (9 m) dia. having 50 rollers of 19.63 in. (498 mm) dia. The pivot piers were 385 ft (117 m) apart. The bobtail counterweights were of masonry. The mid-span locks were simple round lock bars but the draw tails were fastened to the abutments 'by two pairs of jaws attached to the bridge, which, by means of levers, grasp projections of cast iron let into the abutment' (Price, 1879). The bridge was destroyed during World War II.

In what follows some significant tandem swing bridges comprised of symmetric and bobtailed draws having centre bearing, rim bearing and slewing roller bearing pivots will be briefly described.

3.7.1 George P Coleman Bridge

The original George P Coleman Bridge over the York River at Yorktown, Virginia, was a tandem highway swing bridge which was part of a structure 1143 m long. It was opened to traffic in 1952 (Quade, 1954). The overwater fetch across Chesapeake Bay to the east of the bridge is more than 40 km to a barrier beach. On the far side of the barrier beach is the Atlantic Ocean. Figure 3.17 is a photograph taken in 1989. Each draw had equal arms for a total draw length of 500 ft (152.4 m) and weighed 1200 tons (1100 tonnes). The horizontal clearance between pivot piers was 462 ft (140.9 m).

Figure 3.17 Original George P Coleman Bridge. (Photograph by Charles Birnstiel)

Each of the swing spans was centre-bearing with a circular convex bronze disk bearing that slid (rotationally) on a hardened and polished steel concave disk. There were four pairs of balance wheels rolling on a track 34.5 ft (10.5 m) in diameter. Shear locks and movable deck flaps were provided at the deck joints.

In 1994 construction started on a replacement for the original Coleman Bridge superstructure which was completed in 1996 (Abrahams, 1996; Green, 1996). The new bridge is 74 ft (22.6 m) wide compared to the original which was 26 ft (7.9 m) wide. The general configuration of the trusses and control house was not changed from the original. For the river piers, only the tops of the piers were widened. The added weight of the new superstructure had a small effect on the soil bearing pressure at the bottom of the caissons.

The new stabilising machinery is very similar to the original. However, the span drive machinery was changed from electro-mechanical to hydraulic operation. Four hydraulic motors with planetary reducers normally drive each draw, although they were sized to rotate the draw with fewer motors. A vegetable-based hydraulic fluid was selected so as to avoid environmental damage in the event of a spill or leaks.

3.7.2 Harbor Island Bridge across the Duwamish River at Seattle, Washington

Centre bearing swing bridges are built with hydraulic bearings as well as with mechanical bearings, as described in Chapters 1 and 2. Contemporarily, they are, essentially, hydraulic jacks which are used to raise the draw off its central support just prior to, or during rotation of the draw. An air-over-oil pressure intensifier and accumulator was used for operating a swing bridge in Hamburg Harbour, Germany, in about 1905. Modern installations rely on higher through-put pumps, thereby making accumulators unnecessary. An example is the Harbor Island Bridge, which supports Spokane Street

Figure 3.18 Spokane Street Pivot Pier. (Ryall *et al.*, 2000)

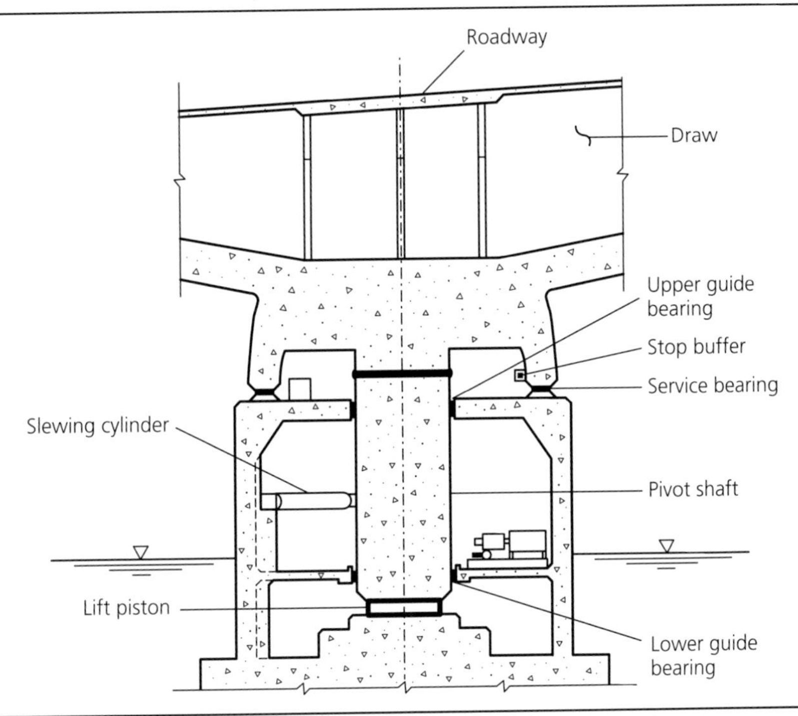

across the Duwamish River. This bridge comprises two bobtail centre bearing draws (Hamilton and Worm, 1990; Birnstiel, 2000). The draws are prestressed concrete box girders with the channel arms 240 ft (73 m)-long and tail arms of 177 ft (54 m). The distance between pivot pier axes is about 480 ft (146 m). Each draw weighs about 6800 tonnes. The hydraulic jack cylinder is 104 in. (2642 mm) in diameter and it raises the draw a distance of 1 in. (25 mm) before the draw is rotated. Each draw can be rotated 45° about the vertical axis by double-acting hydraulic slewing cylinders having a bore diameter of 22 in. (589 mm). The vertical shaft is an 11.8 ft (3.6 m) dia. steel shell filled with reinforced concrete (see Figure 3.18). Circumferential rings were welded to the shells and these were machined to form journals that bear on guide bearings which provide reactions against overturning forces. The bridge was put into operation in 1991.

In March, 2001, after only 10 years in service, one of the lift turn cylinders failed due to cracking of its barrel at one of the hydraulic supply ports (Johnson and Phillips, 2004). A thorough forensic investigation determined the four casual factors as

(*a*) underestimation of stress concentrations
(*b*) over-estimation of material properties
(*c*) high residual stresses due to overlay welding
(*d*) an undetected initial crack.

Some of these topics are discussed in general in Chapter 7. The original cylinders were replaced and the bridge is in operation.

3.7.3 Naestved Swing Bridge

This tandem bobtail swing bridge was built across Naestved Canal near Naestved, Denmark. It crosses the channel at an angle of 80° (Thomsen and Pedersen, 1998). The superstructure comprises a welded steel box with cantilevered floor beams supporting a steel orthotropic deck. Each draw is 49 m long and 14 m wide, shaped as a parallelogram, with the ends parallel to the channel. The distance between pivots is 56.86 m measured along the longitudinal axis of the bridge. The draws are counterweighted so as to minimise dead load bending in the top of the pivot piers. Figure 3.19 shows one draw in the open position.

Each draw is supported at the pivot pier by a standard roller slewing bearing of 5 m diameter having three rows of rollers. The bearing supports actions on the draw: self-weight, live load, impact, wind, etc. The inner periphery of the ring bearing has gear teeth (it is a rack) which are engaged by the hydraulic motors of the span drive. Figure 3.20 is an image of the internal rack on the upper part of the slewing bearing, a pinion, reduction gearing, and a hydraulic motor.

3.7.4 El-Ferdan Bridge

A tandem bobtail swing bridge with a distance between the pivot axes of 550 ft (167.5 m) was completed across the Suez Canal at El-Ferdan, Egypt, in 1963 (Sedlacek, 1965). It was the fourth swing bridge at this location and had the largest opening span in the world. The bridge was designed and built by Krupp but it was extensively damaged during the Six-Day War in June, 1967. It was removed because of the decision to widen the canal and was replaced by tandem bobtail draws having a main span of 1115.5 ft (340 m) and provides a clear channel for shipping of 1050 ft (320 m) (Taha and Buckby, 2000).

The design of the superstructure and machinery was the responsibility of two Krupp entities. The responsibility for the superstructure also involved the design office of Professor Weyer, Dortmund,

Figure 3.19 Naestved Swing Bridge. (Parke and Hewson, 2008)

Figure 3.20 Span drive inside pivot pier of Naestved Swing Bridge. (Parke and Hewson, 2008)

in co-operation with the engineers of the Egyptian National Railway (ENR) and their consultant Halcow, UK (Tomlinson *et al.*, 2000).

The overall length of the bobtailed trusses in the closed position is 2100 ft (620 m) which support a 41.3 ft (12.6 m)-wide orthotropic steel deck that is occupied by a railway track and two carriageways. The trusses vary in height from 197 ft (60 m) at the pivot pier pylon to 49 ft (15 m) at mid-span where the mid-span lock resists positive bending moment when the bridge is closed and subjected to actions such as live load and wind. The moment lock and mechanism is described in Binder *et al.* (2001). The erection of the steel work and quality control is detailed in Mizon *et al.* (2000).

The draws are supported by rim bearings of historic design mainly because of the need to operate in the harsh desert environment. At each pylon there are 112 conical rollers of cast alloy steel with a nominal diameter of 15.75 in. (400 mm), which run on tapered alloy steel tracks that are 56.1 ft (17.1 m) in diameter. There is a central pivot of 1300 mm (51 in.) diameter embedded in the foundation to which all the rotating components are radially connected (Adrian *et al.*, 2000).

Extensive structural and dynamic analyses were performed which were necessary for seismic designs and aerodynamic studies. The behaviour of the structure during various operating conditions was extensively computer-simulated to confirm design assumptions (Schlecht *et al.*, 2000). El-Ferdan is the largest swing bridge in the world and a review of the design procedures could be invaluable to the movable bridge designer embarking on the design of a significant bridge.

REFERENCES

Abrahams MJ (1996) The George P Coleman bridge reconstruction, Yorktown, Virginia. *6th Biennial Symposium, Heavy Movable Structures*, Clearwater Beach, FL, USA, 30 Oct.–1 Nov.

Adrian E, Krüger H and Hess J (2000) Mechanical engineering, drive and control technology of the El-Ferdan Railway Swing Bridge. *Egyptian Society of Engineers Bridge Engineering Conference*, pp. 1–12, Sharm El Sheikh, Egypt, Mar.

Altebrando NJ, Bluni SA, Skelton PM and Moses RS (2002) Woodrow Wilson Bridge bascule span. *Proceedings of the Ninth Biennial Symposium of Heavy Movable Structures*, Daytona Beach, FL, USA, 22–25 Oct.

Anon. (1905) Trunnion bascule bridge at Northwestern Avenue, Chicago. *Engineering News*, **53(3)**: 64–66, 19 Jan.

Arenas de Pablo JJ (2000) La Porta d'Europa bascule bridge in Barcelona, Spain. *Structural Engineering International*, **10(4)**: 218–220, Nov.

Binder B, Pfeiffer M and Weyer U (2001) Die El-Ferdan-Brücke (in German). *Stahlbau*, **70(4)**: 231–244.

Birnstiel C (1996) Bascule bridge machinery rehabilitation at Hutchinson River Parkway Bridge. In *Bridge Management* (Harding JE, Parke GAR and Ryall MJ (eds)), pp. 116–123, 3rd edn. E & FN Spon, London.

Birnstiel C (2000) Movable bridges. In *The Manual of Bridge Engineering* (Ryall MJ, Parke GAR and Harding JE (eds)), ch. 12, pp. 662–698. London: Thomas Telford.

Foerster GA (2006) The support and stabilization machinery for Woodrow Wilson Bridge bascule spans. *Proceedings of the 11th Biennial Symposium, Heavy Movable Structures*, Orlando, FL, USA, 6–9 Nov.

Freudenberg G (1971) The world's largest double-leaf bascule bridge over the Bay of Cadiz (Spain). *Acier-Stahl-Steel*, **971(11)**: 463–472.

Green P (1996) Float out the old, float in the new. *Engineering News Record*, **339(55)**: 24, 8 Jul.

Hamilton WH and Worm L (1990) Design features of a unique swing bridge: hydraulics and control system. *Proceedings of the 3rd Biennial Symposium, Heavy Movable Structures*, St Petersburg, FL, USA, 12–15 Nov.

Hawranek A (1936) *Bewegliche Brücken*. Julius Springer, Berlin, Germany.

Holth N (2012) *Chicago's Bridges*. Shire Publications, Oxford, UK.

Johnson AC and Phillips JM III (2004) Repair and redesign of the Spokane Bridge lift/turn cylinders. *Proceedings of the 10th Biennial Symposium, Heavy Movable Structures*, Orlando, FL, USA, 25–28 Oct.

McBriarly PT (2013) *Chicago River Bridges*. University of Illinois Press, Urbana, IL, USA.

Mizon DH, Mohammed D and Binder B (2000) El-Ferdan Bridge – construction. *Egyptian Society of Engineers Bridge Engineering Conference*, pp. 1–14, Sharm El Sheikh, Egypt, Mar.

Parke G and Hewson N (eds) (2008) *ICE Manual of Bridge Engineering*, 2nd edn. ICE, London, http://dx.doi.org/10.1680/mobe.34525.

Phillips JS (2008) *Two miles – eighteen bridges: a walk along the Chicago River*. See http://www.chicagoloopbridges.com for further details (accessed 08/12/2014).

Price J (1879) Movable bridges. *Minutes of Proceedings, Institution of Civil Engineers*, Session 1878–1879, Part III, London.

Quade MN (1954) Special design features of the Yorktown Bridge. *Transactions*, ASCE 119.

Reusink JH (1996) Bearing the biggest bascule. *Evolution*, **3**: SKF, Göteborg, Sweden.

Ryall MJ, Parke GAR and Harding JE (eds) (2000) *The Manual of Bridge Engineering*. Thomas Telford, London.

Salcedo JM (2003) Bascule bridges in Spain. *Proceedings of the 5th International Symposium on Steel Bridges*, pp. 1–10, Barcelona, Spain.

Saul R and Zellner W (1991) The Galata bascule. *Report of the IABSE Symposium Leningrad*, **64**: 557–562, IABSE, Zurich, Switzerland.

Schlecht B, Wünsch D, Christianhemmers A, Hoffman R and Liesenfeld G (2000) Multibody-simulation of the El-Ferdan swing steel railway bridge. *Egyptian Society of Engineers Bridge Engineering Conference*, pp. 1–12, Sharm El Sheikh, Egypt, Mar.

Sedlacek H (1965) Die neue Drehbrücke über den Suez-Kanal bei El-Ferdan/Agypten (in German). *Der Stahlbau*, **34(10)**: 289–302, Berlin, Germany, Oct.

Taha N and Buckby R (2000) The El-Ferdan Bridge over the Suez Canal rail link to the Sinai. *Egyptian Society of Engineers Bridge Engineering Conference*, pp. 1–11, Sharm El Sheikh, Egypt, Mar.

Thomsen K and Pedersen KE (1998) Swing bridge across a navigational channel, Denmark. *Structural Engineering International, IABSE*, **8(3)**: 201–204, Zurich, Switzerland, Aug., http://dx.doi.org/10.2749/101686698780489225.

Tomlinson GK, Weyer U, Maertens L and Binder B (2000) El-Ferdan Bridge – design. *Egyptian Society of Engineers Bridge Engineering Conference*, pp. 1–18, Sharm El Sheikh, Egypt, Mar.

Wengenroth RH and Mix HA (1975) Fifty-year history of movable bridge construction – Part III. *Journal of the Construction Division, ASCE*, **101: CO3**: 545–557, September.

Wirzberger E (1994) Galata Bridge in Istanbul, the biggest bascule bridge in the world. *Proceedings of the 5th Biennial Symposium, Heavy Movable Structures*, Clearwater, FL, USA, 2–4 Nov.

WisDOT (2002) *State of Wisconsin Structure Inspection Manual*. WisDOT Bureau of Structures, Madison, WI, USA.

Movable Bridge Design
ISBN 978-0-7277-5804-0

http://dx.doi.org/10.1680/mbd.58040.087

Chapter 4
Other movable bridge forms

Charles Birnstiel

4.0. Introduction

This chapter treats some non-traditional movable bridge forms, historical span drives and examples of unusual bridge types that may be suitable for future crossings. Included is a description of the development of the British railway swing bridge.

4.1. Market Street Bridge, Chattanooga, TN

In July 1917 a double-leaf rolling bascule bridge (Scherzer) was completed over the Tennessee River between downtown Chattanooga and the Northshore District (Howard, 1917). It provided for a wide clear horizontal navigation channel having unlimited vertical clearance. Figure 4.1 depicts the closed bridge *c.* 2005. Figure 4.2 shows the bridge partially open during a test opening that same day. Because of the bridge's historical significance the appearance was not altered significantly during the rehabilitation of 2007 (Engstrom, 2008).

Figure 4.1 Market Street Bridge in closed position. (Photograph by Charles Birnstiel)

Figure 4.2 Market Street Bridge in partially open position. (Photograph by Charles Birnstiel)

The bascule was designed by the Scherzer Rolling Lift Bridge Company of Chicago under the overall supervision of the consulting engineer for the project, BH Davis. The Scherzer Company designed the bascule trusses as independent trusses for dead load but as three hinged arches for resisting live loads. The live loads comprised electric street cars, motor vehicles and pedestrians. The three hinges of an arch are the front teeth of the tracks (the arch springings) on each bascule pier and the crown hinge. In addition, perhaps as a redundancy, the standard Scherzer mid-span shear transfer device was installed. The standard mid-span Scherzer automatic jaw and diaphragm shear lock plays an important part in the operation of this bridge. It enables the alignment of the two parts of the crown hinge during the mating manoeuvre to occur almost automatically. The bridge operator does not need to juggle the separate leaves in order to complete the crown hinge joint. Figure 4.3 shows one of the crown hinges just before a test closing is to be completed and Figure 4.4 depicts a mid-span shear lock at about the same time. A springing hinge, a front tooth of a track plate, is shown with the leaf in the fully closed position in Figure 4.5.

The horizontal distance from the channelward face of a front tooth on one side of the channel to a corresponding point on the other side is about 300 ft (91.44 m). The roadway *c.* 2005 is shown in Figure 4.6. Despite the immense weight of the leaves, each leaf was rotated by a single 25 hp (18.6 kw) original equipment motor during the witnessed test opening. Figure 4.7 illustrates a span drive at the time of the author's site visit.

4.2. Armstrong swing bridges

Railways developed first in England, and because that country already had an extensive network of canals and river navigations there was conflict between the competing modes of transportation at waterway crossings. Railways had to cross navigable waterways and many crossings were designed and built by the firm of William E Armstrong and its successors. It came about in the following way.

Figure 4.4 Market Street Bridge – mid-span shear lock in nearly closed position. (Photograph by Charles Birnstiel)

Armstrong served as a solicitor to a company distributing water piped from distant reservoirs throughout the city of Newcastle upon Tyne. At the level of the River Tyne the water pressure was significant and Armstrong suggested that it be utilised to operate cranes in the port. He demonstrated hydraulic crane operation successfully and then after he withdrew from his law firm in 1846 he established a manufactory for hydraulic cranes upstream of Newcastle at Elswich, with the financial backing of his former law partner. He also developed local hydraulic power utilities with which to serve the hydraulic cranes he sold. At first they were low pressure systems but he soon realised the potential advantages of operating at higher pressure. In about 1850 he incorporated gravity accumulators in his systems. In

Figure 4.3 Market Street Bridge – crown hinge in nearly closed position. (Photograph by Charles Birnstiel)

Figure 4.5 Market Street Bridge in closed position – front track tooth serves as springing hinge. (Photograph by Charles Birnstiel)

operation the accumulator in the system would more-or-less continually and slowly be charged with high pressure water from a steam-driven pump. When an application needed energy a throttling valve would be opened to admit high pressure water from the accumulator to the hydraulic engine – usually an oscillating three cylinder type, which powered a gear train by means of a crankshaft.

Figure 4.6 Market Street Bridge – roadway on rolling bascule spans. (Photograph by Charles Birnstiel)

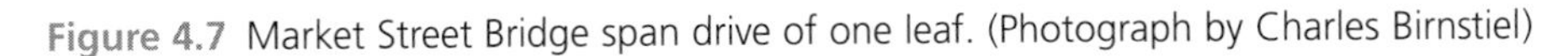

Figure 4.7 Market Street Bridge span drive of one leaf. (Photograph by Charles Birnstiel)

Soon, hydraulically operated material handling machinery became popular in English ports and some outside the UK.

Armstrong conceived the idea of applying his hydraulic power systems and hydraulic machinery to swing bridge operation. Often the swing bridge was remote from a hydraulic utility so he built a miniature hydraulic power system right inside the pivot pier of the swing bridge.

The first hydraulically operated swing bridge was erected in 1852 over the River Severn (Armstrong, 1869). By 1869 there were more than 50 swing bridges (mostly for railways) of two classes:

- the lift and turn kind in which the bridge is lifted off its bearings by a central press and then turned by hydraulic pressure
- the rim bearing type in which the bridge is supported at the pivot pier by a nest of rollers at all times and is turned by hydraulic pressure.

In some cases rim bearing bridges have central presses which are charged prior to turning the draw in order to reduce dead load on the rollers during swinging. The equipment for the second kind of UK swing bridge will be described using as an example photographs of the Newcastle bridge (Figure 4.8).

Figure 4.8 is a lengthwise view of the bridge. The British swing bridges were designed as two-span continuous beams when in the closed position and because the bending moments are a maximum over the pivot the top chords (booms) were given a bowstring shape. Usually, the spanning members were plate girders but in situations where the required flange cross-sectional area for a plate girder was large the truss configuration was selected, as shown in Figure 4.8.

Between bridge openings the draw is supported at the pivot pier by a circular drum girder of box-shaped cross-section having a diameter of about 43 ft (13.1 m). The drum girder rests on 42 cast-iron tapered live rollers of 3 ft (0.91 m) dia. and 14 in. (0.36 m) face width. The rollers run on a cast iron track and are loaded through a cast iron upper path (tread) bolted to the underside of the drum girder. Bolted to the upper roller path is a rack which is engaged by the driving pinion. Figure 4.9 is a

Figure 4.8 Newcastle upon Tyne Swing Bridge – view from a draw end. (Photograph by Charles Birnstiel)

photograph taken within the confines of the drum girder that shows the rollers, the upper path and the live ring framing.

Figure 4.10 shows the upper end of the central press piston. Before the draw is swung, the press is charged from the accumulator, applying an upward force of about 65% of the draw dead load (originally).

Figure 4.9 Newcastle upon Tyne Swing Bridge. Photographed from inside of drum girder. Centre press showing packing gland for piston and also the large-dia. circular bearing for the live ring. Wrought-iron bars are part of live ring. (Photograph by Charles Birnstiel)

Figure 4.10 Rim bearing of Newcastle upon Tyne Swing Bridge. Photographed from inside drum girder. Note tread splice, tapered rollers and axles. (Photograph by Charles Birnstiel)

The accumulators are 17 in. (432 mm) in diameter and 17 ft (5.18 m) long and were installed vertically into the pivot pier and are loaded with cast-iron weights sufficient to develop (originally) a fluid pressure of 700 psi (48.3 bar) (Homfray, 1878). Fluid from the accumulators is throttled to the hydraulic engine shown in Figure 4.11 (it is one of the two original engines). Each engine has three single-acting oscillating cylinders of 4.5 in. (114 mm) dia. and 18 in. (457 mm) stroke that rotate a crankshaft.

Figure 4.11 Newcastle upon Tyne Swing Bridge – hydraulic engine. Three oscillating cylinders turn the crankshaft. (Photograph by Charles Birnstiel)

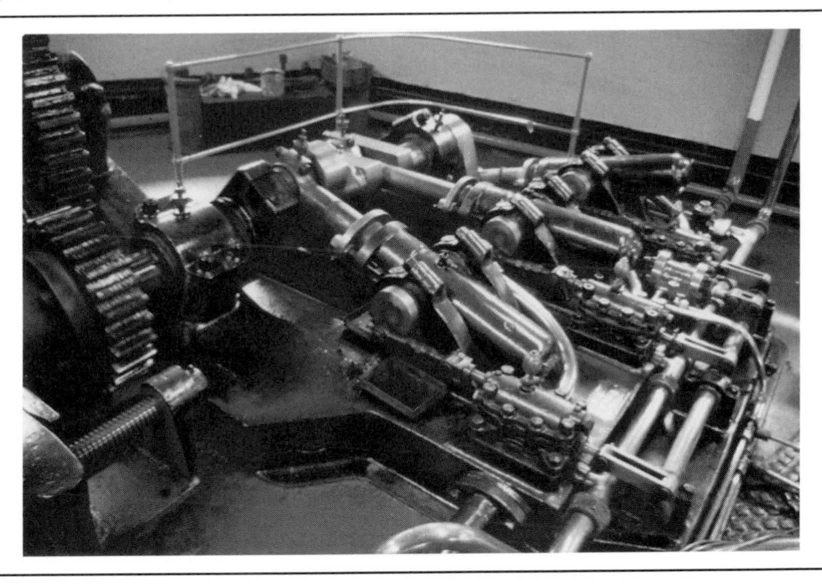

Figure 4.12 Newcastle upon Tyne Swing Bridge – speed reduction gearing. Note gearshift to offer speed change. (Photograph by Charles Birnstiel)

The output torque from the crankshaft is increased by the speed reduction gearing shown in Figure 4.12, the output of which is transmitted to the driving pinion shown in Figure 4.13 which engages the underside of the rack fastened to the drum girder. Speed is controlled by rotating the throttle valve handle atop the cylindrical stand shown on the right side of Figure 4.14.

Figure 4.13 Newcastle upon Tyne Swing Bridge – rack and pinion. Pinion turns about horizontal axes meshing with rack above. (Photograph by Charles Birnstiel)

Figure 4.14 Newcastle upon Tyne Swing Bridge – control station. (Photograph by Charles Birnstiel)

The operation of most swing bridges in the UK was as described above, except that some did not have central presses. The first English swing bridge with an electro-mechanical drive was put into operation in 1899. The difference between UK and US practice is described in Chapters 8, 12 and 13.

4.3. Movable bridges supported by pontoons

Swing bridges have been built that are wholly or partly supported by pontoons, usually across protected waterways with slow currents. Those that were built with one end of the draw supported by a fixed pier are usually on waterways with small fluctuations of water level. Examples of pontoon-supported swing bridges were: the Galata Bridge at Istanbul, Turkey, built in 1912 and replaced in 1990 by a large bascule bridge (Saul and Zellner, 1991) and a railway bridge across the Suez Canal at Kantara, Egypt (Hawranek, 1936). The Galata Bridge pontoon was hinged at one corner to a moored pontoon and originally was swung open or closed by propellers driven by electric motors located in the pontoon at the free end of the draw. The double-decked draw was 262 ft (80 m) long and 82 ft (25 m) wide, and it rotated 180°. The swing bridge across the Suez Canal at El Ferdan, Egypt was supported by a fixed pier at the hinged end and was also swung by a propeller drive. Many of the shorter draws are swung open by pulling the pontoon at the free end by means of a wire rope or chain drive.

4.3.1 Floating bridge at Sargent, TX

An example is the floating bridge at Sargent, TX, depicted in Figure 4.15. It is a steel barge that is pivoted about a vertical pin to clear the channel. The barge is 124 ft (37.8 m) long and supports a 24 ft (7.3 m) wide roadway. Because the bridge floats in a tidal estuary it was necessary to build levelling spans and aprons at each end in order to accommodate the change in water level. In the figure the hinge (vertical axis) about which the barge rotates may be seen just beyond the 'No Fishing' sign and the hauling wire rope is seen connected to the barge at the upper left-hand corner of the photograph. To open, the barge swings counter clockwise. In the photo the bascule aprons have been raised prior

Figure 4.15 Floating bridge, Sargent, TX. (Jim Philips, III, URS Corp.)

to an opening. The levelling spans between the apron hinge and the land pivot about horizontal axes at the abutments and are mechanically adjusted near the apron hinges.

4.3.2 Town Bridge, Northwich, Cheshire

Site conditions strongly influence the choice of movable bridge type and when one first sees Town Bridge, or drawings of it, one wonders what prompted such a peculiar design. In the 19th century an important chemical industry developed in Cheshire due in large part to major subsurface salt deposits in the area. Chemical manufactories drove wells into the ground and injected heated water to dissolve salt and force it to rise above ground level. It became a natural resource for chemical production. So much salt was extracted that the ground surface subsided significantly. The prior bridge at the Town Bridge site, according to Saner (1899/1900), a fixed plate-girder structure of 90 ft (27.4 m) span, had caused

> great inconvenience to the river-traffic through the constantly decreasing headway (vertical clearance). The average yearly loss of headway for about 16 years has been $4\frac{1}{2}$ inches (114 mm) in the locality of the bridge, so that the house property is 6 feet (1.8 m) nearer the water level than formerly. It was imperative to maintain the full headway for present (water) traffic; the author (Saner) ... from time to time raised the girders. The streets in the immediate neighborhood could not, however, be raised in a corresponding manner ... so that the road-gradients became as steep as 1 in 11.
>
> Saner (1899/1900)

Obviously the replacement bridge had to be a movable (opening) bridge which could provide unlimited vertical clearance and the 'subsiding ground necessitated a form of structure easily adjustable'. The design selected was, essentially, as shown in Figure 4.16. In Saner's description

Figure 4.16 Town Bridge, Northwich, Cheshire. (Jenkins, 2000)

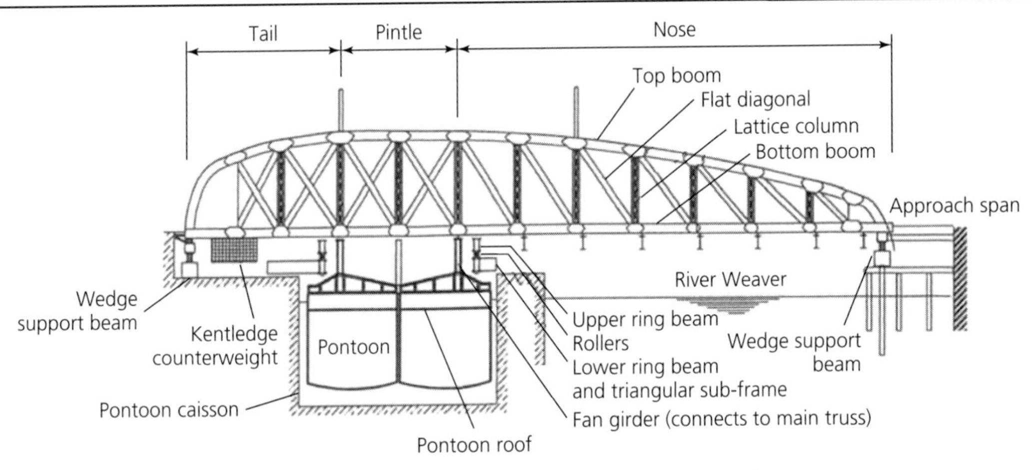

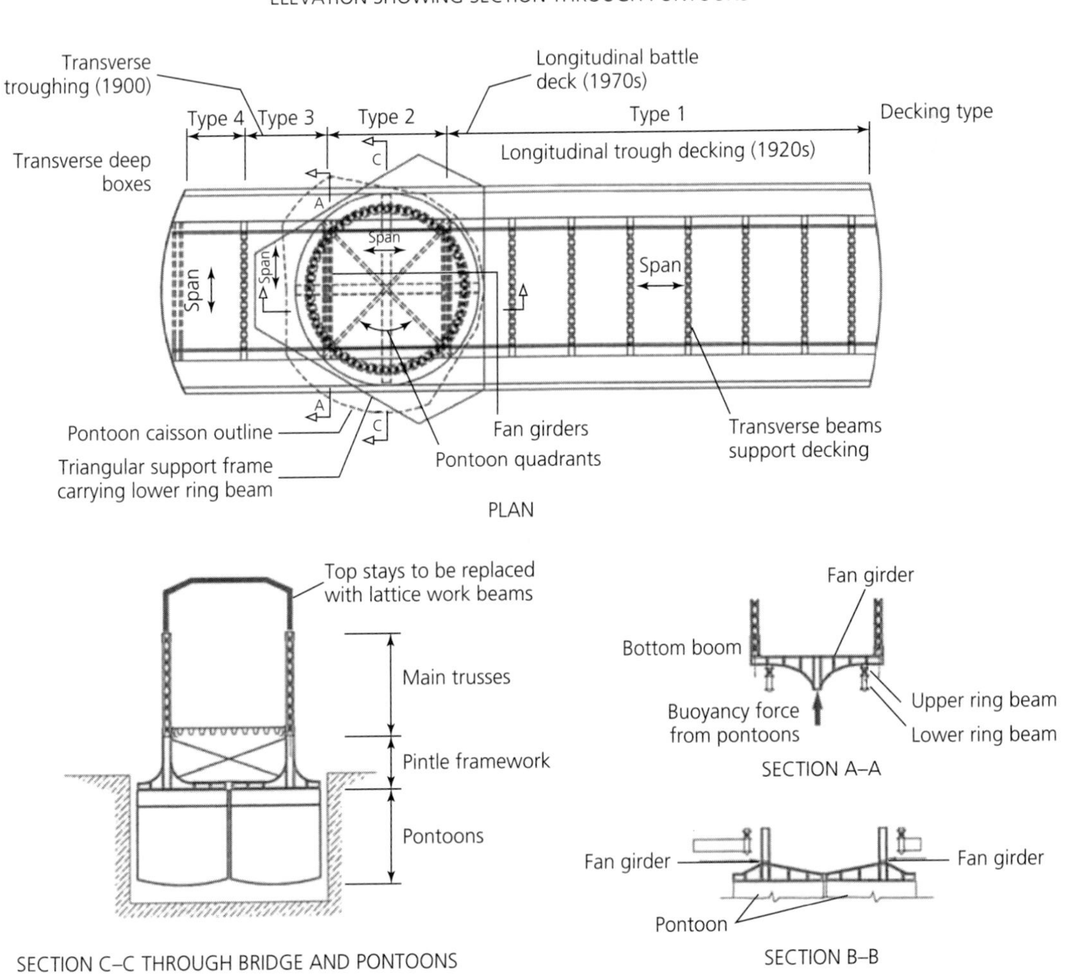

(1) A pair of bow-string lattice girders carrying the road and footways. (2) A sealed circular pontoon or buoy, placed under the center of gravity of the superstructure, and suitably and rigidly connected thereto, in such a manner as to be always submerged in the river, and capable of being turned with the superstructure. (3) A group of cast-iron screw piles surrounding, but clear of, the pontoon and carrying a gridiron girder, which in its turn carries the bottom roller-path; cup-castings and adjustable screws are inserted between the heads of the piles and the girders.

Saner (1899/1900)

The purpose of the adjustable screws between pile-heads and gridiron girder was to enable the girder to be raised or lowered, so 'that a daily adjustment can be effected if required – a by no means improbable contingency'. The pontoon supports a substantial portion of the gross self-weight of the bridge so that the screw pile loads are minimised and load on the rollers, and hence rolling friction, is much reduced.

A drawing included in Saner (1899/1900) shows that the American hand drive was adopted with a circular rack mounted on the fixed structure (onto the lower roller-path) and with the manual span drive machinery mounted on the draw. The 72 conical rollers were hollow castings with flanges on the inward facing ends of 11.25 in. (285 mm) outside diameter. The mean diameter of the finished roller surface does not appear to have been shown on the drawings.

This was the first British swing bridge with an electro-mechanical main span drive. A wire rope of 1.5 in. (38.1 mm) dia. was wrapped around the drum girder of the bridge and a wire rope hoist drum located inside an external cabin located on land adjacent to the tail of the draw. A reversible electric motor with appropriate reduction gearing is used to drive the motor and thereby rotate the draw.

Figure 4.17 is a photo taken after the refurbishment described by Jenkins (2000). Clearly evident is the wire rope wrapped about the drum girder and the absence of a rack attached to the fixed roller-path. Proceeding downward in Figure 4.17, the edge of the footwalk, the drum girder with the span drive wire rope, the conical rollers spaced by axles between the live ring channels, and the top of the pontoon can be seen. It is evident that the span drive is now a cable wrapped about the drum girder which is moved by a hoist drum located in the nearby cabin building.

Figure 4.17 Town Bridge – rim bearing. (Photograph by Charles Birnstiel)

4.3.3 Ford Island Bridge

The Ford Island pontoon bridge at Pearl Harbor, Hawaii, does not swing about a vertical axis in order to create a clear passage for navigation. Instead, it retracts linearly under the approach structure. It is a single 283 m long multi-cell reinforced concrete box which is retracted under the flanking transition and trestle spans by a wire rope haulage system in order to provide a 175 m-wide clear opening for navigation (Abrahams, 1996; Birnstiel and Tang, 1998). The draw span is pulled by 51 mm dia. wire ropes wound on hydraulically powered winches. Two winches are utilised in a counter-torque arrangement with both the ropes that retard and advance the pontoon tensioned so that the draw span is always under control. The rope drive is located on one side of the pontoon thus providing an eccentric towing force which apparently is not objectionable because the pontoon is guided. The time required to open or close the draw is 25 minutes, including vehicular traffic management time.

4.4. Transporter bridges

Transporter bridges comprise an overhead fixed span on which a track is mounted for a carriage (traveller) that can be moved longitudinally. It may be self-propelled or be drawn back and forth by a continuous cable drive. A gondola is suspended from the movable carriage by means of cables or latticed hangers. The gondola transports pedestrians and vehicles back and forth across the channel. The inventor of the overhead transporter is a matter of controversy but the first one built was designed by the Spanish engineer and architect Alberto de Palacio, working with his brother Silvestre, also an engineer, based on a patent granted to Alberto and Ferdinand Arnodin (de Palacio and Arnodin, 1890). It was built in 1893 near Bilbao, Spain, and according to Troyano (2003), Arnodin and other French engineers were involved in the project. Arnodin was contractor or consultant on some 15 transporter bridges built in France, England, Spain, South America and the USA. An Arnodin transporter bridge operates at Newport, South Wales. A promotional dimensioned drawing labelled 'Elévation générale' at the scale 1:400 was signed by Arnodin on 16 February 1901 at Châteauneuf sur Loire, France. A facsimile is on display at the bridge site.

4.4.1 The Newport transporter bridge

The background leading to the construction of the Newport transporter is thoroughly described in Mawson and Lark (2000). As shown in Figure 4.18 the bridge comprises two towers spaced 642 ft (185.6 m) apart from which a boom is suspended from draped cables (as in a suspension bridge) and, near the ends of the span, oblique stays, as for a cable-stayed bridge. The vertical clearance from high-water level to the underside of the deck was originally 177 ft (53.9 m). The height of the towers above the approach roadways was 242 ft (73.8 m) and the distance between the anchorages 1545 ft (471 m).

Figure 4.19 depicts the deck of the boom at about the quarter point of the span. It comprises two walkways alongside the stiffening trusses and an intermediate open space between the track beams. The deck is suspended from a total of 16 spiral round wire strands, four on each side of a truss as shown in Figure 4.20. The suspenders have looped ends through which bars are passed that receive the threaded ends of U-bolts clamped around the strands.

The traveller from which the gondola is suspended rolls on rails mounted on the inside bottom flanges of the track beams. Figure 4.21 depicts a traveller wheel. A continuous wire rope hauls the traveller across the span. The return part of the wire rope is visible in the foreground of the photograph. The drum for the endless wire haul rope is located within the house with the red hipped roof at the far tower in Figure 4.18.

Figure 4.22 shows the gondola approaching a dock during a crossing. The photograph was taken from a stairway landing partway up the tower. The gondola floor is wide enough for three motor vehicles and has protected space for pedestrians and cyclists.

Figure 4.18 Newport transporter – overall view. (Photograph by Charles Birnstiel)

Figure 4.19 Newport transporter – boom. (Photograph by Charles Birnstiel)

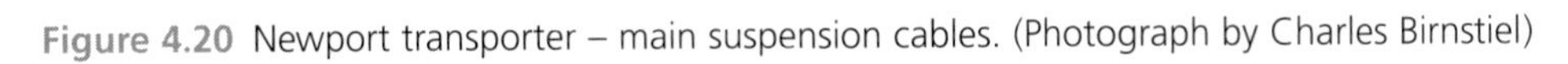

Figure 4.20 Newport transporter – main suspension cables. (Photograph by Charles Birnstiel)

Figure 4.21 Newport transporter – traveller wheel. (Photograph by Charles Birnstiel)

Figure 4.22 Newport transporter – gondola. (Photograph by Charles Birnstiel)

4.4.2 Tees transporter

About five years after the Newport transporter was put into operation, a transporter was completed across the River Tees at Middlesbrough. However, instead of a suspended boom as at Newport, the designer elected to build a rigid trussed cantilever bridge between the towers from which a travelling carriage is supported. The distance between the towers is 590 ft (180 m) – about 55 ft (17 m) less than that at Newport. The gondola has a capacity of 200 persons or 9 automobiles – somewhat greater than at Newport. It has recently been renovated.

4.4.3 Rendsburg transporter

Another operating transporter bridge is that across the North Sea–Baltic Sea Canal at Rendsburg, Germany. The traveller runs underneath a high-level double-track railroad bridge that was opened to railway traffic in 1913. The continuous trusses of the bridge have a main span of 140 m with a vertical clearance of 42 m. As is evident from Figure 4.23, the gondola is suspended from 12 inclined wire ropes fastened to the traveller. The traveller is self-propelled by electric bogies running on tracks outboard of the bridge trusses. The photograph depicts a train at the bridge.

4.5. The Bellmouth passage bridges

While planning the Canary Wharf real estate development in the Docklands area of east London it was found desirable to provide two roadway bridges across the waterway known as Bellmouth Passage, which connects the North and South Docks. The bridges support North Colonnade and South Colonnade which have roadways for one-way vehicular traffic and sidewalks for pedestrians. However, two pedestrian bridges were also planned. All four bridges were to be at the basic elevation of Canary Wharf.

4.5.1 Bellmouth passage road bridges

The architects for the two road bridges, Wilkinson Eyre, were faced with significant restrictions in designing them. The site was 'tight'; no operating mechanism was to project above the roadway when the bridge was open to roadway traffic, and the Bellmouth Promenade was not to be blocked to pedestrians. Because of those restrictions, and the lack of space for a counterweight, the architects

Figure 4.23 Transporter bridge at Rendsburg, Germany. (Courtesy of Jürgen Bätcke)

proposed a single-leaf roller-bearing bascule. Patents were issued in the USA for this type of bridge to John Philo Cowing (1899, 1900, 1901) and others. Cowing was an important engineer in Cleveland, Ohio, and when the need arose for another local movable bridge after his patent was granted, a Cowing style bascule was built. It was opened to road traffic in 1907 and decommissioned in 1957.

Figure 4.24 shows North Colonnade Bridge under construction. It comprises two box girders that terminate in the rings which are supported by rollers below sidewalk level. The deck is orthotropically stiffened steel plate with asphalt surfacing. The circular roller bearing is evident. It is shown close-up in Figure 4.25, with the gear rack and the counterweight on the rear side. Four hydraulic motors located below the sidewalk engage each rack and rotate the ring to move the leaf. The leaf has been reported to be quite span heavy with the centre of gravity of the bridge located about 0.9 m forward of the centre of rotation. Note that the pedestrian walkway passes through the rings and it is said that pedestrians may walk through the rings while the bascule leaf is rotating (Aveni, 2001).

4.5.2 Bellmouth passage pedestrian bridges

The two pedestrian opening bridges that span the passage to facilitate access to the retail establishments opposite the office building, swing to clear the passage. They are steel box beams that support a glazed superstructure. The draw is supported at one end (without a tail span counterweight) by a slewing bearing which is capable of resisting the cantilever bending moment. Figure 4.26 depicts one of the pedestrian swing bridges in the closed position, where the longitudinal box beams essentially behave as simply supported. The pier is of large diameter because it has to support the large-dia. slewing bearing and resist the bending moment about a horizontal axis due to dead load when the free end of the beam is unsupported. Figure 4.27 shows the top of the pier and one of the multistage planetary gear units that rotate the large ring gear (hidden) in order to rotate the draw.

The architect for the swing bridges was Patel Taylor. Techniker were the consulting structural engineers with detailed design by Davy Markham utilising LUSAS software. Mechanical design was by MG Bennett Associates.

Figure 4.24 North Colonnade Bridge over Bellmouth Passage. (Photograph by Charles Birnstiel)

Figure 4.25 Roller bearing and counterweight – Bellmouth Passage Road Bridge. (Photograph by Charles Birnstiel)

Figure 4.26 Bellmouth Passage Pedestrian Swing Bridge. (Photograph by Charles Birnstiel)

Figure 4.27 Pivot pier and span drive. (Photograph by Charles Birnstiel)

4.6. Gateshead Millennium Bridge

Another unusual movable bridge concept that is of current interest is the so-called gyratory bridge. The operating principle is that a movable span rotates about a longitudinal axis, that is about an axis crossing the waterway. In the description of the US patent granted to E Swensson in 1909, the axis is elevated above the closed movable span (see Figure 1.1(f)). At both shores, there is a tower with a trunnion at the elevation of one-half the lift height. The axes of the trunnions are collinear. At each end, a portal frame (a vertical crank arm) extends downward to the movable span, which may be a box truss. To open the bridge for marine traffic, the crank arms are rotated 180° about the trunnion axis. The movable span, which is below the trunnions when the bridge is closed, is then directly overhead when the bridge is open. The waterway clearance is approximately twice the height of the trunnions above the roadway. Counterweights are provided at the free ends of the arms so that the centre of gravity of the moving mass can be made to lie on, or close to, the axis about which it rotates. Waddell referred to the gyratory lift bridge as 'another freak structure – impractical, uneconomic, but exceedingly ingenious' (Waddell, 1916).

The author is unaware (in 2014) of any existing gyratory movable bridge other than the Gateshead Millennium Bridge across the River Tyne near Newcastle upon Tyne. As is evident from Figure 4.28, it is a pedestrian bridge with a movable deck that is close to the water in the closed position. In plan, the deck is symmetrically curved with a central offset of about one-third of the 105 m span. This shallow box deck is supported by cables from a backwardly inclined arch. A large trunnion is connected to each springing of the arch. The trunnion axes are collinear. These trunnions are mounted on massive piers, which also house the operating machinery. Hydraulic cylinders at each trunnion rotate the leaf. Because there is no mechanical connection between the trunnions on opposite sides of the

Figure 4.28 Gateshead Blinking Eye Pedestrian Bridge. (Wikipedia (courtesy of Paul Belchamber, LUSAS))

channel, the strokes of the pistons must be closely matched by electro-hydraulic controls. If the springing of one arch were to rotate without a corresponding rotation of the other springing, the arch (and deck) would be twisted.

Because of the incline of the deck and its pronounced horizontal curvature, a rotation of about 40° is sufficient to create a 25 m vertical waterway clearance at mid-span. Wilkinson Eyre were the architects for the bridge, Gifford and Partners the structural engineers and MG Bennett Associates engineered the mechanisms.

4.7. Belidor bascules

An interesting form of the bascule bridge is that shown in Figures 1.3 and 4.29. One end of the bascule leaf is hinged at A and the other rests on the pier at B. A cable is connected to each side of the leaf near the toe which leads over a sheave C at the top of the tower ending at a rolling counterweight. As the leaf rotates about the hinge A on opening, the counterweight rolls down the curved track. The curve of the track is selected to obtain balance of the rotating leaf with the constant value of the counterweight at all angles of leaf opening. Belidor showed that for the 'balanced' condition, the curve is sinusoid (Belidor, 1729). A cardiod was derived by Fraser and Deakin (2001). The resultant shape depends on the assumptions made in the derivation. The final track profile was determined experimentally at the bridge built by the Erie Railroad over Berry's Creek in the Hackensack Meadows near Jersey City (Anon., 1896). Of course, a truly 'balanced' condition would not be practicable. The bridge should be span heavy with the toe of the closed leaf pressing downward on the rest pier because of

- better vehicle riding quality when the leaf is firmly seated
- the need to resist wind pressure when the leaf is raised
- the need to lower the leaf in the event of operating mechanism failure.

If the operating machinery is independent of the main cable and can resist wind from both the fore and aft directions then a lower amount of span-heavy imbalance is tolerable, which results in lower power draw.

Figure 4.29 Belidor bascule diagram. (Parke and Hewson, 2008)

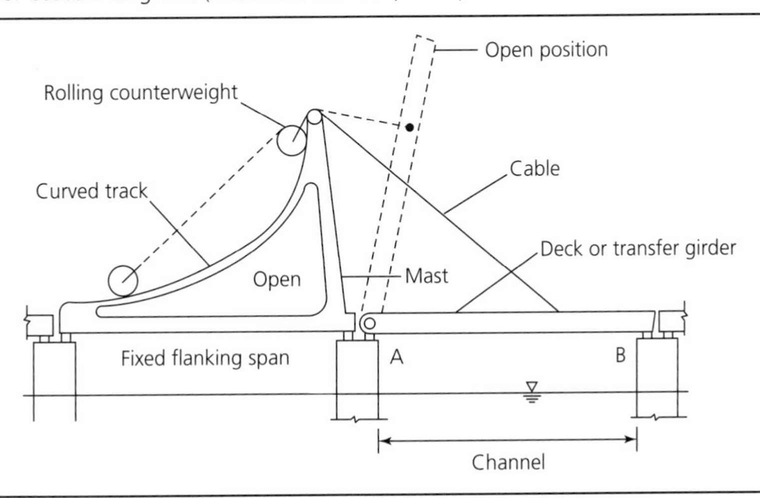

Figure 4.30 Glimmer Glass Bridge – opening. (Photograph by Charles Birnstiel)

A Belidor bascule was constructed to support Brielle Road (two roadway traffic lanes and a sidewalk) over Glimmer Glass Channel near Manasquan, New Jersey in 1898. Figure 4.30 depicts the bridge while it is opening. It is frequently operated during the summer in order to permit passage of recreational boats because of its low vertical waterway clearance when closed. Figure 4.31 is a view of the tower from the flanking span showing the rolling counterweights on each side, the curved tracks on which they roll, the drums about which the lifting wire rope is wrapped, the overhead transverse

Figure 4.31 Glimmer Glass Bridge – tower. (Photograph by Charles Birnstiel)

shaft which forces both drums to rotate in unison (practically), and the electric motor and speed reducer which rotates the assembly. In this machinery arrangement the hoisting cable serves both as a counterweight rope and an operating rope. The tower and track support are of timber; the members are replaced as they deteriorate. The local residents have successfully resisted modernisation of the bridge except for the interior of the bridge operator's house.

More recently, the Belidor bascule form was adopted for the Forton Lake Opening Bridge at Gosport in southeast England. This bicycle and pedestrian bridge has a single leaf of 18 m span and the overall width of the bridge is 6 m. The deck is othotropic steel plate. The single counterweight cylinder is the full width, out-to-out, of the curved tracks (Thorpe, 1999). Each hoist line is a chain passing over a chain wheel (sprocket) that is powered by a 3 kw (4 hp) motor with reduction gearing that can open the bridge in five minutes. The chain was replaced after 12 years of use.

A novel project delivery process was devised for this bridge by Gosport Borough Council and a group at the University of Southampton. Design was performed by Maunsell and Bennett Associates. Aesthetics drove the choice of bridge type to the Belidor bascule because a balance beam bascule was considered unattractive. Actually, both the balance beam bascule and the Belidor bascule have a positive feature in common: the span drive machinery can be located near or above the bridge deck level. The Belidor has the further visual advantage that it does not have an elevated balance beam.

4.8. Folding bridges

A type of movable bridge mentioned in Chapter 1 is the folding bridge. The distinguishing feature of this type is that it is comprised of panels which fold about one or more transverse hinges in order to clear a waterway. The concept is not new. Such bridges were built for roadway traffic in Chicago, IL and Milwaukee, WI from 1890 to 1920 (Birnstiel, 2001).

Figure 1.1(g) is a diagram of the folding bridge in which the top chords of the trusses shorten to coil the bridge. Because the right end of the articulated truss in Figure 1.1(g) is bolted to the foundation, shortening of the top chord of the truss curves the truss in its vertical plane. The top chord length change can be accomplished by using screw or hydraulic actuators as vertical truss members. The 1 m-wide pedestrian bridge at Paddington Basin in London, shown in the closed position in Figure 4.32, has hydraulic cylinders as verticals and pressurisation of the cylinders changes the shape of the geometrical figures of the truss so that the bridge can be folded into a coil (Brown, 2005). However, because the last truss panel at the right is bolted to the foundation, the coiled bridge cannot roll. It is held vertical and stationary in the open position in order to resist horizontal forces.

In December 1997, a folding pedestrian bridge across a fjord on the Baltic Sea in the City of Kiel, Germany, was opened to traffic (Knippers and Schlaich, 2000). Figure 4.33 shows the Hörnbrücke Bridge in the closed position. Based on the congregation of pedestrians, the photograph must have been taken just after a bridge closing. The bridge is used by people travelling between the railway station located at the right and the passenger ferry wharfs located leftward. During the years 1998–1999 the bridge was opened and closed an average of 11.4 times per day in the summer and 8.9 times per day during winter.

The bridge comprises three panels with a total span of 25.6 m and a width of 5 m. Figure 4.34 shows the walkway in the closed position and the frame pylons which rotate about their bases in the longitudinal direction of the bridge. An intermediate open position of the bridge is shown in Figure 4.35.

Figure 4.32 Folding bridge at Paddington Basin – closed. (Photograph by Charles Birnstiel)

The design and analysis of the structural system for the Hörnbrücke Bridge is described in Knippers and Schlaich (2000). The design criteria included the following:

- All cables are to be in tension under all loading conditions without reliance on any mechanical or hydraulic tensioning devices.
- The structural system is statistically determinate in all positions. Temperature change, cable creep and malfunctions shall not result in uncontrollable cable forces.
- The span drive should be simple. All winches are to operate at constant speed or constant torque.

Figure 4.33 Hörnbrücke Bridge – pedestrian bridge at Kiel, Germany in closed position. (Photograph by Jürgen Bätcke)

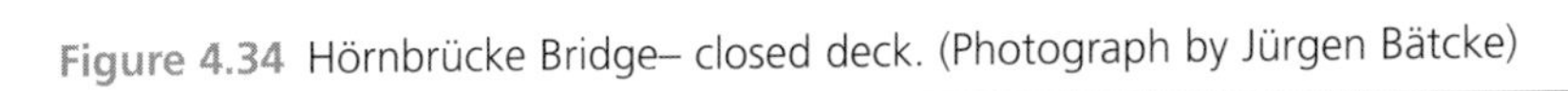

Figure 4.34 Hörnbrücke Bridge– closed deck. (Photograph by Jürgen Bätcke)

Figure 4.35 Hörnbrücke Bridge – intermediate open position. (Photograph by Jürgen Bätcke)

- The bridge shall be held in the open position by mechanical locks in order to perform inspections and maintenance, including replacement of cables.

Behaviour of the bridge under potential wind gusts was simulated during shop tests and the damping, deflections and cable tensions measured. The bridge behaviour under wind was found to be satisfactory during several actual on-site openings at winds of 25 m/s.

The architect was von Gerkan, Marg and Partners. Structural design was by Schlaich, Bergermann and Partners.

4.9. Other movable bridge forms

Many other forms of movable bridges have been proposed and built that offer possibilities for architectural expression. Space is not adequate to present them here but many examples may be found in Hovey (1926), Hawranek (1936), and Hool and Kinne (1943).

REFERENCES

Anon. (1896) A novel lift bridge on the Erie. *The Railroad Gazette*, 27 Nov.

Abrahams MJ (1996) Ford Island Bridge. *Proceedings of the Sixth Symposium on Movable Bridges*, pp. 998–990. Heavy Movable Structures, Inc., Clearwater Beach, FL, USA.

Armstrong WE (1869) Hydraulic swing bridges: The Armstrong Swing Bridge over the Ouse. *Van Nostrands Engineering Magazine*, **1**.

Aveni M (2001) Pedestrians walk through bascule's rolling rings. *Civil Engineering*, **71(4)**: 18.

Belidor BF (1729) *La science des ingenieurs des la condite des travaux de fortification et d'architecture civile*. Chez Claude Jombert, Paris, France.

Birnstiel C (2001) Popular obsolete movable bridges. *Proceedings of the 7th Historic Bridge Conference*, pp. 138–145, 19–22 Sept., Cleveland, OH, USA.

Birnstiel C and Tang M-C (1998) Long span movable bridges. *Structural Engineering World Wide 1998*. SEI-ASCE, Elsevier Science, London, UK.

Brown JL (2005) Rolling London footbridge surprises spectators. *Civil Engineering*, pp. 16–17, Mar.

Cowing JP (1899) Bascule Lift Bridge. US Patent 633,81, Jun.

Cowing JP (1900) Bascule Lift Bridge. US Patent 644,405, Feb.

Cowing JP (1901) Bascule Bridge. US Patent 665,405, Jan.

Engstrom IC (2008) The historic rehabilitation of the Market Street bridge in Chattanooga. In *Historic Bridges: Evaluation, Preservation and Management* (Adeli H (ed.)). CRC Press, Boca Raton, FL, USA, http://dx.doi.org/10.1201/9781420079968.ch14.

Fraser DJ and Deakin AB (2001) Curved track bascule bridges: from castle drawbridge to modern applications. *Proceedings of the 7th Historic Bridge Conference*, pp. 28–36, 19–22 September, Cleveland, OH, USA.

Hawranek A (1936) *Begwegliche Bruecken* (in German). Julius Springer, Berlin, Germany.

Homfray SG (1878) *The Tyne Bridge*. Paper presented before the Tyne Engineering Society on 14 February, 1878. Port of Tyne Authority, South Shields, Tyne and Wear, UK.

Hool GA and Kinne WS (1943) *Movable and Long-Span Bridges*, 2nd edn. McGraw-Hill, New York, USA.

Hovey OE (1926) *Movable Bridges*, vols I and II. Wiley, New York, USA.

Howard WJ (1917) Bascule bridge is erected as a cantilever. *Engineering News-Record*, **79(4)**: 26 July.

Jenkins T (2000) Refurbishment of Town Bridge, Northwich, Cheshire. *Bridge Management 4*, Thomas Telford, London, http://dx.doi.org/10.1680/bm4.28548.

Knippers J and Schlaich J (2000) Folding mechanism of the Kiel Hörn Footbridge. *Structural Engineering International*, **1**: 50–53.

Mawson BR and Lark RJ (2000) Newport Transporter Bridge – an historical perspective. *Proceedings of ICE Civil Engineering 138*, Paper 12086, February, http://dx.doi.org/10.1680/cien.2000.138.1.40.

Parke G and Hewson N (eds) (2008) *ICE Manual of Bridge Engineering*, 2nd edn. Institution of Civil Engineers, London, http://dx.doi.org/10.1680/mobe.34525.

de Palacio MA and Arnodin FJ (1890) Means for transporting loads. US Patent 425,724, Apr.

Saner JA (1899/1900) Swing Bridges over the River Weaver at Northwich. *Minutes of Proceedings*, Institution of Civil Engineers, **140**: Part 2, http://dx.doi.org/10.1680/imotp.1900.18644.

Saul R and Zellner W (1991) The New Galata Bascule Bridge at Istanbul. *Report*, IABSE Symposium Leningrad, IASBE Zurich, Switzerland, pp. 557–562.

Thorpe JE (1999) Forton Lake Opening Bridge, UK. *Structural Engineering International*, **9(3)**: 178–180, Aug.

Troyano LF (2003) *Bridge Engineering: A Global Perspective*. Thomas Telford, London, http://dx.doi.org/10.1680/beagp.32156.

Waddell JAL (1916) *Bridge Engineering*, vols I and II. Wiley, New York, NY, USA.

Wikipedia (2015) *Gateshead Millennium Bridge*. See http://en.wikipedia.org/wiki/Gateshead_Millennium_Bridge (accessed 22/05/2015).

Movable Bridge Design
ISBN 978-0-7277-5804-0

http://dx.doi.org/10.1680/mbd.58040.115

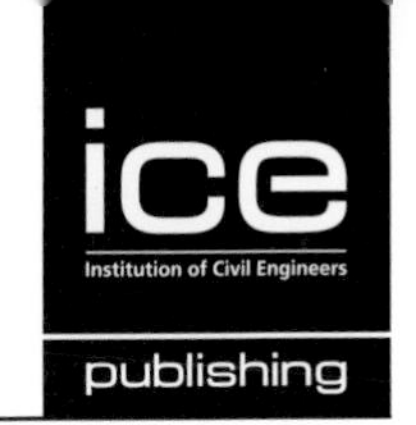

Chapter 5
Movable bridge design standards

Charles Birnstiel

5.0. Introduction
Three notions current in engineering design philosophy for combined structural, mechanical and electrical systems are: safe-life, fail-safe and damage-tolerant design.

Safe-life design is used for systems that are difficult to repair or whose failure could cause personal injury or costly physical damage. *Fail-safe* design recognises that all physical systems have flaws, and hence redundancies are designed into the system so that the negative consequences of failure are minimalised. Fail-safe design is at a level above safe-life, usually at increased cost. *Damage-tolerant design* with its concomitant inspection and maintenance programme recognises that all physical systems have flaws and that they propagate with time, but that the faults can be managed with the aid of a maintenance programme so as to minimise the risk of catastrophic damage.

Bridge design standards and recommended practice are based on safe-life philosophy, although only recently has 'life' been defined. Implicitly, 'life' meant a long time, with appropriate maintenance. However, all bridges have flaws, and some flaws are exacerbated during service. In such cases, if it is necessary to maintain traffic on the bridge, or to maintain operation, a damage-tolerant design approach is required until the damage is repaired.

In the absence of a British Standard or a Eurocode or any other European standard specifically on movable bridge design in the English language, the authors referred to the American Association of State Highway and Transportation Officials (AASHTO) standards and the American Railway Engineering and Maintenance Association (AREMA) *Manual of Railway Engineering* for guidance on engineering design matters. In particular, the current AASHTO load and resistance factor standard (AASHTO, 2007) and the AREMA working stress recommended practice (AREMA, 2014) were selected for all the numerical examples in Chapters 12, 13, 14 and 15. The major reasons leading to this choice are described in Section 5.2, Development of movable bridge design practice.

5.1. Units
Initially the author's intent was to use metric units throughout this book. However, after it became evident that the book had to be based on the AASHTO standard and the AREMA recommended practice in order to be of practical value, the necessity of using both US conventional (or customary) units (USCU) and metric units became obvious. The conversions from USCU to metric, with examples, that were utilised are shown in Table 5.1. Equivalent machined surface finishes are shown in Table 5.2.

5.2. Development of movable bridge design practice
Before the mid-19th century movable bridges were built by construction contractors engaged by the bridge owner – a private venture or a governmental agency – often as toll bridges. In the USA the

Table 5.1 Conversion of units printed in a reference (ref.)

Where reference units are US Customary Units (USCU), the reference value will be listed in USCU, and a conversion to metric will be given alongside in parenthesis.

1	Ref. has metric dimension. Copy without conversion
2	Ref. has dimension in feet and inches. Convert to decimal feet and metres. Example: Ref. says 48 ft–8$\frac{1}{2}$ in or 48′–8$\frac{1}{2}$″ Convert to decimal feet, 48.71 ft and then convert to metres (48.71) (0.3048) = 14.85 m. For book write as 48.71 ft (14.85 m)
3	Ref. had dimension in inches. Convert inches to mm. Example: Ref. says 6″. Convert to mm (6) (25.4) = 152.4 mm. For book write as 6 inches (152.4 mm)
4	Ref. gives weight in pounds (lb) or kilopounds (kips). Convert pounds (lb) to kilograms (kg) or tonnes (metric tons). Example: 55 lb (0.4536) = 24.95 kg and 55 kips (0.4536) = 24.95 tonnes. For the book write as 55 lb (24.95 kg) or 55 kips (24.95 tonnes) for the respective examples
5	Ref. gives torque in ft-lb. Convert to N-m, using factor of 1.3558. Example: 100 ft-lb = (100)(1.3558) = 135.6 N-m. For book write as 100 ft-lb (135.6 N-m)
6	Ref. gives pressure or stress in ksi. Convert to MPa, using factor of 6.89476. Example: 10 ksi = (10) (6.89476) = 68.9 MPa. For book write as 10 ksi (68.9 MPa)
7	Ref. gives velocity as ft/sec. Convert to m/s, using factor of 0.3048. Example: 100 ft/sec = (100)(0.3048) = 30.48 m/s. For book write as 100 ft/sec (30.48 m/s)
8	Ref. gives power in hp. Convert to kW, using factor of 0.746. Example: 50 hp = (50)(0.746) = 37.3 kW. For book write as 50 hp (37.3 kW)
9	Ref. gives energy in ft-lb. Convert to Joules, using factor of 1.3558. Example: 100 ft-lb = (100)(1.3558) = 135.6 Joules. For book write as 100 ft-lb (135.6 Joules)
10	Ref. Temperature in °F. Convert to °C, using formula °C = (°F − 32) × 5/9. Example: 50°F = 10°C. For book write as 50°F (10°C)
11	Ref. roughness surface given in micro-inches. See Table 5.2 for equivalent surface finish in micro-metres. Example: 32 micro-inches = 0.8 micro-metres. For book write as 32 micro-inches (0.8 micro-metres)

Note: All computations have been made and presented on the basis of the AASHTO *LRFD Movable Highway Bridge Design Specifications*, 2nd edn, 2007, in the US Customary Units (USCU). They are: pound force (lbf), inch (in) or foot (ft), second (s), and degree Fahrenheit (f). The unit kilopoundf (kipf) may be used where the lbf number has too many digits

Table 5.2 Equivalent surface roughness finishes

USCU: Micro-inches	Metric: Micro-metres
8	0.2
16	0.4
32	0.8
63	1.6
125	3.2
250	6.3
500	12.5

bridges were usually two-arm swing bridges rotating about a central masonry pier built in the waterway. In Europe, where there were more canals and docks, the proportion of single-arm swing bridges was greater. For them the axis of rotation was located on land. In the Low Countries, balance beam bascules were a preferred solution for canal and other narrow waterway crossings.

Towards the mid-19th century the westward expansion of the US population, industrialisation, and railway growth led to the need for more low-level crossings of navigable waterways. Two types of double-arm swing bridges were developed

- the continuous through truss type which comprised two or more timber trusses, usually of the parallel chord Howe or Pratt patent, that were continuous for the full length of the draw and were supported by a turntable on a pivot pier
- the cable-stayed or 'hog-tied' type which comprised two simple truss bridges of box-shaped cross section, each hinged at one end to a turntable on the pivot pier. Wrought iron 'chains' were draped from the top of a tower mounted on the turntable to far ends of the two trusses. The timber swing bridges were usually rim or combined bearing and were operated manually. Photographs of a 'hog-tied' railway swing bridge built of iron (the Clinton draw), in the open and closed positions, are shown in Birnstiel (2008).

The Clinton draw was built in 1864. It comprised a pair of Bollman patented suspension trusses with inclined stays ('hog-ties') radiating from the top of a cast-iron tower that was mounted on the turntable. The overall length of the draw was 300 ft (91.5 m). (Wendell Bollman patented the first all-metal American railway bridge truss. It had cast-iron compression members and wrought-iron tension members and was detailed so that a member could be replaced without shoring the whole span during the replacement process.) The Clinton turntable was operated by a steam engine that was located at the pivot pier, above the top chords of the trusses.

As wrought iron began to be produced in quantity in the USA, and the country was no longer dependent on importing British iron, many of the 'bridge companies' that had been erecting timber bridges included iron bridges in their offerings. They were vendors to the railways and did the structural engineering design as well as fabricating the ironwork. Because the bridge contracts were based on price, the railways soon learned that it was necessary for them to include design specifications in their purchase orders. These specifications were prepared by the railway's engineering staff or consulting engineers. The consulting engineers helped prepare the design specifications; they did not actually prepare the design drawings: that was left to the iron fabricator. The state bridge engineer for Virginia from 1952–1972, JN Clary, reviewed American specifications for iron and steel bridges (mostly railway) prepared in the last half of the 19th century. The work lists at least 100 railway bridge specifications. That document was discovered in the Virginia Department of Transportation files and was reprinted by Wallace T McKeel, Jr, and Ann B Miller (McKeel and Miller, 2006). However, the railway bridge specifications contained few provisions for movable bridges. An exception was Anon. (1879) which included requirements for swing bridge turntable design (part of the stabilising machinery) but ignored span drive machinery design.

5.2.1 The Price paper

The state-of-the-art of movable bridge practice in the UK and on the Continent (mainly with respect to configuration) was reported by Price (1879). His table of movable bridges lists some US bridges including those over the Mississippi that were described by Warren (1878). The printed discussion of the Price paper contains much information regarding tilting swing bridges and pivot construction as built in the UK, and in France and Germany.

About this time the controversy in America about cable-stayed as opposed to continuous truss construction of double-armed swing bridges appears to have temporarily subsided. Railway swing bridges were to be continuous trusses extending from end-to-end over the pivot piers, primarily because increases in railway loading and speeds required stiffer bridges. Cable-stayed swing bridges such as those built by the King Bridge Company in Ohio were still constructed for streets.

No continuous truss as opposed to cable-stayed swing bridge design issue developed in Britain. From the outset of swing bridge construction, the double-arm swing bridges were built with continuous lattice or plate girders with a bow-shaped upper flange supported by a turntable on the pivot pier. An example is the swing bridge at Newcastle upon Tyne (Figure 1.9).

5.2.2 The Schneider paper and specification

Charles C Schneider published a paper in the American Society of Civil Engineers (ASCE) *Transactions* which included a general specification for movable bridges. Although not limited to them, it was intended primarily for railway bridges (Schneider, 1908). It was a state-of-the-art paper with emphasis on swing bridges, because swing bridges were the most prevalent type, especially among railways. He defined and classified the major configurations of movable bridges and promoted the advantages of the centre bearing swing bridge as opposed to the rim bearing type. Rim bearing swing bridges had been in vogue in the UK and in the Midwest after the civil war. Schneider's paper elicited much written discussion, in which Ralph Modjeski wrote

> In all the so-called standard specifications for bridge superstructures clauses relating to movable bridges seem to have been carefully avoided, and yet such bridges constitute the most expensive part of a railway, . . . Mr Schneider's very clear and concise paper is valuable because it embodies, probably the first general specification on movable bridges, and because it opens the field for a thorough discussion.
>
> Schneider (1908, p. 308)

Schneider had an extensive background in bridge engineering, including movable bridges. From 1900–1903 he was vice president of the American Bridge Company, the firm assembled by the JP Morgan interests in early 1900 from more than two dozen pre-existing independent bridge fabricating companies. One of the acquired companies, located in Chicago, had the name American Bridge Company, and had built double-arm swings in the Midwest in the late 1800s including a 363 ft (110.7 m)-long symmetrical double-arm swing bridge over the Missouri River at Boonville, MO, completed in January, 1874, for the Missouri, Kansas and Texas Railway. Schneider had extensive background experience in proposing specifications for the mechanical and electrical engineering of movable bridges, even though he was a graduate civil engineer.

5.2.3 The Leffler paper and specification

Burton R Leffler published a short paper with an accompanying specification covering railway bridges that move in a vertical plane; that is, bascules and vertical lifts. He presented quite a detailed specification that dealt with general details of design, operating machinery, counter-balancing, wire rope, workmanship, allowable stresses in machinery components, motors (steam engines, gasoline and electric motors) and electrical controls (Leffler, 1913). Leffler was engineer of bridges for the Lake Shore and Michigan Southern Railroad in the early 1900s. He maintained an interest in movable bridges into his senior years, being awarded a patent for a vertical lift bridge with a tower rack drive, jointly with Clifford E Paine (of Strauss and Paine fame) in 1942.

Three discussers criticised Leffler's 1913 work. He accepted some of the suggestions and comments and included a revised specification in the closure to his paper. Many of Leffler's specification provisions may be found today in Chapter 15 of the AREMA *Manual* (AREMA, 2014).

5.2.4 AREA and AASHTO acceptance

The mechanical and electrical specification provisions proposed by Schneider and Leffler were not immediately adopted by AREA (the American Railway Engineering Association, a precursor to AREMA). The AREA *Manual* (AREA, 1921) contains only two pages devoted to mechanical equipment: 'Specifications for bronze bearing metals for turntables and movable railway bridges', which had been adopted by AREA in 1918. However, 'Specifications for movable railway bridges' were distributed by AREA during the following year, 1922. Those specifications dealt only with mechanical machinery and power equipment; they did not address structure. The coverage was approximately that of Section 15.6 in the 2014 *Manual* (AREMA, 2014). So it appears that American specifications for movable railway bridges essentially date from 1922.

The American Association of State Highway Officials (the precursor to AASHTO) adopted a modified version of the AREA movable bridge specification in 1937.

5.3. European movable bridge specifications

Currently there are no specifications specifically for the design of movable bridges in the English language in force in Europe. The Netherlands Standardisation Institute (Nederlands Normalisatie Instituut, NEN) issued *Voorschriften voor het ontwerpen van beweegbare bruggen* [Rules for the design of movable bridges] in 2001 with a supplement in 2002 (NEN, 2002). In 1998 the Deutsches Institute für Normung (DIN) [German Institute for Standardisation] promulgated the three-part standard (in English) DIN 19704-1-2-3 Hydraulic steel structures (DIN, 1998). It is written for canal works including canal bridges, but in the absence of a movable bridge specification designers use it as a reference for movable bridge machinery and control design. In May, 2012 a draft update of DIN 19704-1-2-3 was circulated for comments due 21 September 2012, in anticipation of superseding the 1998 document. The 2012 draft was available only in German as of this writing. In the following subsections the contents of NEN (2002) and DIN (1998) will be very briefly reviewed. Comparisons between those two documents, and between them and AASHTO (2007) and AREMA (2014) are beyond the scope of this work.

Movable bridge engineers in France sometimes refer to *Rules for the Design of Hoisting Appliances* promulgated by the Fédération Européenne de la Manutention (FEM) (FEM, 1998).

5.3.1 The Netherlands' rules for design of movable bridges

NEN 6786:2001 (NEN, 2002) is divided into 11 chapters with two appendices. The chapter headings and contents are as follows

Chapters 1–4	Administrative matters
Chapter 5	General provisions
Chapter 6	Limit states
Chapter 7	Materials and their properties
Chapter 8	Loads
Chapter 9	Component assessment based on material properties, fatigue, contact stresses, stability and testing
Chapter 10	Structural design, machine analysis
Chapter 11	Electrical power and control

Although the norm was not revised in detail, it appears to be more thorough than AREMA (2014) on such topics as mechanical component fatigue, bolting and loading. It even treats the torsion due to Lang's lay wire rope construction.

5.3.2 DIN hydraulic steel structures, May 1998 (DIN, 1998)

The norm for hydraulic steel structures comprises three parts. 19704-1 deals with design analysis, 19704-2 treats details and fabrication, and 19704-3 covers electrical equipment. Parts 1 and 2 both have 10 chapters, Part 3 has seven but has extensive references to other DINs, for example, DIN V 19250 Control technology. Only the chapter headings for Part 1 are listed below.

Chapters 1–3	Administrative matters
Chapter 4	Materials
Chapter 5	Actions (loads) on steel structures
Chapter 6	Friction
Chapter 7	Analysis of steel structures
Chapter 8	Actions (loads) on machinery
Chapter 9	Machinery analysis
Chapter 10	Analysis of special machinery components

It should be noted that most provisions refer to one or more supplementary DIN norms. Of course, many provisions of DIN 19704 would not apply to a movable bridge design, thus simplifying matters.

5.3.3 Fédération Européenne de la Manutention

The 'Rules' are available in English, French and German. FEM 1.001, 3rd edition, revised 1998 (FEM, 1998) has the English title *Rules for Design of Hoisting Appliances* and was published in eight booklets. The booklet titles are:

1. *Object and scope*
2. *Classification and loading on structures and mechanisms*
3. *Calculating the stresses in the structure*
4. *Checking for fatigue and choice of mechanism components*
5. *Electrical equipment*
6. *Stability and safety against movement by wind*
7. *Safety rules*
8. *Test loads and tolerances.*

At the time of writing, Booklet 6 appears to be undergoing revision.

5.4. USA Railroad Bridge Safety Standard

In the USA most railroad track is privately owned. Historically, the design and maintenance of railroad bridges supporting the trackage was subject to little government regulation; local, state, or federal. Partly in response to some serious accidents at movable bridges, the Office of Safety of the Federal Railroad Administration (FRA), a unit of the US Department of Transportation (USDOT), promulgated rules regarding bridges supporting most privately owned active trackage with a guage of 2 ft (0.61 m) or more (USDOT, 2013). Severe civil penalties are prescribed for violations, and in the case of wilful violation, individuals and employers may be assessed. Furthermore, per statute, 'each day a violation continues shall constitute a separate offense' (USDOT, 2013).

Essentially, CFR 49 Part 237 (USDOT, 2013) requires trackage owners with bridges to adopt a bridge management programme to prevent the deterioration of railroad bridges and to reduce the risk of catastrophic bridge failure. The content of bridge management programmes shall include (a) an accurate inventory of railroad bridges; (b) determining and recording the safe load capacity of each bridge; (c) developing a means to retain original design documents and documentation of repairs, modifications, and field inspections, and (d) develop a bridge inspection programme.

The law also deals with qualifications and the designation of responsible persons involved with bridge inspection, review of reports, engineering design, and supervision of repairs and modifications. The law cites the American Railway Engineering and Maintenance Association (AREMA) *Manual for Railway Engineering* as its primary reference document.

REFERENCES

Anon. (1879) General specifications for a wrought-iron railway draw bridge to be erected for the Chicago, Milwaukee, and St Paul Railway Company, over canal in the city of Milwaukee, Wisconsin. *Engineering* News, January 18, **VI**: 21–22.

AASHTO (American Association of State Highway and Transportation Officials) (2007) *Standard Specifications for Movable Highway Bridges*, 2nd edn. AASHTO, Washington, DC, USA.

AASHTO (2007b) *AASHTO LRFD Movable Highway Bridge Design Specifications*, 2nd edn. AASHTO, Washington, DC, USA.

AREA (American Railway Engineering Association) (1921) *Manual of the American Railway Engineering Association*. AREA, Chicago, IL, USA.

AREMA (American Railway Engineering and Maintenance-of-Way Association) (2014) Movable Bridges – Steel Structures. Ch. 15 in *AREMA Manual for Railway Engineering*, Lanham, MD, USA.

Birnstiel C (2008) The Mississippi River Railway Crossing at Clinton, Iowa. In *Historic Bridges: Evaluation, Preservation, and Management* (Adeli H (ed.)). CRC Press, Boca Raton, FL, USA, http://dx.doi.org/10.1201/9781420079968.pt1.

DIN (Deutsches Institute für Normung) (1998) DIN 19704-1-2-3(1998). *Hydraulic Steel Structures*. DIN, Berlin, Germany.

FEM (Fédération Européenne de la Manutention) (1998) FEM 1.001(1998). *Rules for the Design of Hoisting Appliances*. FEM, Brussels, Belgium.

Leffler BR (1913) Specifications for metal railroad bridges movable in a vertical plane. *Transactions*, American Society of Civil Engineers, **76**, Paper No. 1251: 370.

McKeel Jr WT and Miller AB (2006) *History of Early Bridge Specifications: A Reprint of a Paper by JN Clary*, Report No. VTRC 07-R10. Virginia Transportation Research Council, Charlottesville, VA, USA.

NEN (Nederlands Normalisatie Instituut) (2002) NEN 6786:2001(2002). *Voorschriften voor het ontwerpen van beweegbare bruggen (VOBB)* (in Dutch). NEN, Delft, Holland.

Price J (1879) Movable bridges. *Minutes of Proceedings*, Institution of Civil Engineers, Session 1878–1879 Part III, Paper 1537.

Schneider CC (1908) Movable bridges. *Transactions*, American Society of Civil Engineers, **60**, Paper No. 1071: 258–336.

USDOT Federal Railroad Administration Office of Safety (2013) Code of Regulations, Title 49, Bridge Safety Standards, Part 237. The Railway Educational Burea, Omaha, NE, USA.

Warren GK (1878) Bridging the Mississippi River between St Paul, MN, and St Louis, MO. In Appendix X, *Report of the Chief Engineer*, Government Printing Office, Washington, DC, USA.

Movable Bridge Design
ISBN 978-0-7277-5804-0

http://dx.doi.org/10.1680/mbd.58040.123

Chapter 6
Materials of construction

Charles Birnstiel

6.0. Introduction

Materials used to construct the superstructure and machinery of movable bridges will be reviewed in this chapter. The most important are alloys of iron in various forms, but copper alloys and non-metallic materials are also significant. This chapter is not a discourse on material science, although there is reference to elementary metallurgy. The emphasis is on the manufacture of materials. Properties of the materials significant in movable bridge design and material failures are treated in Chapter 7.

6.1. Ferrous materials

Compounds of iron are very prevalent in the Earth's crust in the form of iron ores. Overall, the most common iron ores contain mineral oxides such as magnetite (Fe_3O_4), hematite (Fe_2O_3) and limonite ($FeOOH$, approx.). In Britain the most common iron mineral is a carbonate called siderite ($FeCO_3$) (Sim and Ridge, 2002). Ore minerals are usually mixed with rock or non-metallic matter called gangue. That matter is separated from the ore by dressing, a process that varies with the character of the ore deposit. The gangue residue, non-mineral components of the ore, form a viscous liquid slag when heated above 1000°C.

6.2. Wrought iron

The origin of wrought iron dates back to at least 1400 BC in the Old Hittite Kingdom that was located in Asia Minor. After the collapse of that Kingdom, about 1200 BC, migrants spread the Hittite iron technology through southern Europe and the Middle East. The Early Iron Age in central Europe, from 800 BC to 500 BC, is known as the Hallstatt Period. Hallstatt is a village in Austria. Celtic migrations beginning about 500 BC brought iron technology from Hallstatt to Western Europe and Britain. Iron ore mining, dressing, smelting and forging techniques did not advance much in the next 1000 years. The equipment and processes in place about 1500 AD are shown in Agricola (1558). The smelting process described in Sim and Ridge (2002) was, essentially, still used in backward areas of Germany in the early 1800s, except that the furnace charge may have included limestone.

Smelting of dressed iron ore was accomplished by the direct process until the 18th century. In the case of iron oxide the mineral was heated by burning charcoal, a process that released carbon monoxide (CO) gas. Heating the oxide ore (Fe_2O_3) in an oxygen-starved atmosphere of carbon monoxide (CO) caused oxygen to move from the ore to form carbon dioxide gas (CO_2), leaving iron (Fe). In chemical equation form:

$$Fe_2O_3 + 3CO \rightarrow 2Fe + 3CO_2$$

For a carbonate ore (siderite) the ore was first roasted (Overman, 1850/2011) in order to drive off the carbon (C). The direct process for siderite is:

$$FeCO_3 \rightarrow FeO + CO_2$$

$$2FeO + \tfrac{1}{2}O_2 \rightarrow Fe_2O_3$$

$$Fe_2O_3 + 3CO \rightarrow 2Fe + 3CO_2$$

The significant temperatures while smelting with the direct process were

- bloomery furnace operating temperature about 1200°C (2190°F)
- iron oxide reduces to iron at about 800°C (1470°F)
- melting point of iron at about 1540°C (2800°F)
- slag becomes fluid at approximately 1150°C (2100°F).

So, during operation of the furnace using charcoal as fuel, and often with an air blast, the temperature was hot enough to reduce the mineral and melt the slag, but not hot enough to melt the iron. The product of the furnace was a sponge ball of iron physically containing slag. The slag was in physical association with the iron, but not chemically bound to it. The sponge iron was then hammered to drive out the excess slag. The resulting wrought iron bars were then sent to a forge for further processing. In the 19th century, hammering was largely replaced by rolling.

After 1850, there were many advances in methods of wrought iron manufacture in Britain, the USA, and elsewhere. Henry Cort is credited with developing the puddling process in the 1780s, which revolutionised the economy of iron manufacture. The puddling process and puddling furnaces are described in Ketchum (1924), Moore (1941), Aston and Story (1957), and Paulinyi (1987).

Wrought iron may be considered a composite material: high purity iron with iron silicate (a glass-like slag) dispersed within the iron. The slag is in the form of threads or fibres dispersed throughout the cross-section of the rolled bars. In the 20th-century product there may have been on the order of 250 000 glass-like fibres per square inch (645 mm^2) of cross-section (Aston and Story, 1957). Photomicrographs of wrought iron in longitudinal section are shown in Swain (1924), Moore (1941), Aston and Story (1957) and Cobb (2012).

Although commercial production of wrought iron ceased in 1973, prodigious amounts were produced in the UK and the USA during the 19th century and bridges, some of which are extant, were built from it until the 1890s.

6.3. Steels

Steel is a composition of iron and carbon (a nonmetal), other alloying elements and impurities. The other alloying elements may be metals or nonmetals. However, in most constructional steels, carbon is the principal alloying element, usually varying from 0.10–1.2% by weight. Ferrous alloys with carbon contents from about 2.0–4.5% are conventionally termed cast irons. The 2.0% carbon content dividing line between steel and cast iron is a convenience for definition. It was taken as the maximum carbon content at which the material could be hot-rolled. Actually, iron-carbon alloys with a higher percentage of carbon can be rolled.

The practice of iron making is similar to the smelting operation for wrought iron described previously, but the furnace is immensely larger – it is called a blast furnace. The blast furnace is charged with iron ore, coke, limestone and air. Coke is carbonised coal which serves the same purpose as charcoal did in making wrought iron, to provide heat and a carbon monoxide (CO) atmosphere when burning. The limestone serves as a flux: it facilitates the agglomeration and separation of sand and other undesirable material in the dressed ore from the melt. The basic chemical equation for the blast furnace process is the same as for that for the ancient direct smelting process

$$Fe_2O_3 + 3CO \rightarrow 2Fe + 3CO_2$$

The important difference between the smelter for wrought iron and the blast furnace is that the temperature in the blast furnace is higher so that the iron melts and is discharged as a fluid, rather than as a pasty mass as from an 18th-century wrought-iron bloomery. Liquid iron is discharged (a) into sand moulds in order to solidify into pigs which are stored for later processing, or (b) into ladles for transporting to steel-making furnaces. For about a century Bessemer converters and open-hearth furnaces (the Siemens process) were used for this. The chemistry of the batches in Bessemer or Siemens furnaces was adjusted by adding elements to produce the desired type of steel. The liquid steel was discharged from the furnace into ladles for transporting to ingot moulds into which it was poured.

Currently, the emphasis in the USA is on making steel from scrap metal, rather than iron ore, and the use of continuous casting rather than the batch process for casting ingots. The variability of constituents in the scrap raw material and the reduced amount of hot rolling that was necessary because of the thinner slabs produced by the continuous casting products initially caused the end product to contain more inclusions than in those produced by the older manufacturing method. However, 'clean steel' technology has evolved to minimise this problem at modern plants (Davis, 1998, p. 194). Coarse inclusions are important because they serve as fatigue-cracking initiation sites. The effect of inclusions on mechanical properties has been described by Krauss (2015, pp. 172–174).

Steelmaking processes are thoroughly described in Davis (1998) and Krauss (2015, pp. 9–15).

6.3.1 SAE steel designation system

By the mid-1900s the American Iron and Steel Institute (AISI) and the then Society of Automotive Engineers (SAE) had jointly developed a system for designating carbon and alloy steels, the AISI/SAE steel grades. Since 1996 it has been known as the SAE Designations. It is a general classifier, insufficient for a purchase order, for which a further specification (usually an American Society for Testing and Materials (ASTM)) is needed to define the product. The basic SAE Designation System uses a four-digit code WXYZ, in which W is an integer representing the major (controlling) alloying element(s). The primary and secondary alloying element numerical code is

1 = Carbon
2 = Nickel
3 = Nickel and chromium
4 = Molybdenum
5 = Chromium
6 = Chromium and vanadium
7 = Tungsten and chromium
8 = Nickel, chromium and molybdenum
9 = Silicon and manganese

The second digit, X, represents the concentration of the secondary alloying element(s) in percentage (1 = 1%) by weight. YZ represent the carbon concentration in hundredths (1/100) of a percent. For example: SAE 1015 refers to steel with carbon as the primary alloying element in the amount of 0.15% and with no secondary alloying element. In addition to the above, there are subdivisions to identify alloying elements that improve hardening (e.g. boron) and machining characteristics (e.g. lead). Complete SAE designation tables may be found in Davis (1998). ASM International also maintains several other alloy steel designation systems.

6.3.2 Effect of alloying elements

The effect of adding alloying elements may be briefly, and perhaps simplistically, described as follows.

- Carbon (C) is the most important alloying nonmetallic element although, paradoxically, steel containing carbon is not called an alloy steel in most metallurgical literature. Carbon, a nonmetal, strongly influences the hardness, strength and ductility of steel. As the carbon content percentage increases, so does hardness, but the melting point of the metal and its ductility and weldability decrease. The influence of carbon content in steel on mechanical properties is apparent in Table 6.1.
- Chromium (Cr) combines with the carbon and iron of the steel to increase hardness and toughness and improve high-temperature strength. When the chromium content exceeds 13% the resistance of the steel to wear and corrosion is substantially increased. The upper limit for chromium content is about 35%.

Table 6.1 Mechanical properties of selected carbon and alloy steels in the hot-rolled, normalised and annealed condition

SAE	Treatment	Tensile: MPa	Strength: ksi	Yield: MPa	Strength: ksi	Elongation: %	Reduction in area: %	Hardness: HB
1015	As-rolled	420.6	61.0	313.7	45.5	39.0	61.0	126
	Normalised	424.0	61.5	324.1	47.0	37.0	69.6	121
	Annealed	386.1	56.0	284.4	41.3	37.0	69.7	111
1040	As-rolled	620.5	90.0	413.7	60.0	25.0	50.0	201
	Normalised	589.5	85.5	374.0	54.3	28.0	54.9	170
	Annealed	518.8	75.3	353.4	51.3	30.2	57.2	149
1095	As-rolled	965.3	140.0	572.3	83.0	9.0	18.0	293
	Normalised	1013.5	147.0	499.9	72.5	9.5	13.5	293
	Annealed	656.7	95.3	379.2	55.0	13.0	20.6	192
3140	Normalised	891.5	129.3	599.8	87.0	19.7	57.3	262
	Annealed	689.5	100.0	422.6	61.3	24.5	50.8	197
4140	Normalised	1020.4	148.0	655.0	95.0	17.7	46.8	302
	Annealed	655.0	95.0	417.1	60.5	25.7	56.9	197
4340	Normalised	1279.0	185.5	861.8	125.0	12.2	36.3	363
	Annealed	744.6	108.0	472.3	68.5	22.0	49.9	217
5140	Normalised	792.9	115.0	472.3	68.5	22.7	59.2	229
	Annealed	572.3	83.0	293.0	42.5	28.6	57.3	167
8260	Normalised	632.9	91.8	357.1	51.8	26.3	59.7	183
	Annealed	536.4	77.8	385.4	55.9	31.3	62.1	149
9255	Normalised	932.9	135.3	579.2	84.0	19.7	43.4	269
	Annealed	774.3	112.3	486.1	70.5	21.7	41.1	229

Data were obtained from specimens 12.8 mm (0.505 in.) in diameter that were machined from 25 mm (1 in.) rounds
All grades are fine-grained. Heat-treated specimens were oil-quenched unless otherwise indicated
Abstracted from Table 1 of Davis (1998)

- Manganese (Mn): some manganese is normally present in commercial steels. It contributes to strength and hardness and controls sulphur content. Steel containing 7–15% Mn is very hard and tough and has the ability to harden on the surface with the core below remaining strong.
- Molybdenum (Mo): The effect of adding molybdenum to steel improves deep hardening properties and strength at high temperatures. It also improves toughness and corrosion resistance.
- Nickel (Ni) as an alloying element added to steel improves strength, toughness and fatigue resistance without loss of ductility. Nickel steels are tough and can be readily heat treated. Nickel is used to make stainless and heat-resistant steels. In combination with chromium the benefits of both alloying elements are reinforced.
- Silicon (Si) is a nonmetal element present in all steels. In combination with other alloying elements it adds strength and toughness. It is detrimental to surface quality.
- Tungsten (W): the addition of tungsten improves the hardness, strength and toughness of steel. It is used to make high speed machinable steels because it has greater stability at high temperature. The alloy steel remains hard even when red hot.
- Vanadium (V), when added to steel, refines its grain structure and increases fatigue resistance. Hardenability is increased in some heat-treatment regimes. It is added to iron in order to form high speed and high temperature steels.

The effect of alloying elements on the mechanical properties of selected steels at different heat treatments (hence different microstructures) is evident in Table 6.1, which was abstracted from Davis (1998). For various alloys the tensile and yield strengths, percent elongation and percent reduction in cross-sectional area at tensile fracture are listed corresponding to the as-rolled, normalised, and annealed heat treatment condition. As-rolled means that the final 'heat treatment' was the hot-rolling operation. After passing the last set of rolls the member; bar, plate, or structural shape was placed on a cooling table for atmospheric cooling.

Normalised means that the test specimen was cut from a bar that was heated above the transformation range (defined later in this chapter), and then cooled in air to a temperature well below the transformation range. Annealed means that the member was fully annealed: it was heated to a suitable temperature and was then cooled at a suitable rate to soften the material and produce desired changes in properties or microstructure.

Normalising and annealing processes will be treated in detail later.

Consider the plain carbon steel SAE 1015. The effect of heat treatment on yield strength is

As-rolled	313.7 MPa (45.5 ksi)
Normalised	324.1 MPa (47.0 ksi)
Annealed	284.4 MPa (41.2 ksi)

Full annealing reduces the steel to its basic yield point.

Comparing plain carbon steels SAE 1015 and SAE 1095 in the annealed state, increasing the carbon content from 0.15% to 0.95% resulted in an increase of the yield strength from 284.4 MPa (41.2 ksi) to 379.2 MPa (83.0 ksi), a 33% increase.

Comparing yield strengths of fully annealed alloys with the same carbon content

SAE 1040 Plain carbon	353.4 MPa (51.3 ksi)
SAE 4140 Molybdenum at 1%	417.1 MPa (60.5 ksi)
SAE 4340 Molybdenum at 3%	472.3 MPa (68.5 ksi)
SAE 5140 Chromium	293.0 MPa (42.5 ksi)

Adding 3% molybdenum results in a 34% increase in yield strength over the carbon steel with the same carbon content. Considering the effect of alloying on the ratio of yield strength for normalised against annealed materials in Table 6.1 shows that the ratio increases with alloying, for the same carbon content.

From Table 6.1 it is also evident that for plain carbon steel the increase in yield strength accompanying increased carbon content comes at the expense of significant reduction in elongation at tensile failure. Elongation is a measure of ductility and workability of the material. In the as-rolled condition the elongation for SAE 1015 is 39%. It reduces to 25% for SAE 1040 and to 9% for SAE 1095. Thus the ability of the steel to deform plastically without fracturing is compromised by higher carbon content.

From the data in Table 6.1 it is also obvious that chemical content is not the sole determinant of physical properties of steels. The reason for this lies in transformations within the crystals that form the metal. The properties of steel depend on its microstructure which is linked to the *processing path* as well as the chemical composition. Hence, it becomes desirable to review some elementary basic metallurgy.

It is difficult for a person who has sawn, filed, drilled and tapped a block of steel to accept the fact that small atoms such as hydrogen and carbon can enter into the block. And even worse, the composition of the solid can change with temperature while it is solid! To quote Reardon (2011)

> Convincing the non-metallurgist that atoms move about in solid metals and alloys may be a hard sell, but it is 100% gospel metallurgy. Atoms in a solid are constantly vibrating due to their thermal motion. Although it may be difficult to imagine, diffusion allows them to migrate to different locations within a crystal lattice. The vibrating atom may switch positions with a neighbouring atom in the lattice, or an atom may even move into neighbouring vacancy sites or defect areas.

For a particular steel composition (alloy), most structure-sensitive properties, such as yield strength and hardness, depend on microstructure. Herein the word microstructure relates to the size and shape of the grains comprising the metal and their distribution. Microstructure is determined mainly by examining polished and etched surfaces of metal specimens with the aid of an optical microscope (metallography). All of the important micro-constituents of steel have characteristic appearances on the surface of the polished samples (Verhoeven, 2007). The definition of microstructure (Davis, 1998) is 'the structure of an object, organism, or material as revealed by a microscope at magnifications greater than 25×'. In practical metallography, the optical image enlargement is usually at least three times greater. Metallography is used to identify alloys and predict material properties. It is a means of quality control for heat-treating operations.

6.3.3 Transformations

The discussion about elementary metallurgy which follows is based on two premises: that all matter is made of atoms, and that practically all solids are crystals. A feature of a crystalline solid is that its molecules are packed in space with repetitive regularity. Crystals of many metals have a face-centred cubic (f.c.c.) structure as shown in Figure 6.1 or the body-centred cubic (b.c.c.) structure depicted in Figure 6.2. In the f.c.c. arrangement, each atom has 12 immediate neighbours; in the b.c.c. form the atom has eight immediate neighbours.

Figure 6.1 Face-centred cubic (f.c.c.) crystal

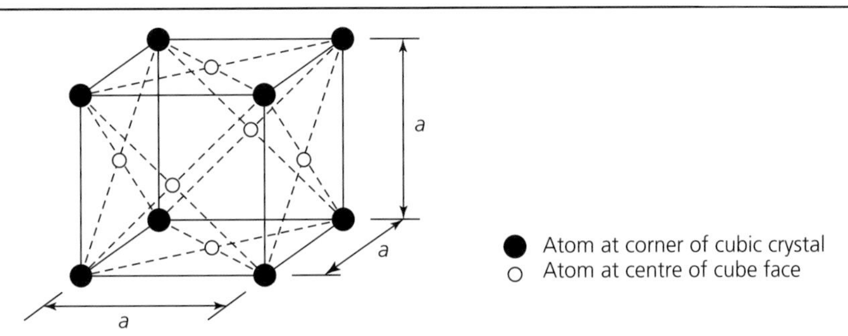

Figures 6.1 and 6.2 portray the centres of atoms. The atoms actually touch each other – they are close-packed. Each diagram shows only a crystal that is repeated in all directions. The crystals are called grains. When a specimen of the metal is ground and polished and then chemically etched, the surface has the appearance shown in Figure 6.3, when viewed and photographed through an optical microscope.

The boundary between two crystals (or grains) is conventionally regarded as a 'crystalline defect' in which a few layers of atoms are uncertain as to which crystal they belong (Holden and Morrison, 1982). The atoms at any boundary of a crystal have a slightly different arrangement from those inside it. Grain boundaries affect the strength of the metal.

A thorough treatment of crystal defects is well beyond the scope of this chapter, but one defect is of such importance in describing metal behaviour that it will be addressed briefly. It is known as a line defect or dislocation. The defect lies along an axis (or line) of atoms in the crystalline lattice. There are two types, the most important one being the edge dislocation shown in Figure 6.4 where a partial plane of extra atoms is present above or below a dislocation line (denoted by the symbol T in the figure). The line defect is not necessarily stationary in the solid metal. Such defects can and do, move through the crystalline structure in response to stress. The formation and movement of dislocations is the basis of slip formation and the mechanism of plastic flow (Reardon, 2011). The presence of dislocations explains why metals can be bent and pounded and rolled and drawn into different shapes without breaking (Holden and Morrison, 1982). If the metal is worked, the grains will align into elongated, pan-caked shapes in the direction of metal flow. Crystal dislocations and distortions of the micostructure,

Figure 6.2 Body-centred cubic (b.c.c.) crystal

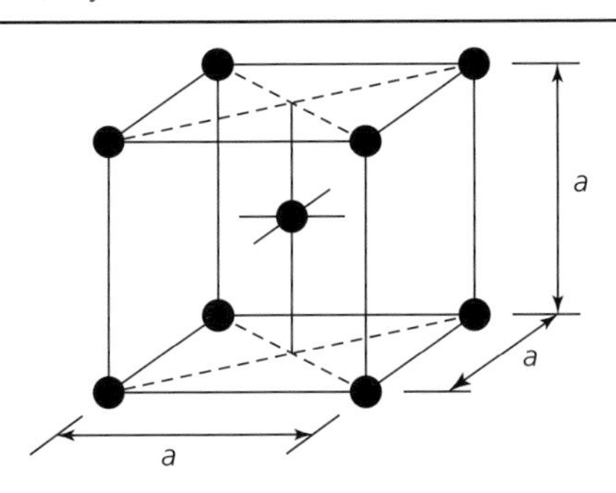

Figure 6.3 Relationship between crystal lattice, grain and grain boundary

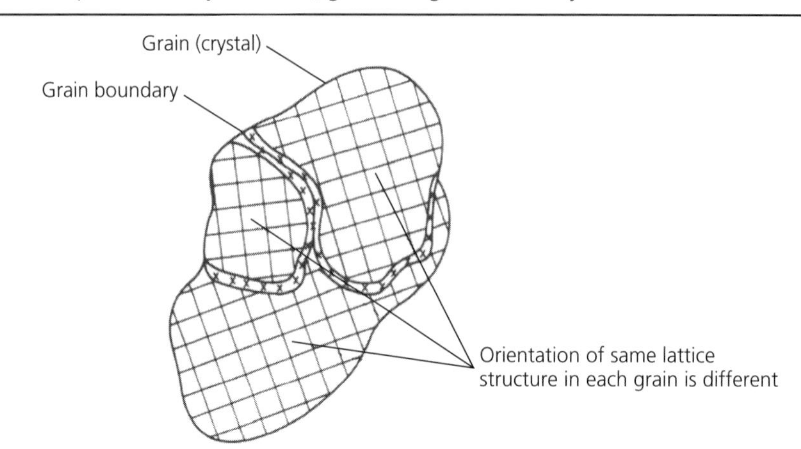

whether present in the original ingot or caused by working the metal, increase the strength of the metal (Verhoeven, 2007).

6.3.3.1 Transformations in pure iron

Iron is an allotropic element; its crystals can exist in the body-centred cubic form (b.c.c.) or the face-centred cubic structure (f.c.c.), depending on temperature. Allotrophy (the property of an element to exist in two or more forms in the same physical state and under the same pressure) is important in metallurgy because it permits the microstructure of solid iron-carbon to be altered using heat. Changes in the structure of pure iron as it transforms from a liquid to a solid when cooling very slowly are depicted in Figure 6.5. At 1540°C (2800°F) the molten iron starts to solidify and the temperature remains constant until all the liquid has 'frozen'. Because this is a pure element it has a definite freezing point. (The freezing point for steel varies depending on the alloying constraints. In general, alloys solidify over a range of temperature.) After the iron has completely solidified it continues to cool uniformly until it reaches 1395°C (2540°F). During this time the solid iron has a b.c.c. structure and the phase is called delta iron or delta ferrite. The temperature is constant at 1395°C (2450°F) for some time while the solid delta iron transforms into gamma iron with an f.c.c. structure. The discontinuities in the temperature-time plot are known as temperature arrests (Digges *et al.*, 1956). They are caused by physical changes in the iron. The arrest at 1395°C is conventionally labelled A_4.

Figure 6.4 Cross-section through crystal dislocation

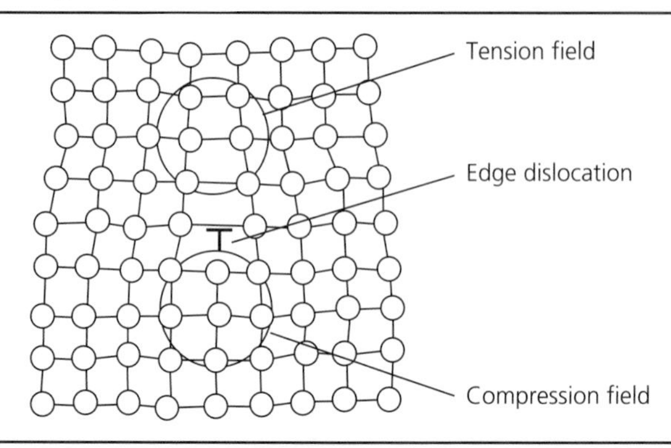

Figure 6.5 Changes in pure iron during cooling from molten state to room temperature

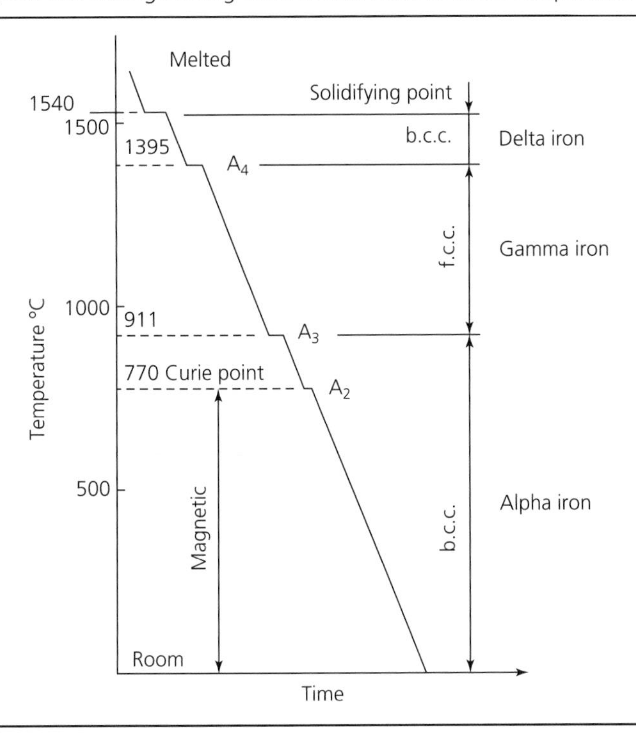

When the transformation to gamma iron has been accomplished, the temperature falls uniformly until it reaches 911°C (1670°F) and it remains constant at A_3 while the gamma iron transforms into alpha iron, which has b.c.c. structure. As the iron cools further, it reaches arrest A_2 at 770°C (1420°F) where the iron becomes magnetic.

The changes in iron structure shown for slow cooling would also occur, in reverse order, for slow heating. The process by which the atomic structure of the iron changes from b.c.c. to f.c.c., or from f.c.c. to b.c.c., is known as a phase transformation. Phase transformations occur in most metal alloys. The iron–carbon system phase transformation is described subsequently.

6.3.3.2 Iron–carbon/cementite phase diagram

Figure 6.6 is a simplified version of the iron–carbon cementite phase diagram at the condition of near equilibrium which is presented in many ASM International publications (Reardon, 2011). The ordinate is temperature (approximately linear scale) and the absisca is percent carbon in the solution by weight (approximately logarithmic scale). The diagram is included herein to show the forms of constituents in steel and cast iron at various temperatures and percent carbon content. The notation herein follows that in ASM publications. Some constituents have two or more names, for example

Austenite = gamma iron (f.c.c.)
Ferrite = alpha iron (b.c.c.) or delta iron (b.c.c.)
Cementite = carbide or Fe_3C

Figure 6.6 shows that there are four curves, labelled liquidus, solidus, eutectic, and eutectoid, that indicate temperatures where phase transformations of iron–carbon–cementite solutions occur. The

Figure 6.6 Simplified iron–carbon cementite near equilibrium phase diagram

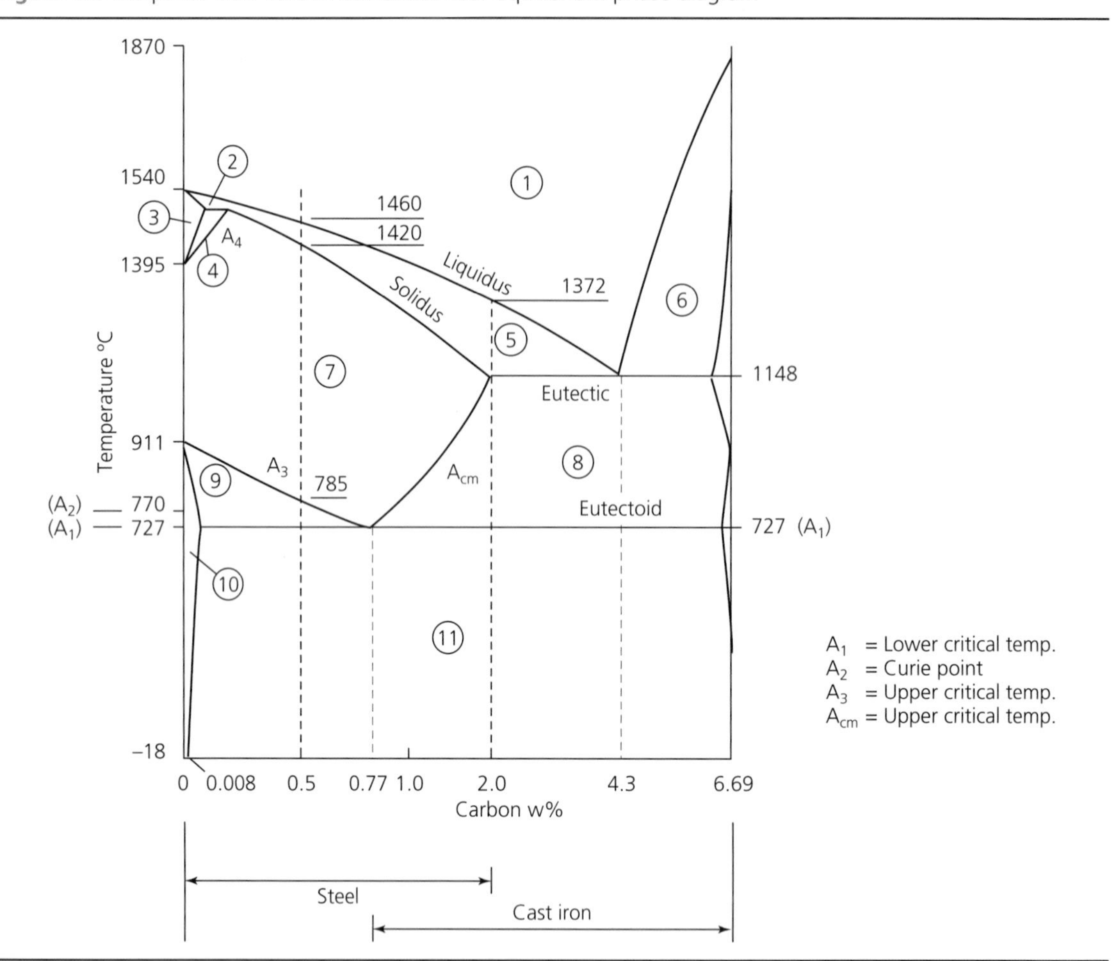

liquidus curve separates the liquid metal from a partially solidified mush. Solidus is the separation between the mush and the solid alloy for steel. Eutectic is where the liquid solution is converted into two or more intimately mixed solids on cooling. Eutectoid is a reaction in which a solid solution is converted into two or more intimately mixed solids on cooling (Reardon, 2011).

The key to the phases in the iron–carbon/cementite phase diagram at near-equilibrium conditions, Figure 6.6, follows

1. Liquid
2. Delta iron + liquid
3. Delta iron
4. Delta iron + gamma iron
5. Gamma iron + liquid
6. Cementite + liquid
7. Gamma iron
8. Gamma iron + cementite
9. Alpha iron + gamma iron
10. Alpha iron
11. Alpha iron + cementite

Significant abscissa are 0.008% which represents the limit of carbon in ferrite at room temperature; 2.0%, which is often taken as the upper limit solubility of carbon in steel; and 6.69% at which the material is all carbide having the chemical composition Fe_3C. Although not absolute, 2% carbon is usually considered the upper limit for an alloy that can be rolled. More carbon makes the alloy brittle and it has to be cast unless there is special pre-heat treatment (Henkel and Pense, 2002).

Considering the temperature scale, the melting point of steel without any carbon content (pure iron) is 1540°C (2800°F). The melting point of steel drops with increasing carbon content to about 1372°C (2500°F) at 2% carbon. It continues to drop to 1148°C (2098°F) at 4.3% carbon, in the cast iron range. The highest temperature at which the steel is solid is 1540°C (2804°F). Another significant ordinate is 727°C (1341°F) at which the solubility of carbon ferrite is 0.02% and it is the lowest temperature at which austenite solely exists, at a carbon content of 0.77%.

A more detailed iron–carbon/cementite phase diagram up to a carbon content of 0.9% is available (Tempil, 2010) which also shows the colour of steel's surface oxide at temperatures above 500°C. This diagram is useful during forging and heat treating operations.

6.4. Heat treatment of steel – general

Heat treatment is defined as controlled heating and cooling of a solid metal or alloy by methods designed to change the microstructure to obtain specific properties (Reardon, 2011). The subject is broad. And much information is closely held. Many consider heat treatment a 'black art'. Only the most elementary aspects of heat treatment will be considered herein and only for the iron–carbon/cementite phases. More thorough discussions of heat treatment of metals may be found in Thelning (1975), Chandler (1995), Henkel and Pense (2002) and Krauss (2005). Heat treatment of steel is possible because of the allotropy of iron and the fact that the carbon atom is much smaller than the iron atom. A table giving approximate mechanical properties of selected carbon and alloy steels in various quenched and tempered conditions is to be found in Davis (1998).

Heat treating of steel occurs at temperatures below the solidus line of Figure 6.6, and the microstructure changes take place inside the solid metal. Microstructure changes are caused by movement of atoms within the crystal lattices at transformation temperatures.

Figure 6.6 illustrates the crystalline constituents of iron–carbon solid solutions with varying carbon content after having slowly cooled from the molten (liquid) state. It is called a near-equilibrium iron–carbon/cementite phase diagram. However, some metallurgists consider 'equilibrium' an inappropriate description because the cementite phase is not really stable. Given sufficient time, iron carbide decomposes to iron and graphite – the steel is said to graphitise. The rate depends on temperature. As an example of the variation in structure, let us consider the structure of the metal for an iron–carbon alloy containing 0.5% carbon by weight. Between 1870°C and 1460°C at the liquidus line, the iron–carbon is in liquid solution. From 1460°C down to 1420°C at the solidus line, the material is a mushy combination of austenite and liquid. From 1420°C down to 785°C (on the A_3 line) it is a fully developed austenite. Between the A_3 line and the eutectoid line at 727°C its structure transforms, it is a combination of austenite and ferrite. Below 727°C the metal structure is a combination of α–ferrite and cementite. In terms of crystals below the solidus line, the structure changed from f.c.c. at the solidus to f.c.c. plus b.c.c. between the A_3 line and the eutectoid and then back to b.c.c. These changes in crystal structure occurred as the metal was slowly cooled from the molten state.

The intent of heat treating is to change the structures from those shown in Figure 6.6 to other microstructures by heating the metal into the transformation range, maintaining it at that temperature for a specific time and then cooling it and maintaining it at the cooled temperature and eventually reducing it

Figure 6.7 Structure of perrite and martensite

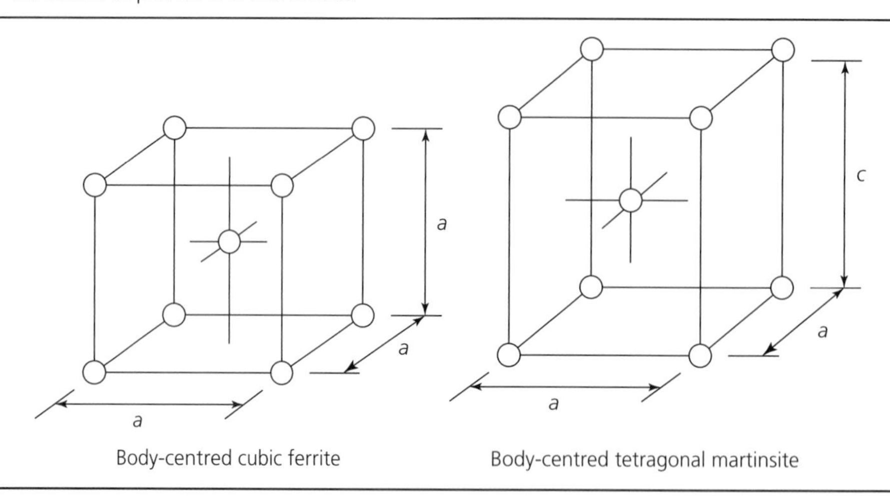

to room temperature. There are many protocols, the selection of which depends on the desired change in microstructure. The subject is treated in detail in Chandler (1995); briefly, however, it should be noted that some protocols are proprietary and the use of computer controlled furnaces has spurred innovations.

As an example of through hardening heat treatment, it is pertinent to return to Figure 6.6 and consider the phase transformations which occur when the steel with 0.5% carbon is heated into the austenitic phase and then rapidly cooled (quenched) from the austenitic phase down to room temperature. During rapid quenching, the time for carbon atoms to diffuse is insufficient and this causes lattice distortion. The alpha iron b.c.c. lattice quickly forms into a new crystal structure called martensite. The martensite cell is a b.c.c. cell with an elongated edge as shown in Figure 6.7. The elongated crystal is called body-centred tetragonal (b.c.t.). The lattice distortion results in volumetric change (usually expansion) which may need to be accounted for in the design of machinery parts, such as through hardened gearing. Martensite is extremely hard. Many steel hardening processes are based on developing this crystal structure (Reardon, 2011).

Heat treating processes include annealing, normalising, tempering and case hardening. Quenching is a cooling process and stress-relieving is a category of annealing. These processes are described briefly below.

6.4.1 Annealing and thermal stress relief

Annealing may be categorised according to the temperature to which the part is raised – seldom above 1000°C (1832°F) and the method of cooling. Annealing temperatures may be classified in the following way (after Reardon (2011)):

- subcritical (below A_1)
- intercritical (above A_1, but below A_3 or A_{cm})
- full annealing (typically 50°C (90°F) above A_3 or A_1).

By convention, use of the word annealing without a modifier implies full annealing.

Subcritical annealing, or in-process annealing, does not involve the formation of austenite. The prior condition of the steel is modified thermally. The history of the steel is important. Thermal stress-relief annealing is an important application of subcritical annealing in movable bridge machinery manufacture. It involves uniform heating of the metallic part, usually to between 550°C and 625°C, holding at that temperature for a predetermined time, followed by uniform cooling. Thermal stress-relief heat treating is used to relieve stresses from the as-rolled condition of the steel and from welding and cold working. Uniform cooling is important; otherwise new residual stresses can arise that are greater than those that the annealing process was intended to relieve. Removal of residual stresses can be important for dimensional stability during subsequent machining processes and to reduce the risk of stress-corrosion cracking (SCC) in some metals. Stress-relief cracking is discussed in Krauss (2015, pp. 449–450).

In the intercritical annealing process, austenite is formed with time at a raised temperature. Because the steel is heated above A_1, it is subjected to slow continuous cooling to permit the development of the desired microstructure.

6.4.1.1 Arc welding of heat-treatable low-alloy steel

According to Davis (1998), the high hardness of many heat treatable low-alloy steels (HTLA) with medium-carbon content in their final hardened and tempered condition often precludes their being welded in that condition because of the strong tendency for 'cold cracking'. This is cracking that develops after weld solidification is complete. Rather than attempting to join the hardened pieces, it is recommended that they be welded in the near annealed condition and then the complete weldment be heated to obtain the desired level of hardness. It should be noted that the medium carbon low-alloy steels SAE 4140 and 4340 – commonly used in machine design – are in the HTLA category. Cold cracking is discussed in more detail in Krauss (2015, pp. 467–468).

It is important to use low-hydrogen consumables and observe recommended preheat and interpass temperatures. Recommended preheat and interpass temperatures for selected HTLA steels are given, as a guide, in Davis (1998). Post-weld heat treatment may be required.

6.4.2 Normalising

Steel is normalised to refine grain size and make the structure more uniform. The steel is typically heated to about 55°C (100°F) above A_3 for steels with $C < 0.77\%$ and above A_{cm} for steels with $C > 0.77\%$, followed by cooling in still or slightly agitated air. Proper normalising requires that the process produce a homogenous austenite phase prior to cooling (Reardon, 2011).

6.4.3 Quenching

Quenching is the process of rapidly cooling metal parts from the austenising temperature to produce controlled amounts of martensite in the microstructure of carbon, low alloy and tool steels. The rate of heat extraction is affected by quenching techniques, of which there are at least six. The most common quenchant media are the liquids and gases listed in Table 6.2.

6.4.4 Tempering

Tempering is a process to reduce hardness and increase toughness. All operations are performed below the lower critical temperature A_1. The temperature is more important than the time at temperature. Steels that temper between 150 and 200°C (300–390°F) will retain much of their hardness and strength with a small improvement in ductility and toughness. This treatment can be used for gears. Tempering at about 425°C (796°F) improves ductility and toughness at the expense of hardness and strength.

Table 6.2 Commonly used quenchants in heat treatment of steel

Liquids	Gases
Water	Natural air
Brine (aqueous)	Exothermic atmospheres
Caustic solutions	Endothermic atmospheres
Polymer solutions	Nitrogen
Oils	Vacuum and pressurised inert gas

6.4.5 Vacuum heat treatment

Heat treatment of steel by means of vacuum processes is of increasing interest (Jones, 1985). Among its advantages is that the charge can be placed into the furnace and run through the complete heat-treatment cycle without further handling and undesirable changes to the surface chemistry. Also, operating costs are often lower because, as the vacuum furnace is fired from an electric power utility, it does not require power when idle and is shut down cold when not in use. Briefly, the batch operation is as follows:

(*a*) The cold furnace is charged and the hatch sealed.
(*b*) The atmosphere in the furnace is extracted to near vacuum.
(*c*) The contents are heated by electric resistance heating elements (similarly to toasting sliced bread).
(*d*) After reaching and being maintained at temperature for a specified duration, the power supply for heating is shut and the vacuum system valved off.
(*e*) The furnace chamber is rapidly filled with nitrogen or argon gas.
(*f*) The gas is violently recirculated over the charge, being sent through a gas-to-water heat exchanger at each pass. The speed of charge quenching using gas can approach 500°C (932°F) per minute.
(*g*) The chamber pressure is reduced to ambient and the hatch opened to enable removal of the hardened product.

Vacuum furnaces are available which can replace oil quenched atmospheric furnaces used for hardening gear alloy steels such as SAE 4130, 4140, 4340 and 8620.

6.4.6 Case hardening

Case hardening is the process of hardening the exterior surface of a steel part while the interior volume retains the core properties. The chemical composition of the surface layer is changed by diffusion treatments or localised heating. Diffusion methods introduce alloying elements into the surface of ferrous parts that assist martensite formation during subsequent quenching. There are at least six commercial diffusion methods of which the following three are the most important

- carburising
- nitriding
- carbonitriding.

Localised case hardening methods include

- flame hardening
- induction hardening
- laser hardening.

There are many case hardening methods in the public and proprietary domains (Reardon, 2011). Only two methods of each class are described below.

6.4.6.1 Carburising

Carburising is a process in which carbon is added into the surface of low-carbon steel by diffusion. It may be done in a gaseous environment (gas carburising), a liquid salt bath (liquid carburising) or with the surface of the part covered with a solid carbonaceous compound (pack carburising). Pack carburising is an ancient method, but still used (Sim and Ridge, 2002; Digges *et al.*, 1956). Carbon is added to the surface of low carbon steels at temperatures at which austenite is stable, typically 850° to 950°C (1560° to 1740°F). The austenite has a high solubility for carbon. The actual hardening occurs when the now high-carbon surface layer is quenched to form martensite. The martensite case, which has good wear and fatigue resistance, covers the tough, low-carbon steel core.

6.4.6.2 Nitriding

In this diffusion surface hardening heat-treatment, nitrogen enters the surface of the steel while it is ferritic. The process differs from carbonisation in that the gas is added to ferrite instead of austenite. The temperature range is typically 500°to 550°C (930°to 1020°F). Nitriting has a very important advantage over carburising – because the process does not require heating the part into the austenite phase and a subsequent quench to form martensite, there is minimal distortion of the part.

6.4.6.3 Flame hardening

Flame hardening is a localised case-hardening method in which the surface of the steel part is heated with an oxy-acetylene torch and immediately quenched. A hard layer of martensite is produced over a softer interior core having a ferrite–pearlite substructure. Because no external carbon is added, the necessary carbon must be present in the steel part before the process is started. Hence, medium-carbon steels are often used in this process.

6.4.6.4 Laser surface-heat treatment

Lasers are used to harden localised areas of steel and cast iron parts. The laser light is absorbed by the metal thereby heating selected areas into the austenite range. The heated material is transformed to martensite by rapid cooling due to conduction of heat into the bulk of the part (a self-quenching process).

6.5. Heat treatment of steel gears

Gears are usually heat treated to improve tooth strength and surface durability. The processes are known as through-hardening and case-hardening. They are adaptions of the procedures described previously.

6.5.1 Through-hardening steel gears

Through-hardening is usually done by the quench and temper process. The gears are heated to form austenite in the range 800°to 900°C (1475°to 1650°F) followed by quenching. The usual quenchant is water for carbon steel and oil for alloy steel. The rapid cooling causes formation of martensite which causes the gear to be harder and stronger. After hardening, the gears are tempered at a temperature below 690°C (1275°F) which improves ductility and toughness, at the expense of a reduction in hardness and strength.

During quenching and tempering, the gears distort because of the martensite transformation (usually a volumetric expansion). Distortion is often sufficient to lower the quality level of the gear, sometimes necessitating a fiishing operation.

The second gear-cutting (finishing) operation may be eliminated by first cutting the gear from a blank that already has been through-hardened. This requires that the gear hardness be less than that for post-cutting treatment, if the usual tooling is to be used.

6.5.2 Case-hardened gears

Gears may be case hardened using the methods described in Section 6.4.6. Gas carburising and nitriding are the most popular methods.

6.5.2.1 Carburising

Carburising is used with low-carbon steels to obtain a hard case with the teeth having high carbon at the surface graduating to a low-carbon core. The steel of the gear is exposed to an environment with sufficient carbon potential to cause absorption of the carbon on the teeth surfaces and then by diffusion move carbon toward the interior of the gear. Gas carbonising is popular. The most commonly used gas-containing carbon is an endothermic gas (endo gas) produced by reacting natural gas with air over a heated catalyst (Rakhit, 2000). Free carbon produced thereby is then dissolved in the austenite teeth when the metal is heated to about 720°C (1330°F) resulting in the formation of iron carbide (Fe_3C). The distance that the carbon penetrates into the steel depends on the carburising temperature, the time held at that temperature and the gas. The holding temperature is normally 40°C (100°F) above the A_3 line in Figure 6.6 where the metal is fully austenite. Carburising temperatures vary between 790°C (1450°F) and 985°C (1800°F), depending on the carbon content of the steel. The most common temperature for carburising is 925°C (1700°F). An equation for anticipating the total case depth for low-carbon carbon and low alloy steels is given by Rakhit (2000) as $d = 0.025\sqrt{t}$ in which d is the total case depth in inches and t is the holding time in hours. Thus, a gear held at 925°C (1700°F) for 16 hours should develop a total case depth of 0.1 in. (2.5 mm). For anticipated total case depths at other holding temperatures, and effective case depths, see Davis (1998). Of course, in production, the actual case depths should be verified by destructive testing.

The carburising process using endo gas causes carbon atoms to move into the gear teeth. As such, it does not influence hardness. After carburising, the gears are quenched in a cooling medium for hardening such as water, oil, special fluids or agitated inert gases such as argon or nitrogen, or sometimes just air. This process is called direct quenching. During the quenching process the gears distort and may crack. For gears with a sensitive geometry, staged quenching is employed to minimise cracking. The quenched gears are tempered after quenching – it is the necessary heat treatment that defines the surface hardness. The normal tempering temperature is between 115° and 175°C (240°and 350°F). The surface hardness decreases as tempering temperature increases. A tempering temperature of 300°F (150°C) should produce a Rockwell surface hardness of between 55 and 58 HRC (see 'Rockwell designations', page 166).

6.5.2.2 Nitriding

Nitriding is a case-hardening process that is used to harden the surface of alloy steel gear teeth. The gears are heated in an air-tight furnace (often vacuum) and heated to below the A_1 line in Figure 6.6. Usually the temperature is held between 480° and 565°C (900°and 1050°F) in an atmosphere of ammonia (NH_3) that is continuously replenished. The NH_3 breaks down into atomic nitrogen and hydrogen. Some of the nitrogen penetrates the steel surface and combines with the base metal and alloying constituents. Because the holding temperature is below the critical temperature, A_1, the grain structure should not change and there is little geometric distortion. This is an important advantage of nitriding. Finishing the gears prior to nitrating is common practice. Nitriding is a slow process at atmospheric pressure and the total case depths developed are less than in the carburising process. Metallographic standards for core structures and nitride cases are shown in Rakhit (2000).

6.6. Cast steel

Castings have been used for manufacturing movable bridge machinery components of complex shape since the early 1800s (see Chapter 1). Early on they were of cast iron; nowadays mostly cast steel is used. The main advantage of cast steel over grey cast iron, in general, is the better tensile properties of the cast steel. Whereas cast irons are perceived as being good in compression and weak (often worthless) in tension, the cast steel physical properties are essentially alike in tension and compression. However, some cast irons have significant tensile strength. Comparisons of some properties of cast steel and grey cast iron are presented in Blair and Stevens (1995). Loads applied to machinery components often reverse in direction of application during operation. Cast steel may be a suitable choice for a component of irregular shape subjected to loading that varies in direction and magnitude. The manufacturing process for cast steel is much like that for cast iron and is described in Blair and Stevens (1995) and Davis (1998).

6.7. Hot-rolled steel and hot forging

Hot-rolling and hot-forging of steel needs to take place above its recrystallisation temperature in order to preclude work hardening during the process. Recrystallisation of steel is a complex mechanism that is still a research subject. However, practically, the recrystallisation temperature of a metal is often taken as 0.6 times its absolute melting temperature. As an example, the recrystallisation temperature T_r of a 0.5% carbon steel may be approximated with the aid of Figure 6.6. At $C = 0.5\%$ the intersection with the liquidus line is 1460°C. Noting that when absolute zero temperature is −273°C, then $T_r = 0.6$ $(273 + 1460) - 273 = 767$°C.

Practically, one would start the hot-work operation at a higher temperature, say 1000°C, because of heat losses (T_r is a minimum temperature) and less work is required at a higher temperature to displace metal. Hot-working of steel by rolling is described in Davis (1998).

Hot forging is the shaping of metal by compression between a hammer or ram and an anvil. It is the oldest ferrous metalworking process. The prior manufacturing step, smelting, is an extractive process leading to the metal. Nearly all forging was performed manually or with trip hammers powered by flowing water until the Industrial Revolution when heavy falling rams were introduced to strike the workpiece that rested on an anvil. The ram was originally raised manually and later by steam engines and dropped. Now the ram is driven downward by steam, hydraulic or pneumatic means. There are at least four types of forging operations. The open-die method, an advance of the old hammer forging, is still used to form shafts and disks of movable bridge machinery. In it the workpiece is compressed between flat dies without lateral constraint to restrain spread of the hot metal during hammering.

The purpose of forging a workpiece, such as a shaft, is to produce a piece that is stronger than would be produced by other shaping methods. During the forging process, the metallic grains deform to follow the general shape of the workpiece and the orientation is continuous, thereby improving the strength properties. The improvement in mechanical properties produced by forging is a function of the reduction in the cross-sectional area of the workpiece due to the process.

6.8. Cold-finished steel

Hot-rolled carbon and alloy steel that is subsequently cold-rolled or cold drawn, sometimes after aging, is commercially available and it can be appropriate metal for movable bridge machinery. The cold rolling occurs with the steel below its recrystallisation temperature (usually rolled at about the shop temperature). The cold rolling or drawing strain hardens the metal thereby increasing its strength. The rolling also improves surfaces finish and the dimensions of the products are held to tighter tolerances during manufacture.

6.9. Cast iron

By convention, alloys of iron that contain more than 2% carbon by weight and from 1 to 3% silicon are called cast irons. Varying the carbon and silicon content and alloying with other metallic and non-metallic elements results in a wide range of possible physical properties. These can be further modified by heat treatment. A characteristic of cast irons is that they have a matrix of steel with excess carbon distributed within the matrix in the form of graphite or combined as iron carbide (Doane, 1987).

- There are four basic types of cast irons: grey, white, ductile and malleable. The type is determined by the form that carbon takes as it precipitates from the melt on cooling. Grey cast iron results when the excess carbon precipitates as flakes of graphite during the slow cooling from the molten state. The grey colour exhibited on fracture is due to the presence of disbursed graphite in the microstructure. Grey iron can easily be machine finished. It was used as a material for sliding bearings.
- White cast iron is formed when the carbon in the molten solution on solidification remains as carbides. This happens during faster cooling of the hot casting. As a result the white cast irons are hard with high compressive strength but little ductility. They are used in applications where wear and abrasion resistance are important.
- Ductile or nodular iron has a composition similar to grey cast iron but the graphite is in the form of spherical particles rather than flakes. Ductile iron has high tensile strength and ductility when compared to grey iron. Up to 18% elongation for some ductile irons compared to 0.6% for grey iron. Ductile iron can be further treated to obtain superior properties. Crankshafts of some high performance sports car motors are ductile iron castings. Malleable iron also contains graphite nodules but they are not spherical because they are a result of heat treatment, not forming from the melt. It can often be used interchangeably with ductile iron but is preferred for thin-section castings.

One advantage of designing with castings is that complex shapes can be produced with less shop machining than when using alternatives such as forgings and weldments.

6.10. Non-ferrous metals

The principal non-ferrous metals used in the manufacture of movable bridge machinery are alloys of copper, zinc, lead and tin. They are often parts of journal bearings for rotating shafts.

6.10.1 Brass

Brass is a copper base alloy with zinc the main alloying element. Other components of brasses, in lesser percentages, are aluminium, arsenic, antimony, iron, lead, manganese, nickel, phosphorous and tin. However, some brasses are called bronzes even though they contain little tin. One of these brasses, termed manganese bronze, has a maximum tin content of 0.20%. It is one of the five copper alloys listed under ASTM B22, the material specification referred to by the American Railway Engineering and Maintenance-of-Way Association (AREMA, 2014) for copper alloys.

Copper has a face-centred cubic crystal structure (f.c.c.) and this structure can dissolve up to about 25% zinc. The f.c.c. is known as alpha structure. At zinc concentrations greater than 35% body-centred cubic crystals (b.c.c.) also form and brasses with both f.c.c. and b.c.c. crystals are termed alpha-beta brasses. The alpha structure is soft and ductile; the beta is harder and less ductile. The ductility improves with increasing zinc content up to about 30%. The 70/30% copper-zinc alloy is called cartridge brass and has the best combination of strength and ductility (Keyser, 1974).

6.10.2 Bronze

Bronze is a copper alloy whose main alloying element is tin. As mentioned in the previous section, the copper-base alloy terminology is confusing. For instance, the brass containing 90% copper and 10% zinc is referred to as 'commercial bronze' although it contains no tin. Bronze is sometimes preferred to the less expensive brass because it is stronger and more corrosion resistant. There are many standard copper alloy compositions listed by the Copper Development Association (CDA), at least 90 for cast products alone. The CDA website gives data on their chemical compositions and properties. Five are of interest for movable bridge machinery: C86300, C90500, C91100, C91300 and C93700. Their mechanical property requirements are discussed in Chapter 7.

6.10.3 Modifying properties of copper alloys

It is sometimes desirable to modify the mechanical properties of copper alloy castings and weldments. The procedure for copper alloys differ from those for ferrous metals mainly because copper is nonallotropic.

6.10.3.1 Hardening

Machine shop folklore has it that copper alloys (brass and bronze) can only be hardened by cold working. This is true in general for the more common copper alloys, but some alloys can be strengthened by heat treatment. The topic is beyond the scope of this book but is treated in Chandler (1996). The website http://www.keytometals.com is a convenient resource (Total Materia, 2014).

6.10.3.2 Annealing

Machine shop folklore also has it that the process of fully annealing copper alloys is exactly the reverse of that used for steel. Brass is annealed when heated and quickly cooled, in contrast to steel which is cooled slowly. However, the matter of cooling rate is not quite that simple.

Cold worked copper alloys are annealed by heating them above the recrystallisation temperature, optimally to a temperature that will promote grain growth. Because of the many variables – temperature, dwell at temperature, furnace atmosphere, residual straining, etc. – it may not be possible to establish a protocol *a priori*.

6.10.3.3 Stress relieving

The intent of stress relieving is to minimise residual stresses in the workpiece due to cold work, welding or from the sudden cooling during annealing because copper alloys are susceptible to SCC, especially in saltwater atmospheres. Besides, all copper alloys are susceptible to more rapid corrosion when in a stressed condition. Workpieces are slowly heated to a temperature below the recrystallisation temperature, held at that temperature and then cooled.

6.10.3.4 Other heat treatments for copper alloys

For information about homogenising and precipitation hardening of copper alloys see Reardon (2011).

6.10.4 Lightweight metals and alloys

Of the four commercial light metals (magnesium, beryllium, aluminium and titanium) only aluminium seems to have application to movable road and railway bridges. For constituents and properties of commercial aluminium alloys see Davis (1998). On a cautionary note, weldments of an aluminium alloy have shown susceptibility to SCC in a seawater atmosphere, with the cracking starting a few years after installation. (See Chapter 7 for a discussion of SCC.)

6.10.5 Babbitt or white metal

Babbitt bearing materials used for lining plain journal bearings are alloys of tin, copper, antimony, lead, and minor percentages of other elements. Eight alloys are listed in ASTM B23 Standard Specification for White Metal Bearing Alloys (known commercially as 'Babbitt Metal'). Grade 2 is recommended in AREMA (AREMA, 2014).

Babbitted bearings will be found on older movable bridges and they will probably be of the traditional poured style whereby the Babbitt was poured into the cavity around the shaft journal that passed through the bearing. This is an expensive procedure nowadays and seldom specified for new movable bridge construction unless there are historical requirements. Babbitted pillow blocks of up to 6 in. (152 mm) shaft diameter are listed in catalogues of machinery component suppliers. The shaft journals are turned to suit the bearing sleeves. See also Chapter 12.

6.11. Non-metallic materials

6.11.1 Plastics

Plastics are used to fabricate electrical components of movable bridges but are not yet very evident in mechanical machinery. They have application to components whose function is to reduce friction, such as the surfaces of sliding wedges and the bushings of journal bearings for rotating shafts. Many types of thermoplastics have been introduced to the journal bearing market and are used in hydroelectric power water control and canal lock machinery, farm equipment, the military, and marine equipment. In order to minimise space requirements herein, only three types will be discussed: nylon, polytetraflorethylene (PTFE) and the propriety thermoplastic product named ThorPlas. A further restriction is that only the self-lubricating versions will be described

- Nylon (MDS-filled) – a nylon into which molybdenum disulphide (MDS) has been mixed that serves to lubricate the bushing and/or shaft surface. Through-thickness nylon shell.
- PTFE (glass-filled) – a woven polytetraflourethylene material that is adhesively bonded to a steel (often stainless) shell backing. Load and chemical resistance ratings exceed that of oil-impregnated bronze bearings. ThorPlas – a proprietary grade, crystalline, lubricated engineering thermoplastic. It can be installed as a full form sleeve, not requiring a metal or other rigid backing. ThorPlas is a homogenous polymer with lubricants formulated into the molecular structure. Plastic bearings and their application limits are treated in Oberg *et al.* (2012).

Nylon (MDS-filled) pads are sometimes attached to plane sliding surfaces, such as the tops of structural reaction end wedges of swing bridges, to reduce the coefficient of friction and to minimise maintenance.

6.11.2 Wood

Timber was used to build movable bridge superstructures, mostly for swing bridges, prior to the 20th century. Especially long swing spans were built in the Midwest USA for railways and road traffic (Simmons, 2001). All were replaced over time because of heavier railway loading and the need for wider roadway traffic lanes. However, shorter timber swing bridges are still extant.

After the 18th century, wood was seldom used to make bridge machinery. Today, wood is likely to be found only in old electrical switch gear and the like where it was used for self-lubricating rotational bearings. The wood was/is a species of lignum vitae (also lignum-vitae) which is native to the Caribbean and northern South America. It is one of the heaviest and hardest woods on the market. Lignum vitae wood is now used chiefly for bearings or bushing blocks for ship propeller shafts (Miller, 2007). It is adaptable for underwater bearing use because of its self-lubricating properties and strength. There were applications of lignum vitae in the railway and electrical power industries.

6.12. Wire rope

Wire rope is not, strictly speaking, a material, but a precision metallic wire product. However, it is included in this chapter because of peculiarities of wire material selection and rope manufacture. Resources for this chapter were wire rope manufacturers catalogues, ASTM and USA Federal specifications, the *Roebling Wire Rope Handbook* (Roebling, 1966), and other technical literature. It includes information on wire material, construction of wire rope and wire rope fatigue.

The first recorded experiment in which iron wire cables were substituted for hemp cables was performed by Claus von Reden (1730–1791) a mining superintendent at the Harz Mountain silver mines in Hannover, a German-speaking member of the German Confederation, then ruled by Britain. WAJ Albert (1787–1846), a successor to von Reden, made wire rope by twisting three or four iron wires together to make a strand and then twisting three or four of the strands together to make a rope (Sayenga, 1999). The stranded ropes were used successfully in Harz mine shafts in 1834. His manufacturing processes are described in Albert (1837). Figure 6.8 depicts a piece of Albert's first rope which was photographed through a display case in Stuttgart, Germany.

6.12.1 Conventional wire rope construction

A wire rope is a plurality of strands twisted about an axis, or about a core, which may be a wire strand or another wire rope, or a nonmetallic rope, or a nonmetallic rod. Each of the strands is a plurality of wires twisted about an axis, usually a central wire. The word 'construction' in wire rope parlance refers to the design of the rope – the number of strands, the quantity of wires per strand, the arrangement of the wires within a strand and the type of core.

6.12.1.1 Standard classification for conventional wire ropes

Wire ropes are classified, nominally, according to the number of strands twisted about a core and the number of wires in each external strand. A Class 6 × 19 rope has six strands, but the number of wires per strand ranges between 16 and 26. For larger-diameter ropes a standard classification is 6 × 36, which has usually been 27 and 49 wires per strand. Bethlehem Wire Rope and Bridon American still refer to this class as 6 × 37. Again, the designation is nominal because none of the strands in this class has 37 wires. In the UK and Europe the use of eight-strand rope is common. It is also used in North America for passenger elevators and in mining applications.

Figure 6.8 Remnant of the first Albert wire rope. (Photograph by Charles Birnstiel)

Figure 6.9 Some common strand constructions. (Adapted from Wire Rope Technical Board, 2005)

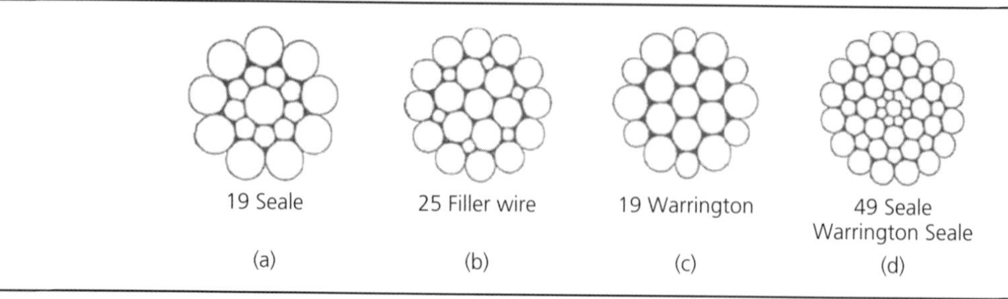

The number of wires per strand is insufficient to define the strand cross-section. It is necessary to select a cross-sectional wire arrangement. Strands have one or more layers of wires twisted about a central core. Single layer strands (as in 6 × 7 rope) are seldom used for movable bridges because the outer wires are relatively large, and hence the strand or rope is less flexible than when strands are of multi-layer construction. The most common conventional strand styles are

- Seale: two layers of wire wrapped about a central wire, as shown in Figure 6.9(a). Each layer has the same number of wires, and all wires in a layer are of the same diameter. The larger outer wires rest in the valleys between the smaller wires of the inner layer.
- Filler wire: two layers of similar-size wires are twisted about a centre wire. The inner layer has half the number of outer wires. Smaller wires (the filler wires) equal to the number of inner wires are laid in the valleys of the inner wires (see Figure 6.9(b). In American practice, classification 6 × 19 with 6 filler wires (also known as 6 × 25 construction) was usually specified for counterweight ropes of vertical lift bridges.
- Warrington: two layers of wires wrapped about a core wire. Inner wires are all of the same diameter. The outer layer comprises wires of two different diameters with the different diameters alternating, as shown in Figure 6.9(c).

When two or more of the basic patterns are combined, the construction is referred to as a combined pattern. Figure 6.9(d) depicts 4-layer construction. The first two layers (the inner two) are Seale, the third is Warrington and the fourth is Seale.

In all the classes described above, the wires in the strands are of circular cross-section, produced by drawing carbon steel rod through a number of successive circular dies of decreasing orifice diameter, thereby cold-working the steel and increasing its strength. During drawing the crystalline structure is altered, which results in increased ultimate strength but at the expense of a loss in ductility. To improve ductility, at some point during the drawing process the wire is heat treated (it is ‘patented’, in wire rope terminology) so that its diameter can be further reduced. The patenting process is discussed in Krauss (2015, pp. 323–324). Wires of different cross-section (triangular, etc.) are used to make strands that are used in other industries.

The round wires in the outer layer of conventional strand have little surface area in contact with sheaves and drums where the rope passes over them to change direction. As a consequence, the outer wires wear because of the relatively small amount of contact area between the wire and sheave. The high-contact stresses between wire and sheave often cause depressions and ridges to form in the sheave grooves because the wire material is usually harder than the sheave material.

Figure 6.10 Round strand before and after compacting. (ASTM, 2009)

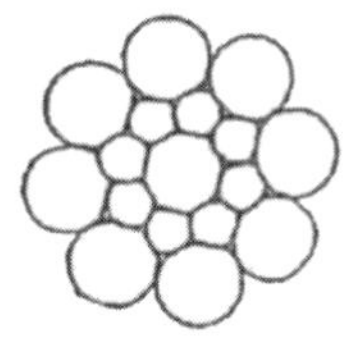

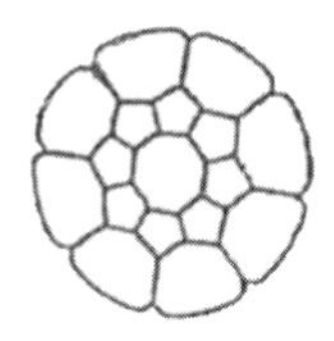

To provide more rope-sheave contact area, some strands made with completely round wires may be compacted by drawing the strands through a die or by roll forming. In these processes the cross-sectional shape of the outer wires of the strand is plastically changed from round to round with a nearly flat outer surface. Figure 6.10 shows a strand before and after compacting. Figure 6.11 depicts a cross-section of a rope closed with compacted strands.

The benefits of using compacted strands are: greater surface area to resist abrasion, greater resistance to lateral crushing and fatigue, and more metallic cross-sectional area as compared to a conventional rope of the same outer diameter. The compaction process smoothes the outer surface of the strands, which then contacts other smooth wire surfaces. Because all the wires within a strand are laid in one operation, the wire contacts are longitudinal – not wire crossing wire. The interstrand wire contacts are of outer wires that have been deformed (see Figure 6.11), with more contact area on the outer side than for the case of round wires in a non-compacted strand.

Ropes may also be swaged after closing. In this process, un-compacted ropes made with strands of round wires are rotary swaged (hammered) to produce ropes with less voids and greater surface area. The effects of multiple passes of swaging on an uncompacted rope is to reduce the diameter of the rope, which is a measure of the increase in density and to smooth the outer surface. The smoother surface produces greater rope-to-sheave contact. But the hammering creates wire-notching where the outer strands contact the inner strands and the core. The notches are not considered to reduce fatigue strength.

There are many classifications of ropes, including ropes with many more strands, triangular strands, flattened strands, etc. ASTM A1023 (ASTM, 2009) has tables of minimum breaking force (MBF) for 25 classifications. Federal Specification RR-W-410G has tables of minimum breaking force for many ropes, including classification 6 × 61, with nominal rope diameters from 2 to 5 in. (50 to 127 mm).

Figure 6.11 Rope closed using compacted strand. (Wire Rope Technical Board, 2005)

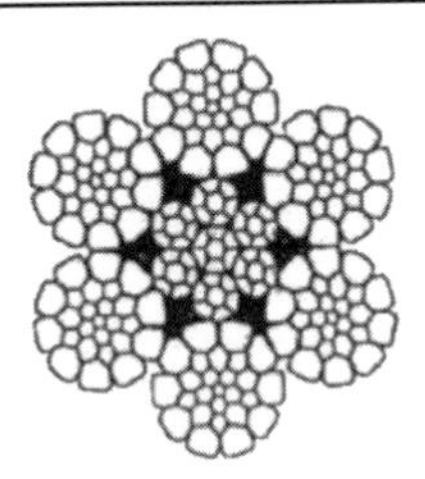

Many special ropes are produced which are not included in ASTM A1023 and Federal Specification RR-W-410G and information about these is available from rope manufacturers.

Because of the helical shape of the strands in a rope, torsion is produced by the tensile force applied to the rope. This torsion rotates the cross-sections of conventional rope unless the rope ends are restrained. The rope length between sheaves or supports also influences this effect. Rotation resistant ropes are manufactured in which the wrapping directions for the strand and wire are arranged so that the torsion induced by axial tension is minimised. To achieve this, rotation resistant ropes have more outer strands (usually 12 or more) and the core is closed in the opposite direction to that of the outer strands. Because most movable bridge ropes are relatively short, and their ends are usually fixed against rotation about the rope axis, rope torsion is not usually an important consideration for movable bridges. Sometimes the effect of rope torsion is minimised by using alternate ropes having left and right lay. Hence, rotation-resistant ropes are not discussed herein.

6.12.1.2 Wire rope cores

The core about which the strands are twisted serves to maintain the round cross-sectional shape during passage of the rope over sheaves and drums. ASTM A1023 states that cores of wire ropes shall be either of steel or fibre composition. Natural fibre (hemp or sisal), polypropylene, solid polymer or an independent wire rope (IWRC) are used as cores. Some bridge engineers consider that hemp and sisal cores can serve as lubricant reservoirs. Of course, this implies that low-viscosity lubricant will be regularly field-applied and will penetrate through the rope into the core because the initial lubrication during manufacture dissipates within a few years. Later in the operating life of the rope, steel cores provide more support for the strands than fibre cores, if the ropes are not relubricated, because steel cores are more resistant to deterioration and crushing.

6.12.1.3 Wire rope lay

The helix formed by the wires of the strand and the strands of the rope is termed lay. The length of the strand spiral in a rope, as it rotates 360°, is termed lay length. It is measured in a straight line parallel to the rope centre. Lay is not part of classification. There are three primary lays: regular, Lang's and alternate.

- Regular lay (also referred to as ordinary lay): the wires of the outer strands are twisted in one direction and the strands in the opposite direction. The wires appear to run nearly parallel to the axis of the rope. If the strands are twisted clockwise, the rope is right regular lay (see Figure 6.12(a)). If the strands are twisted anticlockwise, the rope is left regular lay and appears as in Figure 6.12(b). Right regular lay is normally furnished.

Figure 6.12 Relative service life against D/d ratio based on bending fatigue only. (a) Right regular lay; (b) left regular lay; (c) right lang lay; (d) left lang lay. (Adapted from Wire Rope Technical Board, 2005)

- Lang's lay: the wires in the outer strands and strands themselves are twisted in the same direction and the wires appear to run diagonal to the rope axis. A right Lang's lay rope is depicted in Figure 6.12(c) and a left Lang's lay is Figure 6.12(d).
- Alternate lay: the lays of the outer strands alternate; regular lay followed by Lang lay, such that half of the strands are regular and the other half Lang's lays. Herein, this type will be considered a special rope, beyond the scope of this section.

Each lay type has advantages and disadvantages. Regular lay ropes are less likely to untwist or kink and are more resistant to crushing. Lang's lay ropes are more flexible (in bending) than regular lay and have greater abrasion resistance. Lang's lay ropes have better fatigue resistance (in bending) than regular lay ropes. However, Lang's lay ropes require greater care during installation.

6.12.1.4 Preformed wire rope

Preformed wire rope is rope in which the wires and strands, individually, have been permanently formed during the closing process into the shape they will assume in the completed rope. Preforming reduces internal stresses in the rope during operation and encourages more uniform distribution of axial tensile force between the wires and strands. The rope runs smoother and gives longer service life under bending than non-preformed rope. Also, broken external wires project a shorter distance from the exterior surface during running. Ropes are normally provided preformed, but they can be produced non-preformed on special order. However, not all constructions of the larger wire ropes are supplied preformed (see RR-W-410G). Generally, there is little reason to recommend non-preformed rope over preformed.

6.12.1.5 Wire rope construction materials

Stranded carbon steel wire ropes are traditionally strength-graded as: improved plow steel (IPS), extra improved plow steel (EIP) and extra extra improved plow steel (EEIP). The minimum breaking forces for 25 classes of ropes (up to $2\frac{3}{8}$ in. (60 mm) diameter for some constructions) are given in ASTM A1023 under these grades. Not all classes are listed in all three grades. Federal Specification RR-W-410G gives minimum breaking forces for classification 6×61 for two grades for diameters from 2 to 5 in. (50–127 mm).

The rope grade does not imply that the actual tensile strength of the wires in the rope is of that grade. In fact, the industry has adopted different terms to designate the strength of wires. Table 2 in Section 5 of ASTM A1023 specifies the minimum and maximum wire strength level required for a particular rope grade. The wire rope producer will select the individual wire strength level and diameters to produce the finished rope. Hence, the wire sizes and breaking force levels for a particular class of rope of a certain diameter will vary, depending on the rope manufacturer.

The properties of the wire used for the ropes in ASTM A1023 are defined in ASTM A1007 Standard specification for carbon steel wire for wire rope (ASTM, 2007). ASTM A1007 also specifies all the testing required for the wire. Note that there are five levels of minimum wire breaking force, a force is given for each wire diameter. If one computes the 'engineering' minimum breaking stress (minimum breaking force divided by original wire cross-sectional area) it is evident that it varies with wire diameter.

The European standards for wire rope and wire that correspond to ASTM A1023 and ASTM A1007 are EN 12385–4 Stranded ropes for general lifting applications (DIN (Deutsches Institute für Normung), 2012b) and EN 10264–2 (DIN, 2012a) Cold drawn non-alloy steel wire for ropes for general application, respectively.

The outer wires of wire rope are quite hard. Bridon (2014a, 2014b) gives the following ranges of Brinell hardness: IPS: 388–441, EIP: 415–461 and EEIP: 444–486, for wires of circular cross-section.

ASTM A1007 treats five types of finish for the wires: uncoated, drawn galvanised, final galvanised, drawn zinc-aluminium–mischmetal alloy and final-coated zinc–aluminium–mischmetal alloy. Uncoated, drawn galvanised and drawn zinc–aluminium–mischmetal alloy wires can be furnished in Levels 1 to 5. Final galvanised and final-coated zinc–aluminium–mischmetal alloy may be furnished in Levels 1 to 4. Corrosion tests have shown that the zinc–aluminium–mischmetal alloy coating offers approximately two to three times the amount of corrosion protection compared to standard zinc coated wire. Most use of zinc–aluminium–mischmetal alloy has been in Europe (Klein, 2013). Note that minimum breaking forces for final galvanised ropes are taken 10% lower than the values listed in tables of ASTM A1023 (ASTM, 2009), except for certain rotation resistant ropes.

Wire rope made of stainless steel is also available and is sometimes specified for hydraulic structures where corrosion conditions are severe, such as at navigation locks and dams. These ropes are not covered by ASTM A1023 but are purchased under Federal Specification RR-W-410G. There may be applications of stainless steel for bridges under conditions of severe corrosion. However, their fatigue performance is reportedly poorer than that of carbon steel ropes.

6.12.1.6 Wire size against resistances

The many wire rope designs available result from the fact that operating conditions for ropes vary widely and hence the optimum wire sizes for the ropes vary. The subject is quite complicated but some very general observations follow

- Fatigue resistance – a rope made of many wires will be more resistant to tension-bending fatigue than a rope of the same diameter made with fewer and larger wires. Fatigue resistance should not be equated to bending flexibility. For example, a fibrecore rope is more flexible than an IWRC rope of the same size but the IWRC rope has a greater fatigue resistance, probably because of the firmer support provided for the outer strands by the IWRC and also because of the greater metallic area which results in lower stresses.
- Crushing resistance – is the resistance of the rope to crushing by transverse forces, as occurs when the rope passes over a sheave. Crushing distorts the rope from its nominal circular cross-section and creates internal binding between the wires and strands. IWRC ropes are more crush resistant than fibre core ropes. Regular lay ropes are better than Lang's lay. Six-strand ropes have more resistance than those with a larger number of strands. Compacted ropes and swaged ropes have more resistance than ropes made with only circular wires.
- Resistance to plastic flow and metal loss – because of the high contact stresses between the outer wires of the rope and the sheave (in many cases) the outer wires are flattened if the sheave hardness approaches the wire hardness. The peening action is said to contribute to early wire metal fatigue. Larger wires have lower contact stress for the same rope force and sheave diameter and hence are less subject to plastic deformation. Metal loss is due to abrasion between the sheave grooves and the wires. Corrosion exacerbates this condition.
- Resistance to corrosion – for ropes made from bright carbon steel wires, corrosion resistance is greater for ropes made with larger-diameter wires. The merits of coating wires with zinc or zinc–aluminium–mischmetal alloy are in dispute. However, testing has shown, and expert opinion maintains, that these coatings can provide reliable protection against corrosion. Even if the zinc layer is partly damaged, the steel remains protected because the zinc corrodes first (Feyrer, 2007). It is noted that the movable bridge specification of the

Netherlands requires that all running ropes be made of wires with a drawn-galvanised finish (NEN, 2001).

6.12.1.7 Fatigue

Fatigue, as evidenced by wire breaks, is an important contribution to rope deterioration in some applications. Three situations which lead to fatigue cracking are

- repeated bending
- fluctuating tension
- fluctuating tension with local bendings at fittings.

These phenomena, and how they relate to running ropes of movable bridges, will be described in the following list.

- Bending fatigue: may occur due to repeated changes in direction of a running rope, as over sheaves, while under nominally constant axial tension. Bending fatigue has not been a severe problem for movable bridge ropes in the USA because of the relatively large sheave–rope diameter ratio ($D/d = 80$) specified in the American bridge standards. However, with the trend to less restrictive specifications, and life cycle costing, bending fatigue may become a more important consideration for movable bridges. For example, in the Dutch movable bridge specification, counterweight sheaves are permitted at one-half the diameter per US standards. This would imply that fatigue analyses may be desirable or necessary.
  Despite the tens of thousands of wire rope fatigue tests that have been made since organised laboratory testing was initiated, there is no generally accepted criterion for the fatigue life (even laboratory life) of wire rope. This is partly because the tests have involved so many variables, among which are rope construction, wire properties, cores, lay, lubrication, wire coating, sheave material, axial tensile force, configuration of testing machine and speed of loading. Many rope manufacturers have made and are making fatigue tests on their products. They may offer evaluations to customers on an individual purchase order basis. The writer knows of only one manufacturer that has made public its equation for predicting the number of cycles to bending fatigue failure and also presented sample computations (Verreet, 1998). The equation is based on Feyrer (1995).
  Comprehensive research on the bending fatigue of wire rope has been underway at the Institut für Fördertechnik in Stuttgart, Germany, and results have been published (Feyrer 1995, 2000; 2007). As the fatigue testing is ongoing, the 'Feyrer equation' and the necessary constants are revised from time to time and hence it is suggested that the reader contact the Institut for the current versions.
  A stress analysis method for the bending fatigue design of wire rope was proposed by Zhang and Costello (1996). Their idea was to locate the wires that are most severely stressed and predict their fatigue life utilising a modified Goodman diagram. The stress analysis is based on the theory of wire rope developed by Costello (1987). It was the intent of Zhang and Costello to develop a procedure which could be used as a guide to selecting ropes subjected to severe bending fatigue using only manufacturers' publicised data. This method does not require access to the physical fatigue test results of others. However, because of that simplification, the effects of fretting and rope distortion are not directly accounted for.
  A curve of relative service life against D/d ratio (d = rope diameter, D = sheave diameter that accounts for bending fatigue has been published (Wire Rope Technical Board, 2005) (see Figure 6.13). The curve is intended to give comparative service life for a particular rope, under the same axial (nominally constant) tension, for various ratios of D/d. It is of no value in predicting fatigue life for a specific rope construction without having fatigue experience for a given D.

Figure 6.13 Relative service life against *D/d* ratio based on bending fatigue only. (Adapted from Wire Rope Technical Board, 2005)

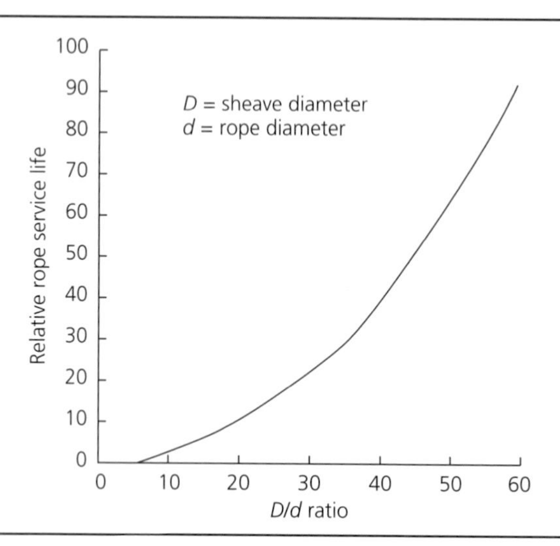

- Fluctuating tension (tension–tension) – variable tension in a rope with only negligible bending at the rope fitting. It seems not to be of importance for running ropes of typical movable bridges. Fewer tension-tension test results have been reported than have results for bending tests. Most research has been done by or for the offshore petroleum production industry, in relation to moored platforms. An empirical equation for the fatigue life of specific cable constructions was proposed by Paton (2001).
  Feyrer made regression analyses of more than 600 tests (mostly tests by others) on six- and eight-strand ropes varying in diameter from 8 to 127 mm in order to determine constants for his equation for each class of specimen. He included a sample life computation for 20 mm Warrington-Seale galvanised ropes with resin sockets. The results are the number of load cycles for a failure rate of 1, 10 and 50% of the ropes.
- Fluctuating tension with local bending at fittings – occurs mostly because of aerodynamic effects and results in fatigue damage at, and within, terminations for cable-stayed bridges and electrical transmission lines. It also occurs where lift bridge counterweight ropes are tangent to the counterweight sheaves. Wire breaks at this location are considered by some owners as sufficient reason for replacing counterweight ropes. The fatigue damage originates mostly from fretting. The economic consequences of this type of fatigue in the electric power transmission industry are enormous. No comprehensive solution appears to have been found to this problem in the bridge or electric power industries. Currently, the favoured approach for power lines and cable-stayed bridges is to install damping devices in order to attenuate the wind-induced vibration. Recently, reported experiments indicate that the fretting fatigue limit of strands coated with zinc is about twice that of strands made with bright wire (Dieng *et al.*, 2007). An extensive research programme on the fatigue of cables under tension that are subjected to small angles of rotation at the anchorages was undertaken by Gourmelon (2002).

6.12.2 Cushioned wire rope

The foregoing discussion dealt with conventional wire rope construction which dates back to Wilhelm Albert's invention of wire rope and fatigue testing thereof about the year 1835. The conventional wire rope comprises strands of round steel wires helically wrapped around a core. The core may be of fibre (natural or synthetic) or an independent wire rope (IWRC) or strand (WSC) construction. The IWRC may, itself, have a steel or fibre core.

In the fibre core rope, the wires of the strands wrapped around the core are supported by the fibre and hence do not have metal-to-metal contact with the core. But there is a metal-to-metal contact between the outer wires of the strands. On the other hand, in the rope having a steel core the wires of the first layer of strands also bear on the wires of the steel core, creating metal-to-metal contact.

Ropes of unconventional construction that have longer fatigue life are available. They have a plastic layer between the steel core and wire strands wrapped around it. They are defined in ASTM A1023 (paragraph 3.17.2.1) as 'cushioned rope – rope on which the inner layers, inner strands or core strands are covered with solid polymers or fibers to form a cushion between adjacent strands or overlying layers'.

Ropes with a plastic layer between the IWRC and the strands wrapped around it combine the advantages of the conventional fibre core and the conventional IWRC ropes. The plastic is a 'soft' bed for the strands and the steel core offers better support for the strands than the fibre core and greater metallic area. Because of the manufacturing process, the plastic layer forms a custom moulded bed for the outer strands, holding their relative positions in the cross-section.

The plastic layer may also seal the steel wire core, trapping the lubricant for the life of the rope. At the same time it seals the core, the plastic layer restricts the entrance of water and abrasives into the rope. Two types of cushioned ropes that have been installed on vertical lift bridges are shown in the Figure 6.14. Stratoplast (Casar trade name) is an eight-strand rope made with conventional round wire strands around a wire rope core. It has a plastic layer between the steel core and the outer strands and is fully lubricated and is used as an operating rope. The largest diameter Stratoplast listed is 72 mm. Turboplast (Casar trade name) is an eight-strand rope made with compacted round wire strands around a wire rope core. Counterweight ropes 85 mm in diameter were installed on a lift bridge in France.

6.12.3 Extension of steel wire rope

Wire ropes lengthen when subjected to tension. The deformations are due to constructional stretch, elastic elongation, and inelastic elongation. The reasons for these three extensions and a means for mitigating their effect will be described subsequently.

6.12.3.1 Constructional stretch

At first application of load on a new rope the peripheries of the helical wires are brought into contact and are 'bedded down' by the circumferential compression created by the tension in the helical wires.

Figure 6.14 Cushioned wire rope. (Casar Rope Works)

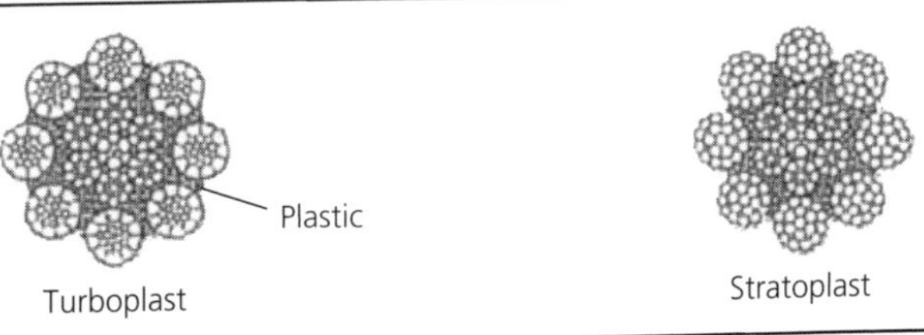

This action causes the overall diameter of the rope to decrease and the helical lay to increase. The extension due to this effect is influenced by the rope construction and other factors including the type of core. The constructional stretch of ropes with fibre cores is about twice that of ropes with independent wire rope cores. Approximate values of constructional stretch of wire rope as a percentage of original length are given in Bridon (2014a). For ropes with a fibre core that are lightly loaded, selected with a design factor (DF) of 8 the constructional stretch (CS) is 0.25%. At heavy loading, DF $\approx$ 3 the approximate CS $\approx$ 0.75%. For ropes with a steel core CS is about half of that for fibre core ropes. Some of the constructional stretch may be removed by prestretching the rope.

6.12.3.2 Prestretching

Wire rope used on movable bridges is prestretched to remove constructional stretch in the range of design operating tensions. For counterweight ropes this is approximately 12% of the minimum breaking force (MBF). The intent is that the rope operate in its elastic range. Prestretching techniques vary among the wire rope fabricators. For bridge design engineers the procedure outlined in Section 6.6.10 'Prestretching' of (AREMA, 2014) may serve as a starting point for writing provisions for the construction contract documents of a particular project. The procedure outlined in AREMA is

1. Tension rope to 0.40 MBF. Hold (dwell) for five minutes.
2. Detension rope to 0.05 MBF. No dwell.
3. Tension rope to 0.40 MBF. Dwell five minutes.
4. Detension rope to 0.05 MBF. No dwell.
5. Tension rope to 0.40 MBF. Dwell five minutes.
6. Detension rope to 0.12 MBF at which rope lengths are marked for cutting counterweight ropes.
7. Release tension.

When the tension on the rope is released some looseness (constructional stretch) will return to the rope. Furthermore, manipulating the rope during spooling, socketing and shipping will also loosen the rope. However, after the rope assembly is installed in its permanent position and run through a few cycles under load the rope cross-section returns to its 'set' or 'bedded down' position that was created during prestretching. In effect, the constructional stretch that would have occurred in an unprestretched rope has been cancelled and the rope now responds to changes in tension elastically.

If the construction contract documents are silent regarding prestretching, then at least one manufacturer uses a modified procedure in which the rope is tensioned to 50% of MBF instead of 40% and the second and third loading cycles are not dwelled.

6.12.3.3 Elastic extension

A prestretched structural strand (and rope) behaves nearly elastically from the stress at about 10% of the MBF to that at 90% of the prestretching force. The slope of the line joining the stress–strain coordinates at those loads is taken as the secant modulus of elasticity, E_s (ASCE, 2010). Operating wire ropes behave somewhat differently. Approximate values of rope modules for prestretched ropes of 'standard or conventional construction' may be found in (Wire Rope Technical Board, 2005) for two stress ranges, zero to 20% of MBF and 21 to 65% of MBF. The rope modulus for rotation-resistant ropes made with compacted strands is greater. See Bridon (2014b) for sample rope modulus values of such ropes at 20% of MBF. It has been observed that stress-strain curves of two ropes of identical catalogue designation differ, presumably because of the use of different machinery during closing of the ropes. ASTM A-1023 has no requirements regarding rope modulus: it requires an MBF.

6.12.3.4 Inelastic elongation

Wire rope tensioned to a stress above that at the dwelled prestretch load is liable to undergo inelastic strains that are not recoverable on unloading. Ropes of movable bridges should be selected such that inelastic strains due to loading will not occur. Inelastic strains due to time-dependent phenomena (creep) are unavoidable. Few test results for prestretched wire ropes tensioned to failure have been made public.

6.12.3.5 Measuring rope length

AREMA-recommended practice for measuring counterweight rope length is that during measuring the rope be supported throughout its length at points not more than 25 ft apart and tensioned to 12% of its MBF.

6.12.4 Rope maintenance

Wire ropes literally wear out, over time. Proper field lubrication during service aids in maximising life. The effect of lubrication on wire rope life is pronounced (Bridon, 2014a). As an example, Bridon gives the relative fatigue life for Bridon Endurance Dyform 34 LR (lubrication reforms) for various lubrication regimes. The life of bright finish rope that is lubricated in service is 160% of that of the same rope that is not attended to after installation. Also that the same rope with galvanised finish and which is lubricated in service has a life 220% of that of the bright finish rope that is not lubricated after installation. The use of drawn-galvanised wire for crane rope is common in Europe. In the USA, typically, bright wire is installed on cranes.

Field inspection is an important part of maintenance. The matter is thoroughly discussed in Wirerope Works (2010), Union Rope (2012) and Wire Rope Technical Board (2005).

REFERENCES

Agricola G (1558) *De Re Metallica*, trans. Hoover HC and Hoover LH (1950) Dover Publications, New York, NY, USA, http://dx.doi.org/10.1063/1.3067253.

Albert WAJ (1837) On the manufacture of whim ropes from iron wire. *Journal of the Franklin Institute* **19(5)**: 369–373, May.

AREMA (American Railway Engineering and Maintenance Association) (2014) Movable bridges – steel structures. Ch. 15 in *Manual of Railway Engineering*. AREMA, Lanham, MD, USA.

ASCE (American Society of Civil Engineers) (2010) ASCE Standard 19–10, Structural Applications of Steel Cables for Buildings. ASCE, Reston, VA, USA.

ASTM (American Society for Testing and Materials) (2007) ASTM A1007 Standard specification for carbon steel wire for wire rope. ASTM International, West Conshohocken, PA, USA.

ASTM (2009) ASTM A1023 Standard specification for stranded carbon steel wire ropes for general purposes. ASTM International, West Conshohocken, PA, USA.

Aston J and Story EB (1957) *Wrought Iron: Its Manufacture, Characteristics, and Applications*. AM Byers, Pittsburgh, PA, USA.

Blair M and Stevens TL (1995) *Steel Castings Handbook*, 6th edn. Steel Founders' Society of America and ASM International, Materials Park, OH, USA.

Bridon (2014a) *Bridon Crane and Industrial Catalog*. Bridon International, Balby Carr Bank, Doncaster, UK.

Bridon (2014b) *Bridon North American Catalog*, Bridon International, Balby Carr Bank, Doncaster, UK.

Chandler H (1995) *Heat Treater's Guide: Practices and Procedures for Irons and Steels*, 2nd edn. ASM International, Materials Park, OH, USA.

Chandler H (1996) *Heat Treater's Guide: Practices and Procedures for Nonferrous Alloys*. ASM International, Materials Park, OH, USA.
Cobb HM (2012) *Dictionary of Metals*. ASM International, Materials Park, OH, USA.
Costello GA (1987) *Theory of Wire Rope*. Dept of Theoretical and Applied Mechanics, University of Illinois at Urbana-Champaign, http://dx.doi.org/10.1115/1.2899552.
Davis JR (1998) *Metals Handbook*, 2nd edn. ASM International, Materials Park, OH, USA.
Dieng L, Urvoy JR, Siegert D, Brevet P, Perier V and Tessier C (2007) Assessment of lubrication and zinc coating on the high cycle fretting fatigue behavior of high strength steel wires. *Oipeec Bulletin 93*: June.
Digges TG, Rosenberg SJ and Geil GW (1956) Heat treatment and properties of iron and steel. *NBS Monograph 88*, US Government Printing Office, Washington, DC, USA.
DIN (Deutsches Institute für Normung) (2012a) DIN EN 10264–2, Cold drawn non-alloy steel wire for ropes for general applications. Beuth Verlag, Berlin, Germany.
DIN (2012b) DIN EN 12385–4, Stranded ropes for general lifting applications. Beuth Verlag, Berlin, Germany.
Doane DV (1987) *Steels and Cast Irons – Application and Metallurgy*. Lecture notes, Course 3, Lesson 4, ASM International, Materials Park, OH, USA.
Feyrer K (1995) *Drahtseile: Bemessung, Betrieb, Sicherheit* (in German). Springer-Verlag, Berlin, Germany.
Feyrer K (2000) *Drahtseile: Bemessung, Betrieb, Sicherheit* (in German). 2nd edn. Springer-Verlag, Berlin, Germany.
Feyrer K (2007) *Wire Rope: Tension, Endurance, Reliability*. Springer-Verlag, Berlin, Germany.
Gourmelon JP (2002) Fatigue of staying cables. *International Journal of Rope Science and Technology, Bulletin* 84. Organisation Internationale Pour L'Etude de L'Endurance des Cables (OIPEEC), Tension Technology Int., Eastbourne, UK.
Holden A and Morrison P (1982) *Crystals and Crystal Growing*. MIT Press, Cambridge, MA, USA.
Henkel D and Pense AW (2002) *Structure and Properties of Engineering Materials*, 5th edn. McGraw-Hill, New York, NY, USA.
Jones WR (1985) Why vacuum? *Heat Treating Magazine*, July: 44–45.
Keyser CA (1974) *Material Science in Engineering*, 2nd edn. Charles E. Merrill Publishing, Columbus, OH, USA.
Ketchum MS (1924) *Structural Engineer's Handbook*, 3rd edn. McGraw-Hill, New York, NY, USA.
Klein TW (2013) Structural cable designs and concepts. *Proceedings of the 7th New York City Bridge Conference*, 26–27 August, New York, NY, USA, http://dx.doi.org/10.1201/b15790-5.
Krauss G (2015) *Steels: Processing, Structure, and Performance*, 2nd edn. ASM International, Materials Park, OH, USA.
Miller RB (2007) Characteristics and availability of commercially important woods. Chapter 1 in *The Encyclopedia of Wood*. Skyhorse Publishing, Inc., New York, NY, USA.
Moore HF (1941) *Textbook of Materials of Engineering*. McGraw-Hill, New York, NY, USA.
NEN (Nederlands Normalisatie Instituut) (2001) NEN 6786:2001 Rules for the Design of Movable Bridges. NEN, Delft, Holland.
Oberg E, Jones FD, Horton HL and Ryffel HH (2012) *Machinery's Handbook*, 29th edn, Industrial Press, New York, NY, USA.
Overman F (1850/2011) *The Manufacture of Iron, in all its Various Branches*. Cambridge University Press, Cambridge, UK, http://dx.doi.org/10.1017/CBO9780511795435.
Paton AG, WM (2001) Advances in the fatigue assessment of wire ropes. *Ocean Engineering* **28**: 491–518, http://dx.doi.org/10.1016/S0029-8018(00)00014-7.

Paulinyi A (1987) *Das Puddeln, Ein Kapitel aus der Geschichte des Eisens in der Industriellen Revolution* (in German). Oldenbourg Verlag, Munich, Germany.

Rakhit AK (2000) *Heat Treatment of Gears*. ASM International, Materials Park, OH, USA.

Reardon AC (2011) *Metallurgy for the Non-Metallurgist*. ASM International, Materials Park, OH, USA.

Roebling (1966) *CF&I – Roebling Wire Rope Handbook*. CF&I Steel Corp. Denver, CO, USA.

Sayenga D (1999) A history of wrought iron wire suspension bridge cables. *Proceedings of an International Conference on Historic Bridges to Celebrate the 150th Anniversary of the Wheeling Suspension Bridge* (Kemp, E (ed)), 21–23 October, West Virginia University Press, Wheeling, WV, USA.

Sim D and Ridge I (2002) *Iron for the Eagles: The Iron Industry of Roman Britain*. Tempus Publishing, Stroud, UK and Charleston, SC, USA.

Simmons DA (2001) Bridging the flats: navigating the streets and railroads of Cleveland. *Proceedings of the 7th Historic Bridge Conference*, pp. 146–155, 19–22 September, Cleveland, OH, USA.

Swain GF (1924) *Structural Engineering, Fundamental Properties of Materials*. McGraw-Hill, New York, NY, USA.

Tempil NI (2010) *Basic Guide to Ferrous Metallurgy*. LA-CO Industries, Elk Grove, IL, USA.

Thelning KE (1975) *Steel and its Heat Treatment*. Buttersworth, London.

Total Materia (2014) , http://www.keytometals.com (accessed 10/12/2014).

Union Rope (2012) *Wire Rope User's Handbook*. Wireco World Group, Kansas City, MO, USA.

Verhoeven JD (2007) *Steel Metallurgy for the Non-Metallurgist*. ASM International, Materials Park, OH, USA.

Verreet R (1998) *Calculating the Service Life of Running Steel Ropes*. CASAR Drahtseilwerk Saar, Kirkel, Germany.

Wire Rope Technical Board (2005) *Wire Rope User's Manual*, 4th edn. Wire Rope Technical Board, Alexandria, VA, USA.

Wirerope Works (2010) *Bethlehem Wire Rope: General Purpose Catalog*. Wirerope Works, Inc., Williamsport, PA, USA.

Zhang Z and Costello GA (1996) Fatigue design of wire rope. *Wire Journal International*, **29(2)**: 106–115.

Movable Bridge Design
ISBN 978-0-7277-5804-0

http://dx.doi.org/10.1680/mbd.58040.157

Chapter 7
Material properties and failure

Charles Birnstiel

7.0. Introduction

The objective of this chapter is to present information on the properties of metals commonly used for building movable bridges, with emphasis on US practice. To that end material data will be extracted from the American Society for Testing and Materials (ASTM) material and testing specifications. This will be followed by descriptions of standard tests utilised to determine mechanical properties. Finally, material failures not explicitly discussed in AASHTO (American Association of State Highway and Transportation Officials) (2007) and AREMA (American Railway Engineering and Maintenance-of-Way Association) (2014) will be described briefly, at an elementary level. Tribology will be introduced (see Section 7.9).

7.1. Material specifications

Most specifications for metals used in movable bridge construction are listed in Table 7.1. They are grouped as structural or mechanical, mostly for tradition, because the structural materials are also used for manufacturing machinery. ASTM is the abbreviation for ASTM (American Society for Testing and Materials) International, the first US testing and materials standards organisation, founded in 1898 in Philadelphia, PA and now located in West Conshohocken, PA.

The principal physical (mechanical) material properties of interest are: tensile strength, yield point or yield strength at a specified percent strain offset, elongation of specimen during test from zero load to complete fracture for a specified gage length, the reduction of cross-sectional area at the fracture location, and the material hardness. Hardness is usually reported according to a Brinell or Rockwell scale. Tables 7.2 and 7.3 list the mechanical requirements for the materials in Table 7.1 as given by ASTM, but in a modified format. The terms in these tables are defined in the descriptions of the standard tensile and hardness tests which are included herein. Figure 7.1 shows the initial engineering stress–strain curves for wrought iron and some steels. A compilation of stress–strain curves for many metals was prepared by Boyer (1987). Further information about steels currently used in American bridge building may be found in FHWA (Federal Highway Association, 2012).

7.2. Tensile properties

Although tensile tests of materials probably had been made in the West before, it is likely that the first systematic study of the mechanical properties of structural materials was that of the Dutch physicist Petrus von Musschenbroek. He invented the tensile testing machine with the horizontal balance lever (Kurrer, 2008). Results of tension tests on various species of timber and kinds of wrought iron made by Musschenbroek were reported in Eytelwein (1808). It is interesting that Eytelwein wrote about the effect of member cross-section size on the strength of wrought iron. Materials testing advanced in the USA after the Civil War, a prominent testing machine manufacturer being Riehle of Philadelphia in Pennsylvania. Industrialisation and railway expansion led to the formation of the American section

Table 7.1 Some ASTM specifications referred to in AREMA (2014)

ASTM designation	Title
Structural	
A36	Standard specification for carbon structural steel
A709	Standard specification for structural steel bridges
A992	Standard specification for structural steel shapes
A572	Standard specification for high-strength low-alloy Columbium-vanadium structural steel
A588	Standard specification for high-strength low-alloy structural steel, up to 50 ksi [345 MPa] Minimum yield point, with atmospheric corrosion resistance
A307	Standard specification for carbon steel bolts, studs, and threaded rod, 60 000 psi tensile strength
A325	Standard specification for structural bolts, steel, heat treated 120/105 ksi minimum tensile strength
A490	Standard specification for structural bolts, alloy steel, heat treated, 150 ksi minimum tensile strength
Machinery	
A27	Standard specification for steel castings, carbon, for general application
A668	Standard specification for steel forgings, carbon and alloy, for general industrial use
A48	Standard specification for gray iron castings
B22	Standard specification for bronze castings for bridges and turntables
B23	Standard specification for white metal bearing alloys (known commercially as 'Babbitt metal')
A449	Standard specification for hex cap screws, bolts and studs, steel, heat treated, 120/105/90 ksi minimum tensile strength, general use

of the International Association for Testing and Materials (ASTM) in 1898. The development of material standards by ASTM made it desirable to develop standardised tests including ASTM E8 Standard Test Methods for Tension Testing of Metallic Materials. Corresponding tests were developed in Great Britain and on the Continent eventually leading to ISO 6892–1 (ISO, 2009).

7.2.1 Some provisions of ASTM E8

The information sought from the standard tension test is shown in Figure 7.2. It is a composite diagram of two curves: an engineering stress–strain curve and a true stress–true strain curve. It is not to any scale and some features are exaggerated for clarity. By convention the term stress–strain curve means engineering stress–strain for which the parameters are computed based on the original dimensions of the specimen. In contrast, for the true stress–strain curve the quantities are computed taking into account the reduction of cross-sectional area and the elongation during necking. The ordinates to the curves in Figures 7.1 and 7.2 are dependent on the rate of straining. The wavy curves of the yield plateau indicate the result of a process of manipulating the control valve of a hydraulic testing machine so as to obtain load readings at which the strain rate was zero or nearly so, in a positive direction. A curve drawn tangent to the valleys represents the lower yield strength.

7.2.1.1 Tensile specimens

Dimensions for preferred standard test specimens are specified for many products in ASTM E8 and ISO 6892–1.

Table 7.2 ASTM-required mechanical properties of structural materials

ASTM spec. no.	Class or grade	Thick./dia. inch: Min.	Max.	Min. tens. strength ksi: MPa	Min. tens yield pt/str. ksi: MPa	Elong. in 2 inch: %	Reduct. of area: %	Brinell hardness HBW: HB	Note/ heat treat
A36	Shapes	–	–	58–80 (400–550)	36 (250)	21	–	–	a
–	Plates	–	–	58–80 (400–550)	36 (250)	23	–	–	b
A572	Gr. 42	–	6	60 (415)	42 (290)	24	–	–	c, f
–	50	–	4	65 (450)	50 (345)	21	–	–	c, e, f
–	55	–	2	70 (485)	55 (380)	20	–	–	c, f
–	60	–	1.25	75 (520)	60 (415)	18	–	–	c, g
–	65	–	1.25	80 (550)	65 (450)	17	–	–	c, g
A588	–		4	70 (485)	50 (345)	21	–	–	plates
–	–	4	5	67 (460)	46 (315)	21	–	–	plates
–	–	5	8	63 (435)	42 (290)	21	–	–	plates
–	–	–	–	70 (485)	50 (345)	21	–	–	shapes
A709	Gr. 36	–	4	58–80 (400–550)	36 (250)	–	–	–	–
–	50	–	4	65 (450)	50 (345)	–	–	–	–
–	50S	–	4	65 (450)	–	–	–	–	–
–	50W	–	4	70 (485)	–	–	–	–	–
–	70W	–	4	85–110 (585–760)	–	19 pl	–	–	–
–	100	–	2.5	110–130 (760–896)	–	18 pl	40–50	235–293	h
–	100W	2.5	4	100–130 (690–896)	90 (620)	18 pl	40–50	–	–
A992	–	–	–	65 (450)	50–65 345–450	21	–	–	–
A307	A	–	–	60 (415)	–	–	–	121– 241	–
–	B	–	–	60–100 (415–690)	–	–	–	121– 212	–
–	C	–		58–80 (400–550)	–	–	–	none	–
A325/	–	0.5	1.0	120 (825)	–	–	–	253–319	–
–	–	1.1	1.5	105 (725)	–	–	–	223–286	–
A490	–	0.5	1.5	150–170 (1035–1170)	–	–	–	311–353	–

Notes
a For W shapes over 426 plf (634 kg/m) the 80 ksi (550 MPa) max tensile strength does not apply
b Yield point 32 ksi (220 MPa) for plates over 8 in. (200 mm) thick
c For W shapes over 426 plf (634 kg/m) min. elongation = 19%
e For grade 50 (345) of thickness $\frac{3}{4}$ in. (20 mm) or less the min. tensile strength = 70 ksi (485 MPa)
f For plates wider than 24 in. (600 mm) the elongation requirement is reduced by 2%.
g Elongation requirement reduced by 3% for plates wider than 24 in. (600 mm).
h For plates wider than 24 in. (600 mm) the reduction in area requirement is reduced by 5% and elongation is reduced by 2%

Table 7.3 ASTM required mechanical properties of machinery materials: part 1

ASTM spec. no.	Class or grade	Thick./dia. inch: Min.	Max.	Min. tens. strength ksi: MPa	Min. tens yield pt/str. ksi: MPa	Elong. in 2 inch: %	Reduct. of area: %	Brinell hardness HBW: HB	Note/heat treat
A27	U60–30			60 (415)	30 (205)	22	30		
	60–30			60 (415)	30 (205)	24	35		
	65–35			65 (450)	35 (240)	24	35		
	70–36			70 (485)	36 (250)	22	30		
	70–40			70 (485)	40 (275)	22	30		
A48	20			20 (138)					
	25			25 (172)					
	30			30 (207)					
	35			35 (241)					
	40			40 (276)					
	45			45 (310)					
	50			50 (345)					
	55			55 (379)					
	60			60 (414)					
A668	A		20	47 (325)				183 max.	
"	B		20	60 (415)	30 (205)	24	36	120–174	A, N or N & T
"	C		12	66 (455)	33 (230)	23	36	137–183	A, N or N & T
		12	20	"	"	22	34		
"	D		8	75 (515)	37.5 (260)	24	40	149–207	N, A or N & T
		8	12	"	"	22	35	"	
		12	20	"	"	20	32	"	
		20		"	"	12	30	"	
"	E		8	85 (585)	44 (305)	25	40	174–217	N & T or NN & T
		8	12	83 (570)	43 (295)	23	37	"	
		12	20	"	"	22	35	"	
"	F		4	90 (620)	55 (380)	20	39	187–235	Q & T or NQ & T
		4	7	85 (585)	50 (345)	"	"	174–217	
		7	10	"	"	19	37	"	
		10	20	82 (565)	48 (330)	"	36	"	
"	G		12	80 (550)	50 (345)	24	40	163–207	A, N or N & T
		12	20	"	"	22	38	"	
A668	H		7	90 (620)	60 (415)	22	44	187–235	N & T
		7	10	"	58 (400	21	42	"	
		10	20	"	58 (400)	18	40	"	
"	J	–	7	95 (655)	70 (485)	20	50	197–255	N & T or NQ & T
		7	10	90 (620)	65 (450)	20	50	187–235	
		10	20	"	"	18	48	"	
"	K	–	7	105 (725)	80 (550)	20	50	212–269	NQ & T
		7	10	100 (690)	75 (515)	19	"		
		10	20	"	"	18	48		
"	L		4	125 (860)	105 (725)	16	50	255–321	NQ & T

Table 7.3 Part 2

ASTM spec. no.	Class or grade	Thick./dia. inch: Min.	Max.	Min. tens. strength ksi: MPa	Min. tens yield pt/str. ksi: MPa	Elong. in 2 inch: %	Reduct. of area: %	Brinell hardness HBW: HB	Note/heat treat
		4	7	115 (795)	95 (655)	"	45	235–302	
		7	10	110 (760)	85 (585)	"	"	223–293	
		10	20	"	"	14	40	"	
"	M		4	145 (1000)	120 (825)	15	45	293–352	NQ & T
		4	7	140 (965)	115 (790)	14	40	285–341	
		7	10	135 (930)	110 (758)	13	"	269–331	
		10	20	"	"	12	38	"	
"	N		4	170 (1175)	140 (965)	13	40	331–401	NQ & T
		4	7	165 (1140)	136 (930)	12	35	"	
		7	10	160 (1100)	130 (900)	11	"	321–388	
		10	20	"	"	"	"	?	
A449		0.25	1.0	120 (825)	92 (635)	14	35	(255)–(321)	
		1.0	105	105 (725)	81 (560)	14	35	(223)–(285)	
		1.5	3	90 (620)	58 (400	14	35	(183)–(235)	
B22	863	55	(380)	110 (760)	60 (415)	12		(223)	j, k
	905			40 (275)	18 (125)	20			j
	937			30 (207)	12 (83)	15			j

j tensile yield strength at 0.05% extension.
k 55 ksi (380 MPa) deformation limit in compression at 0.001 in. (0.025 mm) on specific test specimen.
A annealed
N normalised
Q quenched
T tempered

7.2.1.2 Yield strength

The mechanical requirements of all metallic mechanical specifications require a minimum yield point or yield strength. As is evident in Figure 7.1 the initial shape of the engineering stress against strain diagrams of ferrous metals differs at the knee. Structural carbon steel (mild steel) has a yield plateau (yield point elongation) from the elastic limit to a strain of about 0.014 in. per inch. The start of the plateau (at the left) may be at the upper yield point σ_{uy} or as low as the proportional limit, pl. The yield plateau is a very desirable feature of the steel. The yield plateau is absent for quenched and tempered carbon and alloy steels. For these steels a yield strength is defined by the engineering stress at a tensile strain offset (usually 0.2%). There are other criteria for yield strength in ASTM E8 (ASTM, 2013a) and ISO 6892–1 (ISO, 2009). In the UK, the yield stress is often referred to as the proof stress with strain offset values of either 0.1 or 0.5%.

7.2.1.3 Elongation

The percentage elongation between gauge points marked on the specimen is important because it is an indicator of the ductility of the material – its ability to deform plastically without fracturing. In the USA, the usual gauge lengths are 2 in. (55 mm) for tensile specimens of circular cross-section and 8 in. (20.3 mm) for those of rectangular cross-section.

Figure 7.1 Stress–strain curves of wrought iron and some steels. (Based on Burr (1915) and Galambos (1998))

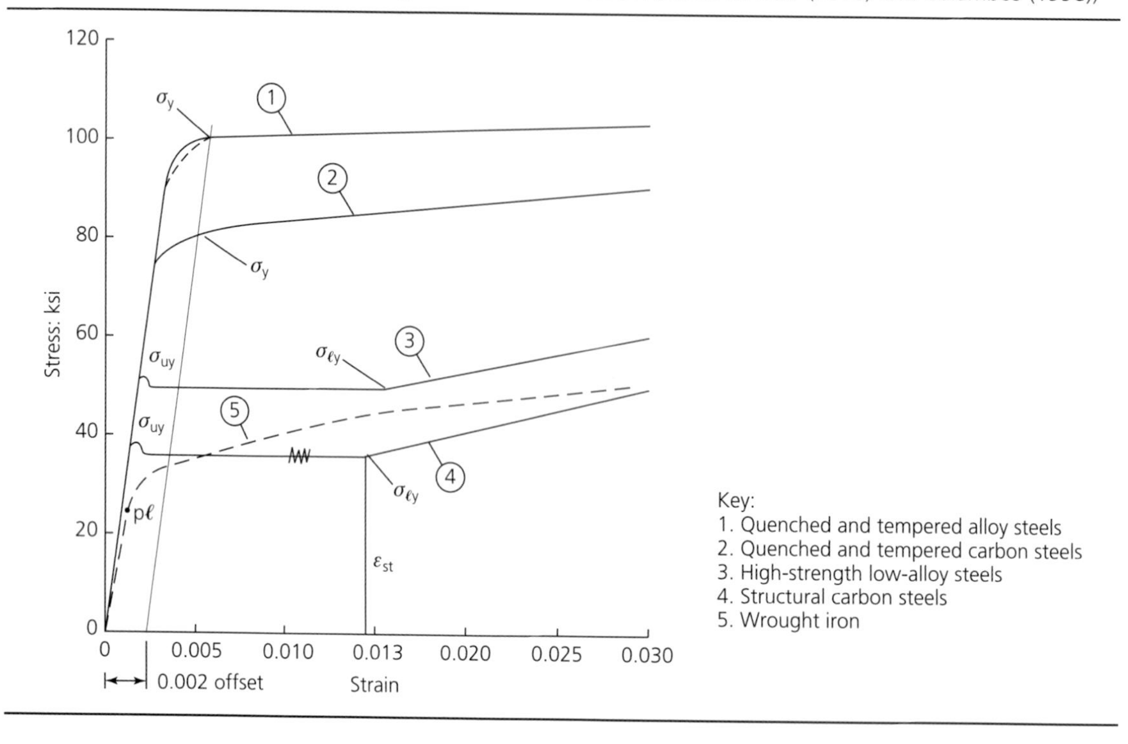

Figure 7.2 Notation for stress–strain curves from tensile tests. (ASTM (2013a) and ISO (2009))

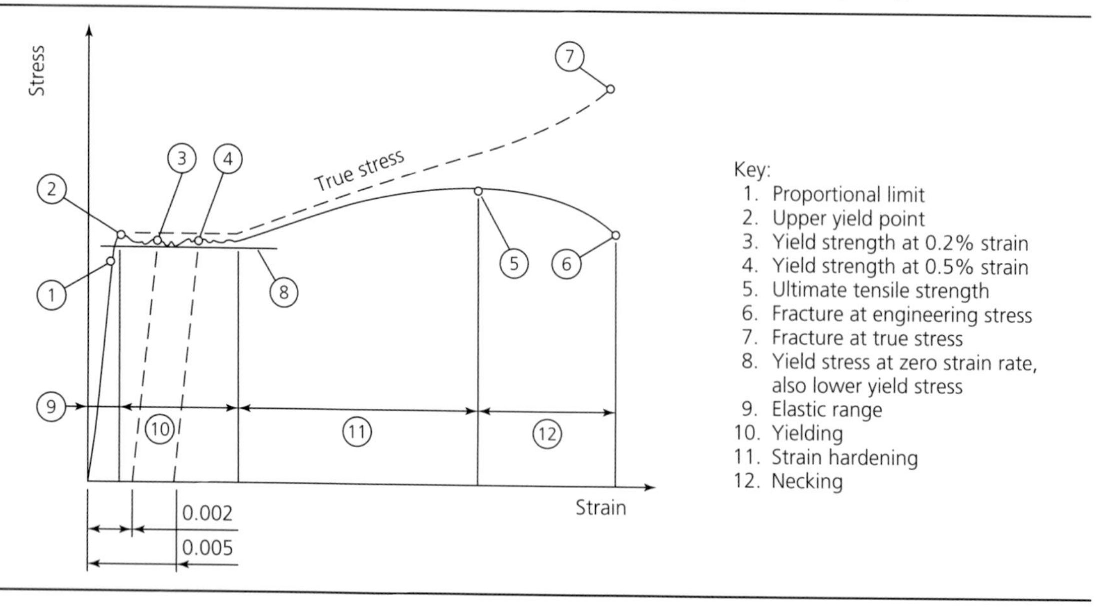

7.2.1.4 Reduction of area
Reduction of area is the difference between the original cross-sectional area of the tension test specimen at the reduced section and the area of the smallest cross-section after fracture, expressed as a percentage of the original area. The quantity is also an indication of the ductility of the material.

7.3. Material hardness
Most ASTM material specifications for metals list requirements for hardness according to a Brinell and/or a Rockwell scale. What is hardness? Material hardness has many meanings. It generally implies a resistance to deformation by mechanical means, usually at room temperature. In metallurgy, hardness is a measure of resistance to localised permanent (plastic) deformation. Commonly, a hard indenter is pressed into the flat surface of the material to be tested under a specific force for a specified time duration (dwell) and the resulting residual indentation, after removal of the force, is measured. A number of such static tests have been developed since the late 19th century. Dynamic hardness tests are also available in which a weight is dropped on the specimen and in other tests an indenter is fired as a projectile onto the subject by a spring-loaded device. In what follows, the more common hardness testers will be briefly described. Details of the equipment and computation of the hardness values may be found in Davis (1998). A state-of-the-art report on conventional hardness measuring methods was presented by Low (2005).

7.3.1 Static hardness tests
7.3.1.1 Brinell hardness testing
The Brinell harness test is conceptually simple. Until recently, a constant force (500 kg, 1500 kg or 3000 kg) was applied on a flat surface of the workpiece through an indenter having as a point a 10 mm dia. ball of hardened steel or of tungsten carbide and the load was maintained for 10–15 seconds before release. The Brinell hardness number is defined as the test force divided by the curved surface area of the indentation (assumed spherical) after removal of the force, all residual deformation being plastic. Obviously, the Brinell hardness number had to be related to the five conditions of test (three forces and two indenters).

In the current ASTM E-10–12 (ASTM, 2012), the test conditions were increased to 25 in number, although only tungsten carbide balls of four diameters at the tips of indenters are now recognised; they are 1, 2.5, 5 and 10 mm. Table 3 of ASTM (2012) lists the 25 valid test force and ball diameter combinations, which are repeated in Table 7.4.

In practice, the Brinell test force is applied to the specimen with a dwell time to permit plastic dislocation of the workpiece material, usually 10–15 seconds. Then the diameter of the indentation at the surface of the workpiece is measured and the hardness number computed from equations given in (Reardon, 2011). Alternatively, one can enter the table of Brinell hardness numbers in the appendix to ASTM (2012) with the measured diameter and read off the hardness number. The table covers all 25 Brinell test conditions.

The notation for reporting Brinell hardness has varied over the years. As per ASTM (2012), the current Brinell hardness numbers shall be followed by the symbol HBW and be supplemented by an index indicating the test conditions in the following order: ball diameter in mm, test force in kg and applied force dwell time in seconds (if other than 10–15 seconds). For example, if a Brinell number of 211 were to be obtained using a 750 kg force applied for 30 seconds by a 5 mm dia. tungsten carbide ball, the Brinell value would be reported as 211 HBW 5/750/30. The exception to this scheme is that the test conditions need not be listed for one case – when a 10 mm tungsten carbide ball is used with a 3000 kg force and

Table 7.4 Brinell hardness scales and test conditions (From ASTM (2014a))

Brinell scale	Ball diameter D: mm	Test force F: kgf
HBW 10/3000	10	3000
HBW 10/1500	10	1500
HBW 10/1000	10	1000
HBW 10/500	10	500
HBW 10/250	10	250
HBW 10/125	10	125
HBW 10/100	10	100
HBW 5/750	5	750
HBW 5/250	5	250
HBW 5/125	5	125
HBW 5/62.5	5	62.5
HBW 5/31.25	5	31.25
HBW 5/25	5	25
HBW 205/187.5	2.5	187.5
HBW 2.5/62.5	2.5	62.5
HBW 205/31.25	2.5	31.25
HBW 205/15.625	2.5	15.675
HBW 2.5/7.8125	2.5	7.8125
HBW 2.5/6.25	2.5	6.25
HBW 1/30	1	30
HBW 1/10	1	10
HBW 1/5	1	5
HBW 1/2.5	1	2.5
HBW 1/1.25	1	1.25
HBW 1/1	1	1

All balls to be tungsten carbide

the dwell is between 10 and 15 seconds. ASTM requires that Brinell values obtained using a hardened steel ball be reported as xxxHB and those obtained with a carbide ball be reported as yyyHBW.

7.3.1.2 Rockwell hardness testing

Rockwell hardness testing is a popular method for determining metal hardness in America. By means of different loads and indenters the Rockwell tester can be used for determining the hardness of a wide range of materials. The hardness number is determined by the depth of the indentation made by an indenter under a prescribed constant load. This is in contrast to Brinell in which the surface area of the residual impression determines the Brinell number.

There are two types of Rockwell hardness tests that, paradoxically, use the same physical machine. For historical reasons they are called the Regular Rockwell Test (Regular R.) and the Superficial Rockwell Test (Superficial R.) Both use the same basic machine but different loading conditions; the loads differ in magnitude and the indenter points differ in material and size. The test conditions for Regular R. are listed in Table 7.5 which is based on ASTM E18 – 14a standard test methods for Rockwell hardness of metallic materials (ASTM, 2014b). Table 7.6 lists the test conditions for Superficial R.

Table 7.5 Regular Rockwell hardness scales and test conditions (ASTM, 2014b)

Scale symbol	Indenter	Total test force: kgf
A	Diamond	60
B	1/16 in. (1.588 mm) ball	100
C	Diamond	150
D	Diamond	100
E	$\frac{1}{8}$ in. (3.175 mm) ball	100
F	$\frac{1}{16}$ in. (1.588 mm) ball	60
G	$\frac{1}{16}$ in. (1.588 mm) ball	150
H	$\frac{1}{8}$ in. (3.175 mm) ball	60
K	$\frac{1}{8}$ in. (3.175 mm) ball	150
L	$\frac{1}{4}$ in. (6.350 mm) ball	60
M	$\frac{1}{4}$ in. (6.350 mm) ball	100
P	$\frac{1}{4}$ in. (6.350 mm) ball	150
R	$\frac{1}{2}$ in. (12.70 mm) ball	60
S	$\frac{1}{2}$ in. (12.70 mm) ball	100
V	$\frac{1}{2}$ in. (12.70 mm) ball	150

Table 7.6 Superficial Rockwell hardness scales and test conditions (ASTM, 2014b)

Scale symbol	Indenter	Total test force: kgf
15N	Diamond	15
30N	Diamond	30
45N	Diamond	45
15T	$\frac{1}{16}$ in. (1.588 mm) ball	15
30T	$\frac{1}{16}$ in. (1.588 mm) ball	30
45T	$\frac{1}{16}$ in. (1.588 mm) ball	45
15W	$\frac{1}{8}$ in. (3.175 mm) ball	15
30W	$\frac{1}{8}$ in. (3.175 mm) ball	30
45W	$\frac{1}{8}$ in. (3.175 mm) ball	45
15X	$\frac{1}{4}$ in. (6.350 mm) ball	15
30X	$\frac{1}{4}$ in. (6.350 mm) ball	30
45X	$\frac{1}{4}$ in. (6.350 mm) ball	45
15Y	$\frac{1}{2}$ in. (12.70 mm) ball	15
30Y	$\frac{1}{2}$ in. (12.70 mm) ball	30
45Y	$\frac{1}{2}$ in. (12.70 mm) ball	45

7.3.1.2.1 ROCKWELL TEST PROCEDURE

The hardness test is divided into four steps of force application and removal as per ASTM E18 – 14a (ASTM, 2014b).

(*a*) Step 1 – bring indenter into contact with the workpiece and apply the preliminary test force F_o. After holding F_o for the specified dwell, measure depth of indentation. This is called the baseline depth of indentation.
(*b*) Step 2 – Increase force on the indenter by adding test force F_1, to achieve a total test force $F_o + F_1 = F$. Hold F for specified dwell.
(*c*) Step 3 – Remove F_1 and hold F_o for specified time and measure the final indentation.
(*d*) Step 4 – Remove F_o and the indenter from the workpiece.

Rockwell hardness relates inversely to the difference between final and baseline indentations while under load F_o.

7.3.1.2.2 ROCKWELL HARDNESS TEST CONDITIONS (REGULAR R.)

For Regular R. the preliminary test force $F_o = 10$ kgf and the total test forces F are 60, 100, or 150 kgf.

7.3.1.2.3 ROCKWELL SUPERFICIAL HARDNESS TEST CONDITIONS (SUPERFICIAL R.)

For Superficial R. the preliminary test force $F_o = 3$ kgf and the total test forces F are 15, 30, or 45 kgf.

7.3.1.2.4 ROCKWELL DESIGNATIONS

The hardness number is followed by the symbol HR and the scale designation. When a ball indenter is used the scale designation is followed by the letter W if it is tungsten carbide or S if the ball is hardened steel. For example, 64 HRC means hardness 64 on the Rockwell C scale made with a diamond indenter (there is no W or S with a diamond indenter).

7.3.1.2.5 CONVERSION TO OTHER HARDNESS SCALES OR TENSILE STRENGTH VALUES

Rockwell hardness numbers on any scale cannot be accurately converted to Rockwell hardness numbers on another scale, or to other types of hardness numbers, or to tensile strength values per ASTM E18 – 14a (ASTM, 2014b). However, the traditional Standard hardness conversion tables for metals of ASTM E140 (2012) are approximate conversions which are still accepted by industry. The Rockwell hardnesses in E140 were determined using ball indenters. Excerpts from ASTM E140 are given in Table 7.7. Several other types of static hardness testers are available, including Vickers, Knoop and Tukon, which lend themselves to microhardness testing, that is testing using small indentations (Davis, 1998).

7.3.2 Dynamic hardness tests

7.3.2.1 Shore sceleroscope

The Shore sceleroscope was developed in the early 1900s. A diamond-tipped hammer is dropped from a fixed height onto the workpiece and the rebound distance is measured. By this means there is only one condition of test for a wide range of hardness, compared to the Brinell and Rockwell machines which utilise 25 and 30 test conditions, respectively. To perform a test, the hammer is raised to the elevated position and then released. The height to which the hammer rebounds on the first bounce is a measure of the material hardness. ASTM E448 Standard practice for sceleroscope hardness testing of metallic materials suggests operating details and precautions.

Table 7.7 Approximate hardness conversion numbers for steel based on ASTM E140. (Adapted from Davis, 1998, p. 72)

Brinell	Rockwell			Vickers	Scleroscope
HBW Note (1)	A scale Note (2)	C scale Note (3)	D scale Note (4)	Note (5)	Note (6)
	85.6	68	76.9	940	97.3
	84.5	66	75.4	865	92.7
	83.4	64	73.8	800	88.5
	82.3	62	72.2	746	84.5
	81.2	60	70.7	697	80.8
615	80.1	58	69.2	653	77.3
577	79.0	56	67.7	613	74.0
543	78.0	54	66.1	577	70.9
512	76.8	52	64.6	544	67.9
481	75.9	50	63.1	513	65.1
455	74.7	48	61.4	484	62.4
432	73.6	46	60.0	458	59.8
409	72.5	44	58.5	434	57.3
390	71.5	42	56.9	412	54.9
371	70.4	40	55.4	392	52.6
353	69.4	38	53.8	372	50.4
336	68.4	36	52.3	354	48.2
319	67.4	34	50.8	336	46.1
301	66.3	32	49.2	318	44.1
286	65.3	30	47.7	302	42.2
271	64.3	28	46.1	286	40.4
258	63.3	26	44.6	272	38.7
247	65.4	24	43.1	260	37.0
237	61.5	22	41.6	248	35.3
226	60.5	20	40.1	238	35.2

Key: (1) 3000 kgf load, 10 mm carbide ball; (2) 60 kgf load, diamond indenter; (3) 150 kgf load, diamond indenter; (4) 100 kgf load, diamond indenter; (5) Load varies; (6) Dynamic test, falling weight

7.3.2.2 Leeb rebound hardness test

A portable rebound tester was developed by Leeb and Brandestini about 1975. With this machine the hardness value is determined from the energy loss of an indenter projectile. The instrument impacts a spherical tungsten carbide, silicon nitride or diamond-tipped projectile, with a velocity generated by a spring force, onto the surface of the workpiece. The ratio of the rebound velocity to the impact velocity is a measure of hardness. There are eight types of projectiles used in Leeb testing, descriptions of which may be found in ASTM A956–12, Standard test method for Leeb hardness testing of steel products (ASTM, 2014c). The impact device is calibrated for the downward vertical impact direction. For other orientations, compensation values need to be applied to the readings. These are listed in tables in

ASTM (2014c) for each of the six impact devices. The EQUOTP, Leeb's original instrument with later improvements, is thoroughly discussed in Kompatscher (2005).

7.3.2.3 Some modern portable hardness testers

Advanced electronics have made possible the production of new portable hardness testers and data loggers which greatly simplify hardness testing. Four instruments that are available from GE Inspection Technologies and are suitable for metals testing are Krautkramer MIC10, Krautkramer DynaMIC, DynaPocket and Krautkramer MIC20. The methods that are utilised by these instruments are briefly described in what follows.

- Ultrasonic contact impedance method (UCI): in the UCI method, a resonating rod with a Vickers diamond tip is pressed into the surface of the specimen under a specific test force produced by a spring. Before contacting the specimen, the rod is excited into longitudinal oscillation at approximately 70 kHz by piezoelectric transducers. As the diamond penetrates into the workpiece under the impact force a frequency shift occurs. The frequency shift is proportional to the indentation of the diamond into the workpiece. The hardness value is not visually determined by the diagonals of the indent (as in Vickers) but by an electronic measurement of the frequency shift. The Krautkramer MIC10 and MIC20 portable testers utilise this principle.
- Rebound method: the modern portable rebound hardness testers operate similarly to the Shore Scleroscope and the Leeb testers. A mass is accelerated onto the surface of the workpiece and strikes it at a defined speed or kinetic energy. The impact deforms the workpiece plastically. The velocities before and after impact are measured electronically. The Krautkramer instruments have an autobalancing feature which obviates the need for corrections to readings because of impact direction. Krautkramer DynaMIC, DynaPocket and MIC20 operate on this principal (it is an alternate built into the MIC20).

The use of advanced electronics has enabled the development of efficient mobile hardness testers which make possible inspections that were impractical or impossible heretofore. However, it is unlikely that they will completely replace conventional bench-top machines.

7.4. Fracture of metals

Structural and mechanical systems of bridges have initial flaws and develop other flaws during service. The initial flaws may or may not develop into cracks large enough to compromise the strength of the system to the extent that safety becomes a concern. New cracks may form due to fatigue in critical components of the span-stabilising system, for example, counterweight trunnions of vertical lift bridges. With so many American movable bridges approaching 70 years of service, such damage is bound to occur frequently. For economic reasons, it may not be feasible to take the bridge out of service immediately to effect repairs. The bridge owner must 'live with the situation'. Sometimes loading can be decreased, but often the only politically viable approach is to adopt the notion of damage-tolerant design. This involves analysing the crack using fracture mechanics theory and the concomitant field inspections to monitor crack growth. The rate of required inspections depends on the computed speed of crack growth in terms of loading cycles and time in a corrosive environment.

Although the subject of fracture mechanics has been studied for almost a century, it has only been within the past 50 years that it has become of major engineering interest, primarily because of the adoption of damage-tolerant design in aeronautical engineering (Wood and Engle Jr, 1979). It is still a specialist form of analysis among civil and mechanical engineers. Recommended texts on the subject of fracture mechanics are Anderson (1991) and Barsom and Rolfe (1999).

7.5. Residual stresses

Residual stresses are created in all metals during processing – casting, hot-rolling, welding, cold working and machining. Castings have been known to break during handling on the foundry floor. Hot-rolled I-shapes have split along the centreline of the web while lying in the fabricating shop. Residual stress is partly responsible for stress corrosion cracking and hydrogen embrittlement failures in metals, and the overall reduction in strength of hot-rolled steel shapes and weldments (Totten *et al.*, 2002). In weldments, residual stresses have been measured as high as the base material yield stress (Gurney, 1968).

An impetus for research into residual stresses in metals was the sudden failure of welded ships during and after World War II (Osgood, 1954). In the late 1950s the effect of thermal residual stresses on the load capacity of centrally loaded hot-rolled and welded W-shaped columns was investigated at Lehigh University, PA (Tall and Rao, 1961; Beedle and Tall, 1962). Researchers measured residual strains corresponding to residual compressive stresses at the tips of W-shape flanges on the order of one-third of the material yield stress. About the same time, residual stresses were measured in as-rolled wide flange shapes and H-shapes hogged from stress-relieved weldments in connection with experiments on columns subjected to biaxial bending (Birnstiel *et al.*, 1967; Birnstiel, 1968).

However, residual stresses can be either beneficial or detrimental to the behaviour of a member. Structural steel members are straightened or curved by mechanically inducing residual stresses in the member. In the rotarising process, the member is repeatedly subjected to reverse bending as a means of straightening. Rotarising changes the residual stress pattern from that originally produced by cooling after hot-rolling (Birnstiel *et al.*, 1967). Sometimes compressive residual stresses are induced in the surfaces of workpieces in order to reduce the risk of stress corrosion cracking. High residual tensile stresses are a factor in environmental embrittlement cracking.

Four methods for measuring the residual strains (which are considered to be the residual stresses divided by the modulus of elasticity) are

- Sectioning – residual strains are determined from measurements of the distance between small diameter holes drilled through a plate, flange or web before and after slicing using a 10 in. (254 mm) Whittemore gauge. A slice pattern for specimens from W shapes appears in Birnstiel *et al.* (1967). The measurements give the average residual strain through the thickness of the slice. It is a destructive test.
- Blind hole drilling – is a semi-destructive test for measuring residual strains on the surface of the material. A special electric resistance strain gauge that has three foil resistance grids arranged radially in a circular pattern is affixed to the specimen and the strain relaxation is measured as a hole is drilled into the workpiece at the centre of the strain gauge group (Micro-Measurements, 2010). To obtain data on the variation of residual strain through the thickness of the material it is necessary to drill the hole in small increments of depth. After each increment of hole depth, the observed strains are recorded and the validity of some coefficients in theory is confirmed. The problem is that the theory in TN503 is based on a small hole drilled completely through a thin, wide, flat plate subjected to plane stress. When residual stress varies with depth, the stress computed from the strain gauge readings is lower than the actual residual stress. Much strain gauge data reduction is required for each blind-hole measurement. Software for reducing this data in accordance with ASTM E837 (ASTM, 2013b) is available commercially.
- X-ray diffraction (XRD) – a non-destructive technique for measuring residual strains to a depth of less than 0.001 in. (0.025 mm) that has been proven reliable and generally applicable. The

method is most suitable for the laboratory but a small instrument, 114 mm × 248 mm × 111 mm, weighing 3 kg is commercially available. According to Ruud in Totten *et al.* (2002) XRD residual strain measurements should be performed by trained technologists using x-ray instrumentation designed for such measurements, not conventional scanning diffractometers.

- Neutron diffraction (ND) – a non-destructive method for measuring elastic residual strains throughout the volume of relatively thick steel components. The ND techniques measure the spacing between crystallographic planes in the workpiece, which spacing is affected by residual and applied stress. Unfortunately, the unstressed cyrstallographic plane spacing for steels and cast iron is not easily measured, because of variations in the composition of the workpiece. ND measurements are still very expensive.
- Ultrasonic measurement – a non-destructive method based on the fact that the speed of ultrasonic waves travelling through a material are affected by the direction and magnitude of stresses present in the material: accoustoelasticity. Measurement depths to at least 100 mm are possible. An ultrasonic computerised complex (UCC) was developed in Ontario, Canada, for the measurement of residual and applied stresses. The equipment enables the determination of magnitudes and signs of uni- and biaxial stresses. The overall dimensions of the UCC are 300 mm × 200 mm × 150 mm. The error of residual stress determination in steel is reported to be in the order of 10% of the yield strength (Kudryavstev and Kleiman, 2011).

As mentioned previously, residual strains can be beneficial and hence may be purposely induced in a workpiece to enhance resistance to stress–corrosion cracking (SCC) and to lengthen fatigue life. Löhe *et al.* in Totten *et al.* (2002) presented stress–life (S–N) curves (Wöhler diagrams) from rotating beam fatigue tests showing that the endurance limit of steel specimens in air that had been shot peened was 20–25% higher than those of smooth specimens. In corrosive environments, the S–N curve for steel takes on the shape of that for nonferrous metal (the endurance, or fatigue limit, disappears) but there is still benefit by introducing compression into the surface of the specimens by shot peening.

Residual stresses may be removed from members and weldments by thermal annealing, as discussed in Chapter 6. They may also be reduced in magnitude by vibrating the workpiece, at room temperature. In the vibratory stress relief (VSR) process, the workpiece is set into flexure by resonance achieved through electro-mechanical exciters temporarily attached to it. The VSR process has been used successfully to achieve dimensional stability in welded and cast fabrications. The degree of residual stress relaxation achieved depends on many factors but a 70–80% reduction of peak residual stress is not uncommon. VSR can be used on materials intolerant of thermal treatment or of large size. A rectangular welded mild steel fabrication 10 m × 5 m × 0.6 m, intended as a base for an aerospace milling machine, was stress relieved by VSR in order to improve dimensional stability prior to final machining. Because a very large furnace was not required, the processing cost was much less than if thermal stress relief had been specified. However, VSR is dependent on the equipment used and the operator (TWI, 2015).

Finally, it should be noted that the residual stresses at any section of a member are in equilibrium. The added stresses due to applied loads are in equilibrium. Hence, the total stresses are also in equilibrium.

7.6. Corrosion

Corrosion is a time-related phenomenon that affects movable bridges by loss of cross-sectional area of components, involvement in SCC and corrosion fatigue, and the need for cosmetic maintenance. Uniform, or general, corrosion occurs when numerous small electro-chemical cells reside on a metal surface. It is relatively easy to evaluate and monitor. Localised corrosion can also occur that creates pitting, crevice corrosion, intergranular corrosion and contributes to SCC in steel and aluminium.

Corrosion may be retarded by alloying, coating with an anodic material, or covering with an impermeable surface coating.

Alloying can improve the corrosion resistance of carbon steel. For example, the addition of a small amount of copper decreases the atmospheric corrosion rate by a factor of at least two. A plot of thickness loss against time in Reardon (2011) shows that atmospheric corrosion for copper-bearing steels is about half of that of structural carbon steel after 20 years' exposure. The addition of other elements to the copper bearing steel resulted in the high-strength–low-alloy (HSLA) steel that corrodes at less than one-sixth the atmospheric rate of structural carbon steel, over a 20-year time span.

Between 1904 and 1912, researchers in France, Germany, Britain and the USA independently experimented with the addition of chromium (Cr) and nickel (Ni) to carbon steel to produce so-called stainless steels. There are scores of standard compositions, the AISI alloy designations, and many proprietary variations, and special alloys for specific applications. For a brief description of modern stainless steel alloys see Reardon (2011). The history of stainless steel is related in Cobb (2010)

7.6.1 Forms of corrosion

Some researchers, including Fontana and Greene (1978) have identified eight forms of corrosion, all interrelated to some degree: uniform or general, galvanic, crevice corrosion, pitting, corrosion fatigue, selective leaching, erosion corrosion, and stress corrosion. Brief discussions of each form follow.

7.6.1.1 Uniform corrosion

Uniform corrosion is the most common form. It is due to chemical or electro-chemical action over the entire exposed surface, such as that of a truss, beam, or plate girder. There is reduction of cross-sectional area, which results in loss of structural capacity. However, this form of corrosion can be predicted for the specific environmental conditions. It can be mitigated by coatings and cathodic protection. Unexpected failures or premature failures are unlikely.

7.6.1.2 Galvanic corrosion

If two dissimilar metals are placed in a conductive liquid and are in contact or otherwise electrically connected, electrons will flow between them. Pure water is not an electrolyte, but rainwater and ground water contain contaminates which make them so. For any combination of two metals, the less noble (less resistant to attack) becomes anodic and the more noble becomes cathodic in the electrolyte cell. Table 7.8 lists some metals and alloys arranged according to increasing nobleness. The list is termed a galvanic series for a particular electrolyte, in this case seawater. The order of the table entries would vary for other electrolytes. However, the seawater series is usually used for general engineering design.

For example, consider an uncoated steel bridge truss structural connection made with zinc-coated steel bolts. Referring to Table 7.8, zinc is less noble than structural steel so that the bolts will be the anode and give up the zinc to the more noble mild steel. The voltage between metals in cells is greater the greater the distance between the metals in Table 7.8. Thus, zinc-galvanised bolts connecting stainless steel will lose their coating faster than the same bolts connecting carbon steel.

Type 304 and 316 stainless steel are listed twice, far apart, in Table 7.8. When the steel is in the so-called active state, it is far less noble than when it is in the passive state. The state is dependent on the oxidising power of the solution. On a plot of oxidising power against corrosion rate, the corrosion rate increases as the oxidising power increases, up to a point where increasing oxidising power decreases the corrosion

Table 7.8 Ranking of some metals and alloys in a galvanic series for seawater environment. (Adapted from Reardon, 2011)

Most anodic and susceptible to corrosion
Magnesium and its alloys
Zinc
Aluminium 1100 and 6053
Cadmium
Aluminium 2024
Mild steel (carbon) (structural)
Grey and ductile cast irons
Type 400 series stainless, 13% Cr (active)
Nickel cast irons
Type 304 and 316 stainless steels (active)
Lead
Tin
Muntz metal and manganese bronze
Naval brass
Nickel (active)
Yellow and red brasses, aluminum bronze, and silicon bronzes
Copper and copper-nickel alloys
Nickel (passive)
Titanium
Type 304 and 316 stainless steels (passive)
Silver, gold, platinum
Most cathodic (noble) and least susceptible to corrosion

rate. Metals that possess an active-passive transition become passive in moderately to strongly oxidising environments (Reardon, 2011).

The corrosion resistant properties of stainless steel are due to a self-healing passive surface film. It is not necessary to chemically treat a stainless steel to obtain the passive film: the latter forms in the presence of oxygen.

7.6.1.3 Crevice corrosion

This type of corrosion is common on the edges of gusset plates on riveted truss connections, often due to excessive rivet spacing or edge distance. Foreign deposits in the lap joint create pockets for stagnant solution.

7.6.1.4 Pitting corrosion

Pitting corrosion can cause failure by perforating the structure, often through a horizontal surface. The pit develops at an anodic site on the surface and continues to grow because the availability of the surrounding cathodic area. Defects in surface coating such as paints lead to pitting corrosion.

Modern methods of testing and statistical modelling of uniform and pitting corrosion in high-strength steels are discussed in Li *et al.* (2014). The paper is focused on suspension bridge wires but the techniques are more general.

7.6.1.5 Stress-corrosion cracking (SCC)

Stress corrosion cracking is cracking that occurs under the combined action of corrosion and tensile stress. The stress may be residual from the manufacturing process or due to active loading on the structure. See Section 7.7.

7.6.1.6 Corrosion fatigue

Corrosion fatigue is fatigue of a member in a corrosive environment. The effect of corrosive environment is especially important for some steels because it changes the shape of their S–N curves (Wöhler diagrams). The S–N curves obtained from the rotating beam fatigue tests of low and medium carbon steels conducted in air have a knee at less than 10^6 cycles of rotation leading to the concept of a fatigue limit (or an endurance limit) defining a stress at which the material has an infinite life. However, when conducted in a synthetic seawater environment, the S–N curve lacks a knee, indicating that the material does not have an infinite fatigue life in that environment.

Figure 7.3 shows S–N curves resulting from rotating beam tests on notched low-carbon steel specimens in air and in 3% salt solution (to simulate seawater) at two temperatures. The notch was a v-shaped circumferential groove 2.5 mm deep with a 0.05 mm root radius. Note that the 'air curve' has a knee at about 7×10^6 cycles. The curves obtained in the saline environment are below the 'air curve' and do not have a knee. At about 5×10^7 cycles the ordinate to the 13°C (55°F) saline curve is only about 30% of that for the 'air curve'. The difference in nominal stress between the 'air curve' and that produced for the 33°C (91°F) saline solution is even greater, illustrating the importance of electrolyte temperature on corrosion.

The marked effect of corrosive environment on the fatigue life of notched rotating beam specimen of stainless steel (13% Cr) and the loss of the endurance limit plateau was reported by Phull (2003a). In those tests the environment was 'only' a 1% solution of sodium chloride.

Figure 7.3 S–N curves for V-notched low-carbon steel specimens tested in brine. (Based on McEvily, 1990)

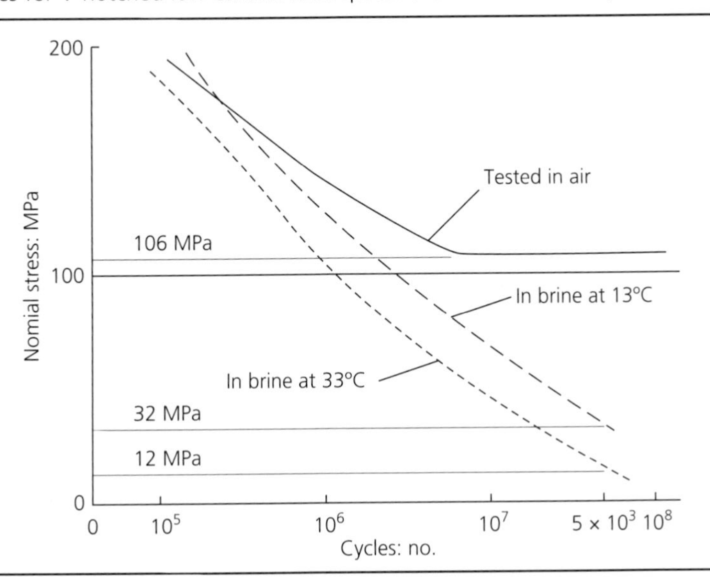

Besides corrosive environment other factors that reduce the value of the endurance limit are: tensile mean stress, large shape size, rough surface finish, chrome and nickel plating, decarburisation and severe grinding (Bannantine *et al.*, 1990. See also Battelle Memorial Institute, 1941).

7.6.1.7 Selective leaching

Selective leaching is the removal of one element from a solid alloy by a corrosion process. An example is the removal of zinc from brass alloys. Common yellow brass is comprised of 35% zinc and 65% copper. If the workpiece colour changes to red or orange, in contrast to the original yellow, dezincification has occurred.

7.6.1.8 Erosion corrosion

Erosion corrosion occurs when the electrolyte moves over the metal surface, as inside a pipe or tube. See Fontana and Greene (1978).

7.7. Stress corrosion cracking

SCC is cracking due to the interaction of corrosion and tensile stress (Fontana and Greene, 1967; Dieter, 1986; Jones, 1992; Reardon, 2011). The stress may be applied externally (mechanical stress due to loading) or be residual, remaining in the workpiece from arc welding or cold work. Cracks may be either transgranular or intergranular and are perpendicular to the tensile stress. Usually there is little evidence of surface corrosion. The underlying mechanism for SCC has not been established: it may occur by a number of mechanisms. When cracking results from the action of hydrogen, it is termed hydrogen embrittlement cracking and is discussed, for convenience, in Section 7.8.

SCC is a progressive type of fracture. Over time, cracks grow gradually until they reach a critical size after which the stress concentration causes a sudden brittle fracture. In order to occur, SCC simultaneously requires:

- a susceptible material (herein, a metal)
- a specific environment that can produce SCC for that material
- sufficient tensile stress to induce SCC.

Nearly all metal alloys are susceptible to SCC in the presence of specific environments. They include carbon and alloy steels, copper alloys, stainless steels, aluminium alloys and others (McEvily Jr, 1990). It seems that pure metals are rarely subjected to SCC. A few of the known SCC systems are listed in Table 7.9.

The only requirements for SCC of a susceptible metal is the combination of a tensile stress above a threshold value at the surface of the metal, and a critical environment. The stress may be a service stress (due to dead, live or wind load) but often it is due to residual effects of the manufacturing process – casting, rolling, forging and arc welding for steels and manganese bronzes, casting for most copper alloys, and cold work for aluminium alloys, brasses and stainless steel. Photographs of SCC cracking in stainless steel may be found in Reardon (2011).

In order for SCC to occur, a specific corrosive environment is a necessary condition but the environment does not have to be really very chemically corrosive. Usually, products of chemical corrosion are not found on SCC fracture surfaces.

As most movable bridges are located in a moist environment, many in seawater atmosphere, some suggested measures to avoid SCC are

Table 7.9 Some common stress corrosion cracking systems. (Data from Reardon, 2011)

Metal	Environment
Aluminium alloys	Air, water vapour Sea water NaCl solutions
Copper alloys	Ammonia vapour and solution Water and water vapour
Carbon and alloy steels	Seawater Moist H_2S gas NaOH solutions Ammonium solutions
Stainless steels	Seawater H_2S gas Acid chloride solutions

- select metals that are inherently more resistant to SCC
- design to avoid sharp notches, crevices and rough finishes
- use low design stresses under service loads
- stress-relieve weldments and castings
- limit metal hardness.

There are two approaches to studying the SCC susceptibility of a metal. The traditional time-to-failure approach which is similar to the S–N approach for fatigue and the Barsom and Rolfe fracture-mechanics approach which has been accepted by ASTM (Barsom and Rolfe, 1999). More information on SCC may be found in Davis (1998).

7.8. Hydrogen embrittlement

The phrase hydrogen embrittlement (HE) commonly refers to the fracture of metal that has been exposed to hydrogen. That definition is of the 'effect' of HE. Strictly, hydrogen embrittlement is a 'process': one form of hydrogen damage to susceptible metals. The process causes a decrease in toughness or ductility of the metal because of the presence of atomic hydrogen. As little as 0.000001% by weight of hydrogen can cause cracking in steel (Dieter, 1986). HE has been subdivided into two types based on the hydrogen source.

- Internal hydrogen embrittlement occurs when the hydrogen enters the molten metal, which becomes supersaturated with hydrogen immediately after solidification.
- Environmental hydrogen embrittlement: in this case, the hydrogen is absorbed by the solid metal. This can occur during pickling or other aspects of galvanising or electroplating, contact with certain chemicals, and during galvanic corrosion. When associated with corrosion, HE is sometimes considered a special case of SCC (Davis, 1998).

7.8.1 Internal hydrogen embrittlement

When high-strength steel containing hydrogen is subjected to tensile stress, it may fracture in a brittle manner, even if the applied stress is less than the yield strength of the steel, and do so without warning.

Unlike the stress–corrosion cracking (SCC) described previously, cracks caused by hydrogen embrittlement usually do not branch, and the crack path can be either transgranular or intergranular (Kim, 1986). The susceptibility of metal to HE generally increases with increasing tensile strength. Carbon steels with tensile strengths greater than about 1034 MPa (150 ksi) are susceptible to HE. High-strength low-alloy steels such as SAE 4130 and SAE 4340 are susceptible to HE cracking in a marine atmosphere when the sum of residual and applied tensile stresses exceed a threshold value. Cracking usually occurs as a delayed fracture. Steels with tensile strengths less than 690 MPa (100 ksi) seem resistant to HE (Kim, 1986).

7.8.2 Environmental hydrogen embrittlement

The most common sources of environmental hydrogen are manufacturing operations such as arc welding, pickling, galvanising and electroplating. The workpieces should be baked immediately after pickling to remove some of the hydrogen before applying electroplate. ASTM has promulgated specifications dealing with preventing hydrogen embrittlement cracking as follows

- A143: Standard practice for safeguarding against embrittlement of hot-dipped galvanised structural steel products and procedure for detecting embrittlement.
- F519: Standard test method for mechanical hydrogen embrittlement evaluation of plating/coating processes and service environments.
- F1459: Standard test method for determination of the susceptibility of metallic materials to hydrogen gas embrittlement (HGE).
- F1624: Standard test method for measurement of hydrogen embrittlement threshold in steel by the incremental step loading procedure.
- F1940: Standard test method for process control verification to prevent hydrogen embrittlement in plated or coated fasteners.

Hydrogen embrittlement of fasteners has been an industry problem to such an extent that ASTM has promulgated a standard for manufacturing process control verification to prevent hydrogen embrittlement in plated or coated fasteners (ASTM F1940). The story is told of manufactured parts that were assembled in a production setting with high-strength plated cap screws. The screws were each torqued to a specified preload by means of calibrated wrenches. After assembly, the parts were placed into wooden boxes and nailed shut for shipment the next afternoon. The following day, loud noises were heard emanating from the boxes. On opening the boxes, it was discovered that many of the cap screw heads had been separated from the bodies where the head joins the body, a zone where the stress concentration factor is high-valued.

From IFI (Industrial Fastener Institute) (2001) and Greenslade (2005, pp. 10, 92) hydrogen embrittlement bolt failures have the following characteristics

- Fasteners have core hardness exceeding 32 HRC.
- Parts have been in contact with an acid during manufacture.
- Failure occurs sometime after preloading the fastener assembly, usually between one and 24 hours.
- Parts have a non-porous finish. The most common finish associated with HE failures is electroplated zinc.
- Parts are subjected to stress at failure.
- The failure location is either where the fastener's head joins the bolt body or in the threads within two pitches of the mating threads.

A photograph of the HE failure of a cap screw appears in Greenslade (2005, pp. 10, 92) which shows intergranular fracture. Bibliographic reviews and recent research on hydrogen embrittlement are to be found in Woodtli and Kieselbach (2000), Phull (2003b), Barnoush (2011) and Krauss (2015, pp. 466–476).

7.9. Tribology

Tribology is the study of friction, wear and lubrication of interacting surfaces (in contact) that are in relative motion. In movable bridge design friction; of shafts in bearings, between gear teeth, and the sliding of span guides and wedge supports, has to be accounted for in sizing the prime mover and auxiliary machinery, as discussed in Chapter 12. Those frictional resistances to motion need to be minimised by adequate lubrication. Some effects of improper lubrication are shown in photographs reproduced in Chapter 13. The subject of tribology, as applied to movable bridges, is broad and space considerations preclude treating the subject in depth herein. The most that can be done is to cite sources in the literature and present the briefest of overviews of the major aspects of the phenomena. Friction, in general, is discussed in Rabinowicz (1995) and Ludema (1996). Rotational bearing design and lubrication are treated in Wilcock and Booser (1957), Shigley and Miscke (2001), Spotts *et al.* (2004) and Oberg *et al.* (2012).

The history leading to the current adhesion theory of friction is reported by Ludema (1996). According to this theory, the interface between contacting solids comprises two regions. One region consists of a number of small highly stressed spots where the surfaces are in actual intimate contact and another region (all of the rest of the 'contacting area') where there is no, or little interaction between the mating surfaces. The action of the asperities is explained thoroughly in Ashby and Jones (1996).

Sliding between two components of a machine or structure sometimes occurs at a nearly constant velocity and at other times the velocity varies, often jerkily. The phenomenon is explained by Rabinowicz (1995). It is commonly called stick–slip. The two main conditions under which irregular stick–slip are observed are the sliding of clean metals and metals covered by a solid film lubricant that has been partially worn away. Stick–slip can often be cured by proper lubrication. On a plot of starting friction as ordinate against sliding velocity as abscissa, a curve for a well lubricated surface will show decreased friction with decreased speed. A poorly lubricated, or unlubricated curve will show increased friction with decreased speed. Avallone *et al.* (2007) shows a friction against a logarithmic velocity curve for lightly loaded lubricated cast iron on cast iron that illustrates stick–slip at velocities less than 0.5 in. (13 mm) per minute.

As a practical matter, designers need estimates of the likely friction coefficients for various combinations of metals and lubrication. Such information may be found in Oberg *et al.* (2012) and Avallone *et al.* (2007).

REFERENCES

AASHTO (American Association of State Highway and Transportation Officials) (2007) Standard Specifications for Movable Highway Bridges, 2nd edn. AASHTO, Washington, DC, USA.

Anderson TL (1991) *Fracture Mechanics: Fundamentals and Applications.* CRC Press, Boca Raton, FL, USA.

AREMA (American Railway Engineering and Maintenance-of-Way Association) (2014) Movable bridges – steel structures. Chapter 15 in *AREMA Manual for Railway Engineering.* AREMA, Lanham, MD, USA.

Ashby MF and Jones DRH (1996) *Engineering Materials One*, 2nd edn. Butterworth–Heinemann, Oxford, UK.

ASTM (American Society for Testing and Materials) (2012) E140-12be1 Standard hardness conversion tables for metals relationship among Brinell hardness, Vickers hardness, Rockwell hardness, superficial hardness, Knoop hardness, scleroscope hardness, and Leeb hardness. ASTM International, West Conshohocken, PA, USA, http://dx.doi.org/10.1520/E0140.

ASTM (2013a) E8-13a Standard test methods for tension testing of metallic materials. ASTM International, West Conshohocken, PA, USA, http://dx.doi.org/ 10.1520/E0008.

ASTM (2013b) E837 Determining residual stresses by the hole-drilling strain-gage method. ASTM International, West Conshohocken, PA, USA, http://dx.doi.org/10.1520/E0837.

ASTM (2014a) E10-14 Standard test method for Brinell hardness of metallic materials. ASTM International, West Conshohocken, PA, USA.

ASTM (2014b) E18-14a Standard test methods for Rockwell hardness of metallic materials. ASTM International, West Conshohocken, PA, USA, http://dx.doi.org/10.1520/E0018-14A.

ASTM (2014c) A956-12 Standard test method for Leeb hardness testing of steel products active standard. ASTM International, West Conshohocken, PA, USA.

Avallone EA, Baumeister III T and Sadegh AM (2007) *Mark's Standard Handbook for Mechanical Engineers*, 11th edn. McGraw-Hill, New York, NY, USA.

Bannantine JA, Comer JJ and Handrock JC (1990) *Fundamentals of Metal Fatigue Analysis*. Prentice-Hall, Upper Saddle River, NJ, USA.

Barnoush A (2011) *Hydrogen Embrittlement*. PhD Thesis, Saarland University, Saarbrücken, Germany.

Barsom JM and Rolfe ST (1999) *Fracture and Fatigue Control in Structures*, 3rd edn. ASTM, West Conshohocken, PA, USA.

Battelle Memorial Institute (1941) *Prevention of the Failure of Metals Under Repeated Stress*. Wiley, New York, NY, USA.

Beedle LS and Tall L (1962) Basic column strength. Column Research Council Symposium on Metal Compression Members, *ASCE Transactions*, **127(II)**: Paper No. 3366.

Birnstiel C (1968) Experiments on H-columns under biaxial bending. *Journal of the Structural Division, Proceedings of ASCE*, **94(ST10)**: October.

Birnstiel C, Leu K-C, Tesoro JA and Tomasetti RL (1967) *Experiments on H-Columns Under Biaxial Bending*, AISI Project 114. New York University School of Engineering and Science, Research Division, New York, NY, USA.

Boyer HE (1987) *Atlas of Stress–Strain Curves*. ASM International, Materials Park, OH, USA.

Burr WH (1915) *Elasticity and Resistance of the Materials of Engineering*. Wiley, New York, NY, USA.

Cobb HM (2010) *The History of Stainless Steel*. ASM International, Materials Park, OH, USA.

Davis JR (1998) *Metals Handbook, Desk Edition*, 2nd edn. ASM International, Materials Park, OH, USA.

Dieter GA (1986) *Mechanical Metallurgy*, 3rd edn. McGraw-Hill, New York, NY, USA, http://dx.doi.org/10.5962/bhl.title.35895.

Eytelwein JA (1808) *Handbuch der Statik Fester Körper* (in German), **2**. Realschulbuchhandlung, Berlin, Germany.

FHWA (Federal Highway Association) (2012) *Steel Bridge Design Handbook*. Publication No. FHWA-IF-12-052. Federal Highway Administration, US Department of Transportation. Washington, DC, USA.

Fontana MG and Greene ND (1978) *Corrosion Engineering*. McGraw-Hill, New York, NY, USA, http://dx.doi.org/10.1149/1.2129187.

Galambos TV (1998) *Stability Design Criteria for Metal Structures*, 5th edn. Wiley, New York, NY, USA.

Greenslade J (2005) Here is what a hydrogen embrittlement failure really looks like. *Distributor's Link Magazine*, pp. 10, 92.

Gurney TR (1968) *Fatigue of Welded Structures*. Cambridge University Press, London.
IFI (Industrial Fastener Institute) (2001) Identifying Hydrogen Embrittlement Failures. *IFI Technical Bulletin*. IFI, Independence, OH, USA.
ISO (International Organization for Standardization) (2009) ISO 6892-1 *Metallic Materials – Tensile Testing – Part 1: Method of Test at Room Temperature*. ISO Copyright Office, Geneva, Switzerland.
Jones RH (1992) *Stress-Corrosion Cracking: Materials Performance and Evaluation*. ASM International, Materials Park, OH, USA.
Kim CD (1986) Hydrogen – damage failures. *Metals Handbook*, vol. 11. ASM International, Materials Park, OH, USA.
Kompatscher M (2005) Dynamic hardness measurements. *Journal of the Metrology Society of India* **20(1)**: 25–36.
Krauss G (2015) *Steels: Processing, Structure, and Performance*, 2nd edn. ASM International, Materials Park, OH, USA.
Kudryavstev Y and Kleiman J (2011) Ultrasonic technique and equipment for residual stresses measurement. *Residual Stress* **8**: 55–65. Springer, Berlin.
Kurrer KE (2008) *The History of the Theory of Structures*. Ernest & Sohn, Berlin, Germany, http://dx.doi.org/10.1002/9783433600160.
Li S, Xu Y and Guan X (2014) Uniform and pitting corrosion modeling for high strength bridge wires. *Journal of Bridge Engineering* **19(7)**: July. ASCE, Reston, VA, USA, http://dx.doi.org/10.1061/(ASCE)BE.1943-5592.0000598.
Low S (2005) State of the art of conventional hardness measuring methods: Rockwell, Brinell, and Vickers. *Journal of the Metrology Society of India* **20(1)**: 15–24.
Ludema KC (1996) *Friction, Wear, Lubrication: A Textbook in Tribology*. CRC Press, Boca Raton, FL, USA, http://dx.doi.org/10.1201/9781439821893.
McEvily Jr AJ (1990) *Atlas of Stress-Corrosion and Corrosion Fatigue Curves*. ASM International, Materials Park, OH, USA.
Micro-Measurements (2010) Measurement of residual stresses by the hole-drilling strain gage method. *Technical Note TN-503*. Vishay Precision Group, Raleigh, NC, USA.
Oberg E, Jones FD, Horton HL and Ryffel HH (2012) *Machinery's Handbook*, 29th edn. Industrial Press, New York, NY, USA.
Osgood WR (1954) *Residual Stresses in Metals and Metal Construction*. Reinhold, New York, NY, USA.
Phull B (2003a) Evaluating corrosion fatigue. In *Metals Handbook*, vol. 13A. ASM International, Materials Park, OH, USA.
Phull B (2003b) Evaluating hydrogen embrittlement. In *Metals Handbook*, vol. 13A. ASM International, Materials Park, OH, USA.
Rabinowicz E (1995) *Friction and Wear of Materials*, 2nd edn. Wiley, New York, NY, USA, http://dx.doi.org/10.1115/1.3625110.
Reardon AC (2011) *Metallurgy for the Non-Metallurgist*. ASM International, Materials Park, OH, USA.
Shigley JE and Miscke CR (2001) *Mechanical Engineering Design*, 6th edn. McGraw-Hill, New York, NY, USA.
Spotts MF, Shoup TE and Hornberger LE (2004) *Design of Machine Elements*, 8th edn. Pearson/Prentice-Hall, Upper Saddle River, NJ, USA.
Tall L and Rao NRN (1961) Residual stresses in welded plates. *The Welding Journal* **40**: October.
Totten G, Howes M and Inove T (2002) *Handbook of Residual Stress and Deformation of Steel*. ASM International, Materials Park, OH, USA.

TWI (2015) FAQ: *Is Vibratory Stress Relief as Effective as Thermal Stress Relief?* http://www.twi-global.com/technical-knowledge/Faqs/structural-integrity (accessed 25/2/2015).

Wilcock DF and Booser ER (1957) *Bearing Design and Application*. McGraw-Hill, New York, NY, USA.

Wood HA and Engle Jr RM (1979) *Damage Tolerant Design Handbook: Guidelines for Analysis and Design of Damage Tolerant Aircraft Structures*. Air Force Flight Dynamics Laboratory (AFFDL/FBE) Wright-Patterson AFB, OH, USA.

Woodtli J and Kieselbach R (2000) Damage due to hydrogen embrittlement and stress corrosion cracking. *Engineering Failure Analysis* **7**. Pergamon, Elsevier Science, London, http://dx.doi.org/10.1016/S1350-6307(99)00033-3.

Movable Bridge Design
ISBN 978-0-7277-5804-0

http://dx.doi.org/10.1680/mbd.58040.181

Chapter 8
Span drive arrangements

Robert J Tosolt

8.0. Introduction

As is evident from Chapters 1 to 4 the configuration of movable bridges is quite diverse, but within that community the diversity of electro-mechanical span drive arrangements is even greater. Machinery design objectives encompass functionality, simplicity, reliability, efficiency, novelty and uniqueness. However, there are common mechanical systems which have proven their effectiveness through an extensive usage history. This chapter seeks to describe some systems which are widespread in the industry, especially in the Americas.

One point that warrants consideration when reviewing the movable bridge drive systems presented in this chapter is that in each case the drive system has been designed to operate a 'balanced structure' – that is, the action of the counterweight is almost equal to the action of the movable span self-weight, but of opposite sign. While machinery drives have been designed to operate unbalanced structures (for example, 8th Street Bridge, Sheboygan, WI), the power requirements and size of the machinery is far larger so that one or more of the options presented subsequently might be impractical or unachievable. Most of the components of the span drives illustrated in this chapter are described in Chapter 9.

8.1. The simple trunnion bascule

The trunnion bascule is regarded as the simplest of movable bridges. The operating principle of the simple trunnion bascule is that the bascule leaf rotates about a transverse axis, the trunnion shaft, as shown in Figure 1.1(a). The trunnion shaft carries the entire self-weight of the leaf and other loads, and it is supported in trunnion bearings mounted on piers. The purpose of the drive machinery is to impart force to the leaf and cause it to rotate open or closed about the trunnion bearings.

The simplest form of trunnion bascule has the span drive machinery located below deck level on the bascule piers. All machinery is anchored to the bascule piers with the exception of the racks, which are the final driven component. They are mounted to the primary load carrying members of the bascule leaf typically the underside of the bascule girders, as shown in Figure 2.1, or are framed into the bascule trusses, dependent upon the bridge's structural configuration.

Figure 8.1 illustrates a typical span drive machinery arrangement for a smaller bascule leaf 61 ft (18.6 m) in length from trunnion to toe, that has bascule girders as the primary structural members. The normal power for each span drive is obtained from one 20 hp, 850 rpm, electric motor with a double extended shaft. The motor is mounted to one side of an enclosed differential reducer with double extended input, intermediate and output shafts. The motor is connected to one end of the reducer input shaft by a double-engagement grid-type coupling.

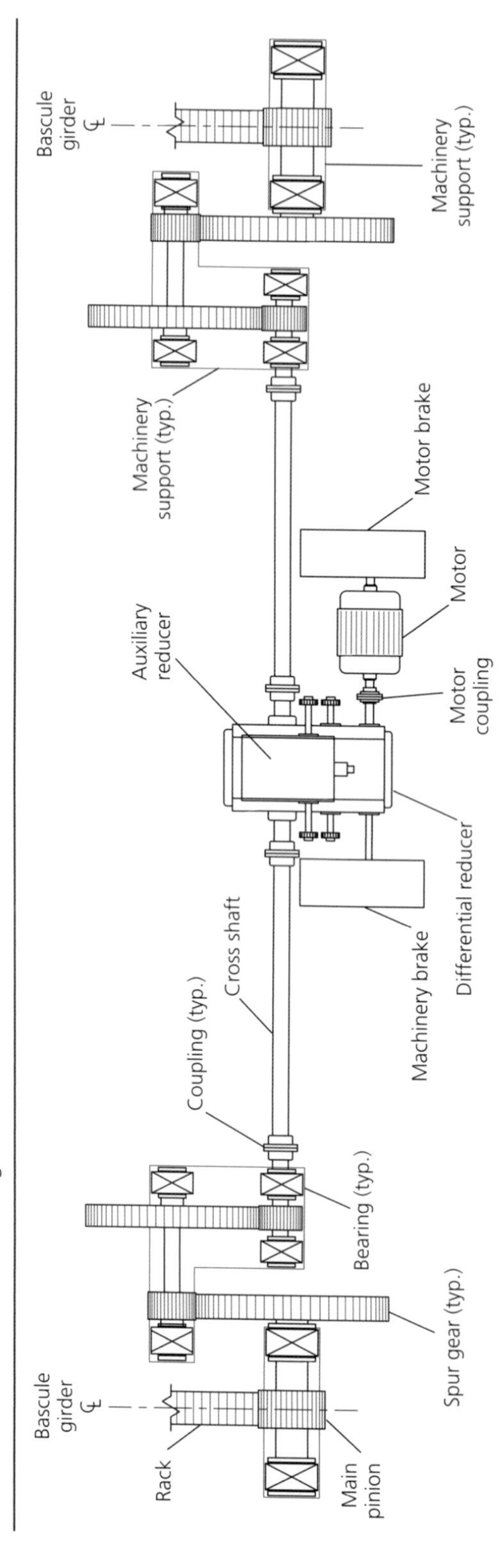

Figure 8.1 Trunnion bascule bridge machinery layout. All drive machinery with the exception of the racks is mounted on the pier. The racks are mounted onto the underside of the bascule girder

The span drive has two brakes: a motor brake and a machinery brake. Both brakes are spring-set thrustor-released shoe-type brakes (see Chapter 15). The motor brake is mounted on the shaft extension at the non-driving end of the motor. The machinery brake is mounted on the other input shaft extension, opposite the motor. The intermediate shaft extensions from the reducer are provided for auxiliary operation, which is addressed below. Machinery driven by the double extended output shaft of the differential reducer is symmetrical about the differential centreline.

Transverse 'floating' shafts are utilised to transmit power output from the reducer to the first of three open gearsets. The inboard end of each floating shaft is coupled to a reducer output shaft extension and the outboard end of each floating shaft is coupled to a short shaft (jack shaft) which is simply supported in two pillow blocks. The floating shaft ends are coupled by single engagement gear-type couplings. These couplings can accommodate small amounts of angular and offset misalignment which may exist between the reducer output shaft and the mating shafts due to errors made during installation or by structural movement during bridge operation. A pinion, the first gear in the open gear train, is straddle mounted on the jack shaft.

All open gears, with the exception of the rack and rack pinion which comprise the final gearset in the gear train, are interference-fitted (shrink or force-fitted) and keyed to their shafts. Shafts are supported in sleeve-type split bronze bearings mounted in pillow block housings. The rack pinion is usually forged integrally with the rack pinion shaft. The rack is mounted onto a frame at the underside of the bascule girder and aligned radially about the trunnion axis.

During normal operation, the electric motors operate the bridge. Rotation of the electric motors causes the bascule leaves to rotate about the trunnions, providing an open channel for marine traffic.

In the event of power disruption, provision was made for auxiliary operation of the leaf. A right-angle speed reducer mounted atop the differential reducer has a square input shaft extension and a double extended output shaft. Sprockets are mounted on both output shaft extensions. The sprockets for the right-angle reducer are mounted vertically in-line with sprockets that are installed on the intermediate shaft extensions of the differential reducer. The sprockets on opposite sides of the reducers provide for two speeds of manual operation. During auxiliary operation, a roller chain is used to connect the pair of sprockets for the desired speed of manual operation. Only one pair of sprockets is used at a given time. The roller chain is not present during normal operation. Because the auxiliary drive is connected into the normal span drive system, all components of the normal span drive are common to the auxiliary drive. Prior to auxiliary operation the motor and machinery thrustor brakes must be hand released. A hand crank or power drill mounted onto the right-angle reducer input shaft extension may then be used to operate the bridge.

A cautionary note: many trunnion bascule bridges are in service with the span drive arrangement shown in Figure 8.1. This arrangement was popular primarily for its economy of engineering and construction. However, the arrangement does not satisfy the current AASHTO specification or the AREMA recommended practice because of the location of the machinery brake. If a failure occurred in the differential reducer (the most sensitive component of the mechanical drive), control of leaf movement would be lost. A single-point failure analysis (Sööt, 1990) will readily confirm this. For the scheme in Figure 8.1, the machinery brake should be removed from the input shaft and a machinery brake installed at each jack shaft. (It is necessary to have a minimum of the two machinery brakes because the primary reducer has a differential.)

8.2. The rolling bascule

The rolling bascule bridge shown in Figure 1.1(b) is an old concept which was studied by William Scherzer (1893), who submitted a patent for the bridge type in 1893 before his untimely demise, and was posthumously granted US Patent 511,713 in 1894. The bridge type was subsequently marketed and advanced by Scherzer's brother Albert Scherzer to great effect so that it was one of the most widely adopted bridge types at the beginning of the 20th century (see Chapter 1).

The operating principle of the rolling bascule is that as the movable leaf rolls open the leaf both rotates and translates in a vertical plane. This characteristic results in an increased channel opening for a given-length leaf and opening angle when compared to a trunnion bascule bridge (see Chapter 2).

The movable leaf is equipped with curved tread plates which mate with and roll upon flat track plates that are mounted to the pier. The span drive machinery is often located at, or connected to, the bascule leaf at the centre of the curved tread radius, which is denoted the centre of roll. When such is the case, the drive machinery imparts a horizontal force at the centre of roll in order to roll the leaf open or closed. The horizontal force is produced by rotating pinions engaged to straight racks mounted on dedicated frames that are anchored to the bascule pier.

Figure 8.2 illustrates a typical span drive machinery configuration for a Scherzer bridge. This arrangement is for a large bascule leaf, 159 ft (48.5 m) in length centre of roll to toe that utilises a through trussed structure with an overhead counterweight as shown in Figure 2.18(c). All machinery is mounted on the movable leaf above deck level, except the racks.

Normal power for the span drive is obtained from two 75 hp, 650 rpm, electric motors. Each motor is mounted to one side of a central differential reducer with double extended input and output shafts being connected by a double engagement grid-type coupling to one end of the differential reducer input shaft. The differential reducer is a retrofit for the original open gear differential.

The span drive has four brakes; two motor brakes and two machinery brakes. All brakes are thrustor operated (spring set, thrustor released) shoe-type brakes. One motor brake is mounted on each reducer input shaft extension. One machinery brake is mounted on each output shaft extension.

The differential reducer is totally enclosed and oil splash-lubricated. The machinery driven by the double extended output shaft of the differential reducer is symmetrical about the differential centreline.

Transverse floating shafts transmit power from the reducer to the first of three open gearsets. The inboard end of each floating shaft is coupled to a reducer output shaft extension and the outboard end is coupled to an intermediate shaft that is simply supported in two bearings. The floating shafts are coupled at either end with single engagement gear -type couplings. These couplings accommodate small amounts of angular and offset misalignment, which may exist between the reducer output shaft and the mating shafts, or develop during operation. A pinion, the first gear in the open gear train, is cantilever-mounted to the end of the intermediate shaft, outboard of the bearings.

The shafting for the three open gearsets are supported in bronze bushed plain bearings in pillow block housings. All bearing housings, with the exception of the bearing adjacent to the main pinion, are foot mounted; the bearing adjacent to the main bearing is flange mounted to the end vertical of the truss. Because the pinion shaft could not be installed if the pinion were shop-mounted, the main pinions are field-installed on the pinion shafts with gib head keys.

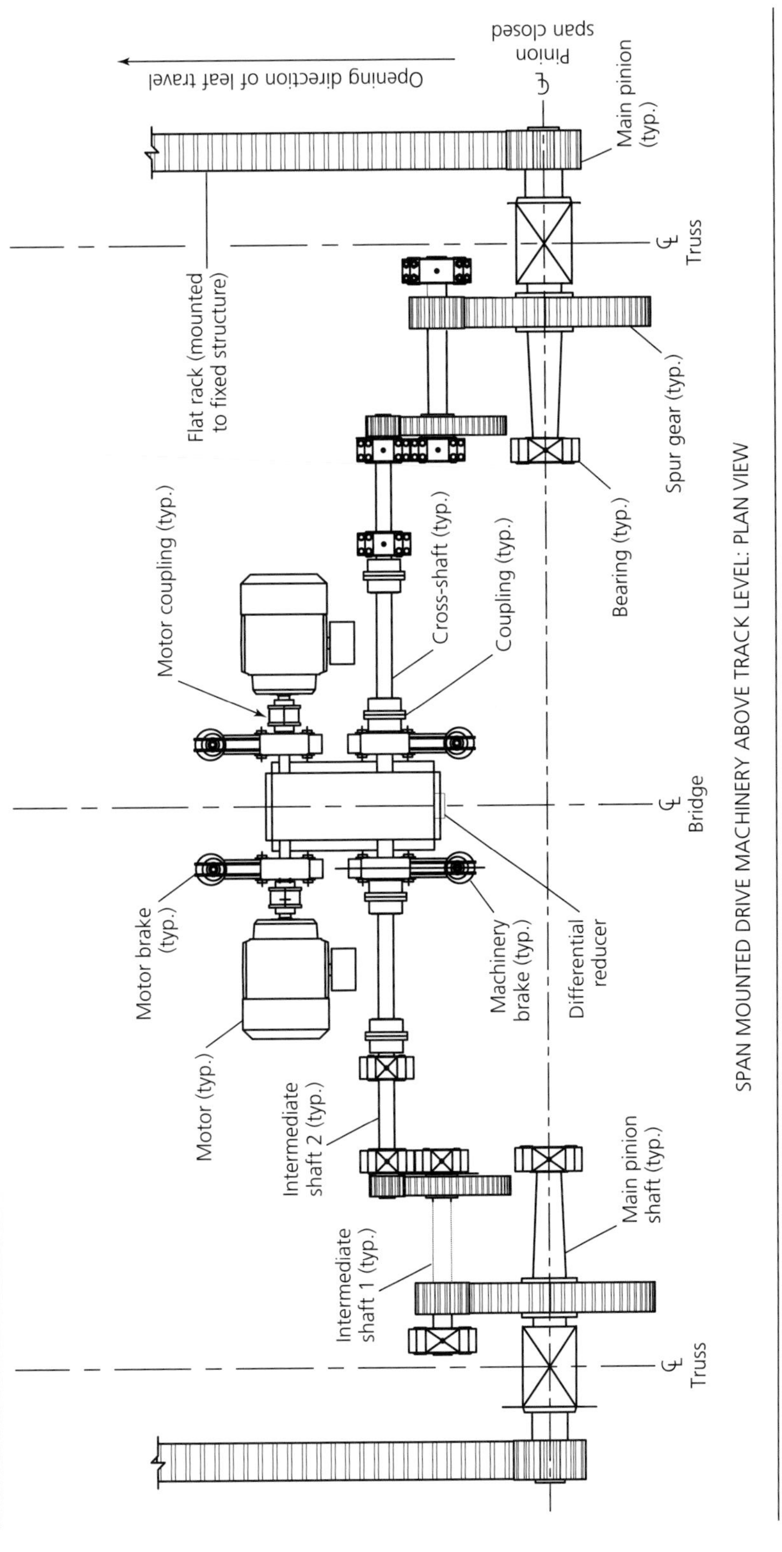

Figure 8.2 Rolling lift bascule bridge machinery layout. All drive machinery with the exception of the racks is mounted on the movable leaf. The racks are mounted onto a structural frame anchored to the pier

With the exception of the racks, all of the machinery rotates with the movable structure. The racks are supported on structural steel frames which are anchored to the piers. Rotation of the motors causes the rack pinions to rotate and travel along the racks with the result that the movable leaf rolls on the curved treads. The span opens as a result of rotary and translational motion.

8.3. The heel trunnion bascule

The heel trunnion bascule bridge was brought to prominence and widespread usage at the beginning of the 20th century by Joseph B Strauss, a contemporary and competitor of the Scherzer Company. Strauss provided an alternative to the rolling lift bridge for low-grade (elevation) crossings where it was impractical or undesirable to build water-tight pits in the bascule piers that would be required for a conventional simple trunnion bridge. Strauss was granted a series of patents at the beginning of the 20th century for his work in advancing the development of the heel trunnion bascule bridge. In this bridge type the counterweight is located remotely from the leaf. The bascule bridge was maintained essentially balanced in all position as described in Chapter 2, and illustrated in Figure 2.13. This design provision resulted in the parallelogram geometry that is synonymous with the Strauss heel trunnion. With the leaf effectively balanced, Strauss achieved operation by pulling the leaf open or closed with an operating strut. The operating machinery could be mounted either on the movable leaf or on the fixed tower structure, as shown in Figure 2.13.

Figure 8.3 illustrates a typical span drive machinery arrangement where the machinery is mounted on the movable leaf. This configuration is for a bascule leaf with a length of 159 ft (48.5 m) heel trunnion to toe. The span drive machinery dates to original construction *c.* 1910 and utilises a standard power train for that time, comprising open spur gearsets supported in straight sleeve bearings mounted in pillow block or flange mounted housings.

The machinery is powered by two 65 hp, 465 rpm, electric motors. Three brakes, one at each motor and one additional holding (machinery) brake were provided. All span drive machinery with the exception of the racks and operating struts is mounted to the heel end of the movable structure above track level. The racks are mounted to the underside of operating struts, which are supported through a bogey wheel carrier on the rack pinion shaft at one end and are pin-connected to the counterweight support truss at the opposite end. Span operation is accomplished through rotation of the rack pinions, which generate longitudinal travel along the straight racks and result in the span-mounted machinery pulling the leaf open or closed along the pinned operating struts.

The original design provided for emergency operation in the event that both motors fail through a manual chainwheel system. The system required the sprockets on the auxiliary pinion shafts to be manually engaged through cut-out couplings. The auxiliary pinion shaft sprockets were then chain-connected to sprockets on the chainwheel shaft. Then, by the installation of chains on the four chain-wheels, personnel at track level could provide the power necessary to operate the leaf.

8.4. The Hopkins trunnion bascule

The Hopkins trunnion bascule is a variation on the simple trunnion bascule that was pioneered by Leonard O Hopkins in 1934 and granted US Patent 2066110A (Hopkins, 1936). The Hopkins bascule sought to improve the constructability of bascule bridges. Hopkins observed that the proper alignment of the machinery to the movable leaf was crucial to proper operation and that having multiple components that needed to be individually aligned in the field increased the time and complexity for field installation. The Hopkins construction sought to address this problem by mounting all of the machinery onto a common frame which could be shop aligned and then field installed utilising a connection to

Figure 8.3 Strauss heel trunnion bridge machinery layout. All drive machinery with the exception of the racks is mounted on the movable leaf. The racks are mounted onto operating struts which each have one end pinned to the counterweight tower

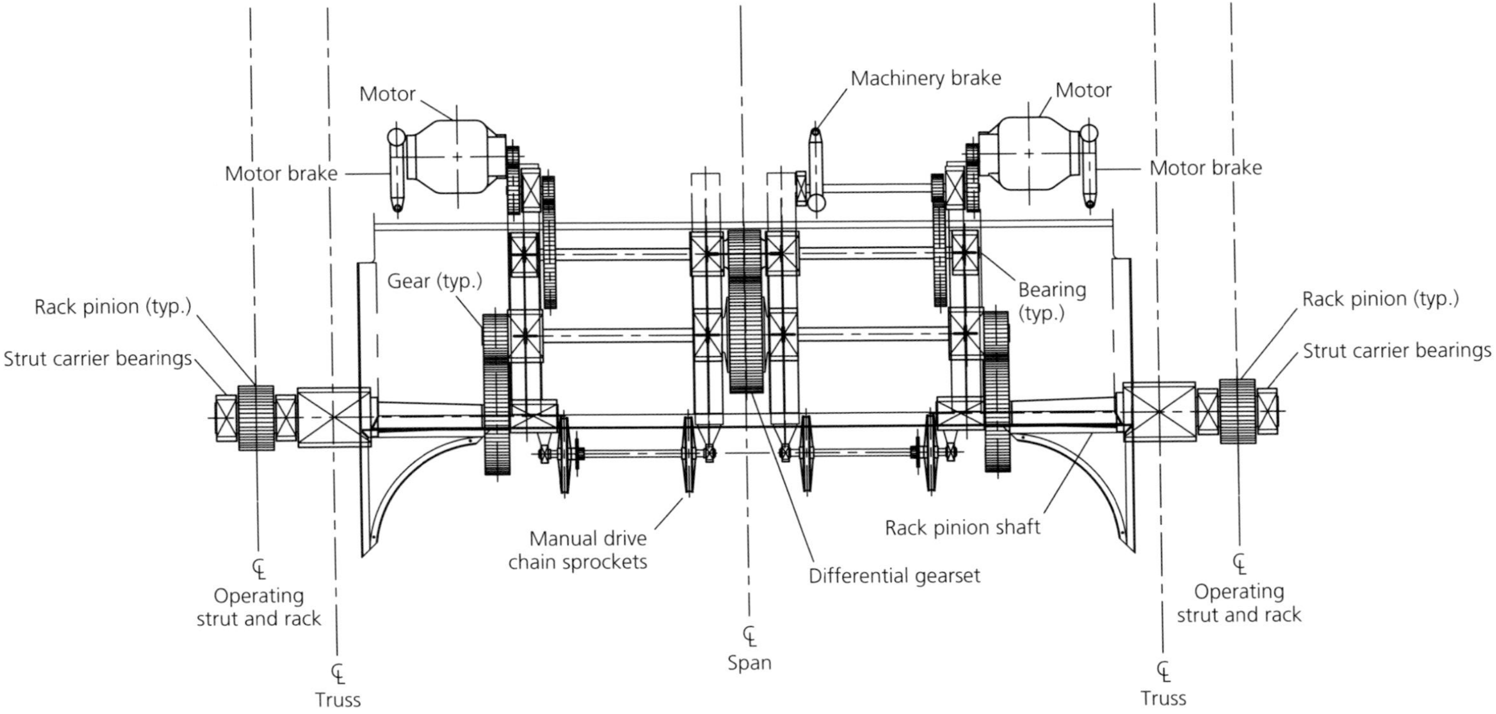

MACHINERY ON MOVABLE LEAF ABOVE TRACK LEVEL: PLAN VIEW

the leaf wherein small inaccuracies in the positioning of the machinery frame might be compensated for so that proper engagement of the trunnions and the machinery might be maintained.

Figures 8.4 and 8.5 illustrate the Hopkins frame configuration as applied to a bascule leaf that is approximately 80 ft (24.2 m) long from centre of trunnion to toe of leaf. Figure 8.4 depicts the arrangement of the drive machinery mounted on the common frame, which is referred to as the Hopkins frame, viewed from the channel.

Normal power for each span drive is obtained from two 30 hp, 850 rpm, electric motors with double extended shafts. The motors are mounted to either side of an enclosed differential reducer with double extended input and output shafts and a single extended intermediate shaft. The motors are connected by double engagement grid-type couplings to the input shafts of the differential reducer.

Each leaf is provided with three brakes: two motor brakes and one machinery brake. All brakes are thrustor-operated shoe-type brakes. The motor brakes are mounted on the non-driving ends of the motor shafts. The machinery brake is mounted on the input shaft extension of the differential reducer between one motor and the reducer. An intermediate shaft extension from the differential reducer is provided for auxiliary operation, which is addressed below.

The machinery driven by the double extended output shaft of the differential reducer is symmetrical about the differential reducer centreline. The outboard end of each reducer output shaft extension is supported in a sleeve-type bronze bearing mounted in a split housing. A pinion is mounted on each of the output shaft extensions between the reducer and the respective bearing. The pinion mates with a gear to form the first of two open gearsets. The rack and rack pinion form the second open gearset.

All open gears with the exception of the rack, which is the final gear in the gear train, are keyed to their shafts. All shafts are supported in sleeve-type split bronze bearings mounted in split housings. Each rack is supported in a frame that is mounted to the underside of a longitudinal girder (i.e. the rack girder) that is framed into the leaf. The racks and rack frames are aligned radially about the trunnion so that the rack pitch circle is centred on the trunnion. Figure 8.5 depicts the relation between the Hopkins frame, the rack support frame and the movable leaf and illustrates how alignment is maintained.

The frames are supported by two vertical legs that are mounted on pins supported in a clevis joint connection anchored to the bascule pier. The clevis pin joint not only facilitates installation of the frame-mounted machinery, but also allows slight rotation of the frame to accommodate small misalignment errors during installation, or leaf deflections during bridge operation. To maintain the proper centre distance for the rack and rack pinion, a link arm connects the rack pinion shaft, which is supported off the Hopkins frame, to a rear stabilising pipe, which is supported in the rack frame and located at the centre of the rack pitch circle (on the trunnion axis). The link arm is equipped with bearings at each connection to allow for the rotational movement of the arm relative to the pipe and the pinion shaft as the bridge opens. During normal operation rotation of the electric motors causes the bascule leaves to rotate about the trunnions, providing an open channel for marine traffic.

Provision was made for auxiliary operation of each leaf in the event of power disruption or failure. The intermediate shaft extension from the differential reducer is provided for auxiliary operation. The intermediate shaft extension has one hub of a jaw-type coupling mounted on its outboard end. The mating hub of the jaw-type coupling is mounted on the output shaft extension of an auxiliary enclosed worm gear reducer. The auxiliary reducer has a square input shaft extension. Because the jaw-type coupling

Figure 8.4 Hopkins frame machinery layout. Front elevation, viewed from channel. All drive machinery with the exception of the racks is mounted onto a frame anchored to the pier. The racks are supported in a frame that is connected to the movable leaf

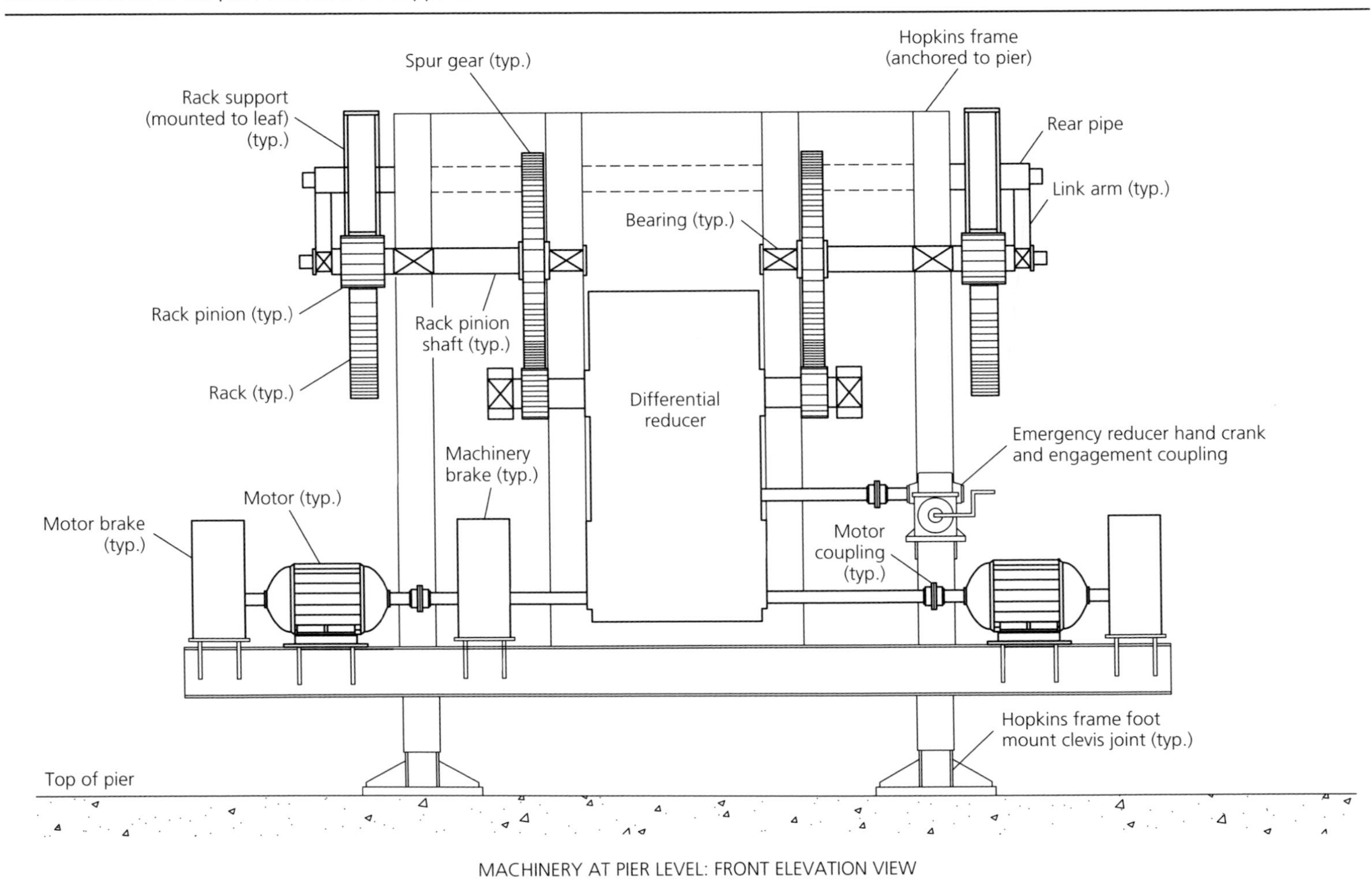

Figure 8.5 Hopkins frame machinery layout. Side elevation depicting relationship between Hopkins frame, rack support frame, link arm, rear stabilising pipe and rack girder

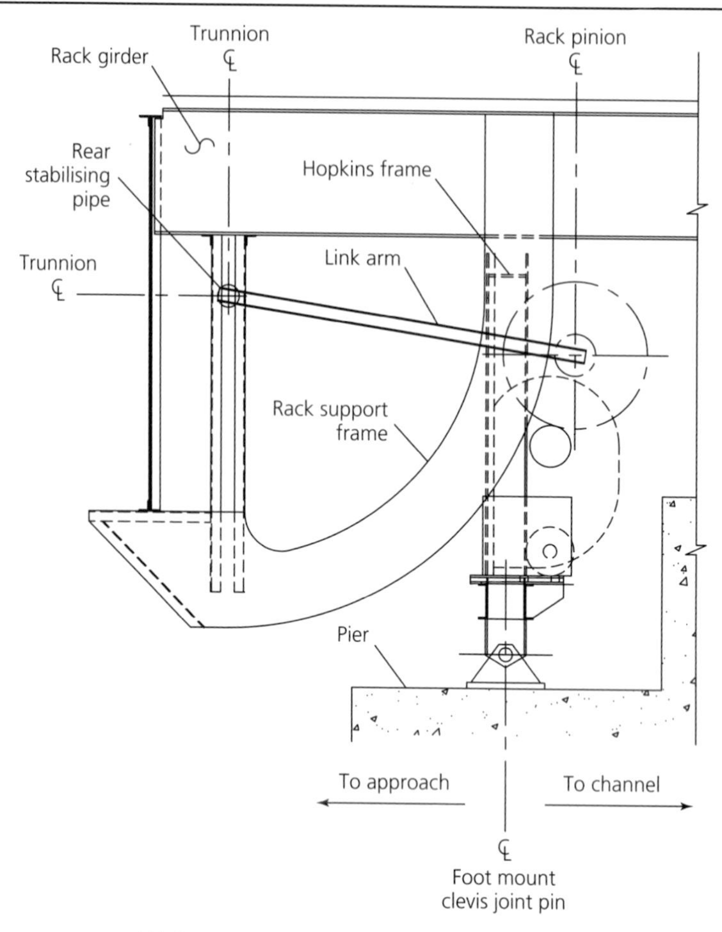

forms a rigid connection with the intermediate shaft to the differential reducer when engaged, all components of the normal span drive are common to the auxiliary drive. The jaw-type coupling is disengaged during normal operation. Prior to auxiliary operation, the jaw-type coupling is engaged by way of a screw mechanism and the motor and machinery brake are hand released. A hand crank or power drill mounted to the square input shaft extension of the auxiliary reducer will operate the bridge.

8.5. The swing span

Swing bridges were popular during the 19th century and beginning of the 20th century in the UK and in the Americas. The operating principle of the swing span differs from the bascule bridge in that the bridge rotates in a horizontal plane as opposed to a vertical plane to affect a bridge opening. The most common type of swing bridge in America has a symmetric draw and a pivot pier that is usually centrally located within the navigation channel. The pivot pier is the foundation on which the swing span is supported. The terminal machinery component, the ring gear, is also mounted on the pivot pier. It is the only piece of drive machinery that is not mounted on the movable span of most American swing bridges. As illustrated in Chapter 2, pier-mounted span drive machinery with the rack mounted on the swing span is often used in the UK and in mainland Europe.

Figure 8.6 illustrates a typical drive machinery configuration for a symmetrical swing bridge with the machinery located above the roadway. This arrangement is for a through truss swing span that is 297 ft (90.6 m) long overall and dates to 1900. The span drive operating machinery is comprised of motors, a brake, open gearing, cross-shafting bearings and couplings. The prime mover and high-speed gear reductions are located in the operator's control room at the top of the truss above track level. The machinery is powered by two 25 hp, 580 rpm, electric motors. The motors are located at opposite ends, and on opposite sides, of a gear frame which is at the centre of the operator's control room. A pinion mounted on each motor shaft meshes with a mating gear to form the first of two gear reductions in the frame which terminate at a central secondary gear that is common to both drive paths. The central gear incorporates differential gearing to equalise the torque going to the north and south transverse output shafts that are driven by the centre gear. The central gear also backdrives a third pinion at its top which is on a common shaft with the sole brake in the system. The brake is manually actuated by the bridge operator through a linkage and is intended to serve as a holding brake. All shafts in the frame are simply supported.

The output shafts (cross-shafts) from the central gear frame run below the control room floor and transmit torque to the north and south sides of the control room. These transverse shafts terminate at bevel gearsets which turn vertical shafts mounted to the exterior face of each truss. Each vertical shaft extends the full depth of each truss and turns an open gearset mounted onto the exterior of the drum girder (see Chapter 2). The driven shaft of each gearset is the final drive pinion shaft at that side of the span. The final drive pinions are located 180° apart and mate with the full circumference ring gear that is anchored to the pivot pier. Rotation of the motor causes rotation of the swing span as the pinions travel about the rack. The full circumference rack (360°) allows the swing span to open in either direction of rotation.

Two hand-operated gear trains (not shown in Figure 8.6) are provided for emergency span drive operation; one drive train is located to the east of the rack and one to the west of the rack. The emergency gear trains are identical; each gear train utilises a capstan inserted at track level to turn a vertical shaft which connects to a gear pair that then connects to a pinion which meshes with the rack. During normal operation, the emergency gear trains are back-driven.

8.6. The span drive vertical lift bridge

The vertical lift bridge is arguably regarded as the most robust of movable bridge types. It functions by raising a horizontal span vertically upward to provide clearance for marine traffic to pass below. The most common type of vertical lift has a tower at each approach to the lift span. The movable span is counterbalanced by two main counterweights, one supported off the tower at either end of the bridge. Wire ropes connect to the span, pass up and over sheaves at the top of each tower, and connect to the counterweights. To effect operation, the drive machinery must either transfer force directly from the span to the towers (span drive) or indirectly transfer the force by turning the counterweight sheaves (tower drive). The tower drive is dependent on friction between the counterweight ropes and the sheave grooves.

Figure 8.7 illustrates a typical span drive machinery arrangement for a span drive lift bridge. This configuration is for a through-truss vertical lift bridge which has a lift span 195 ft long (59.5 m) and

Figure 8.6 Swing span machinery layout. All drive machinery with the exception of the racks is mounted on the swing span. The racks are anchored to the pivot pier

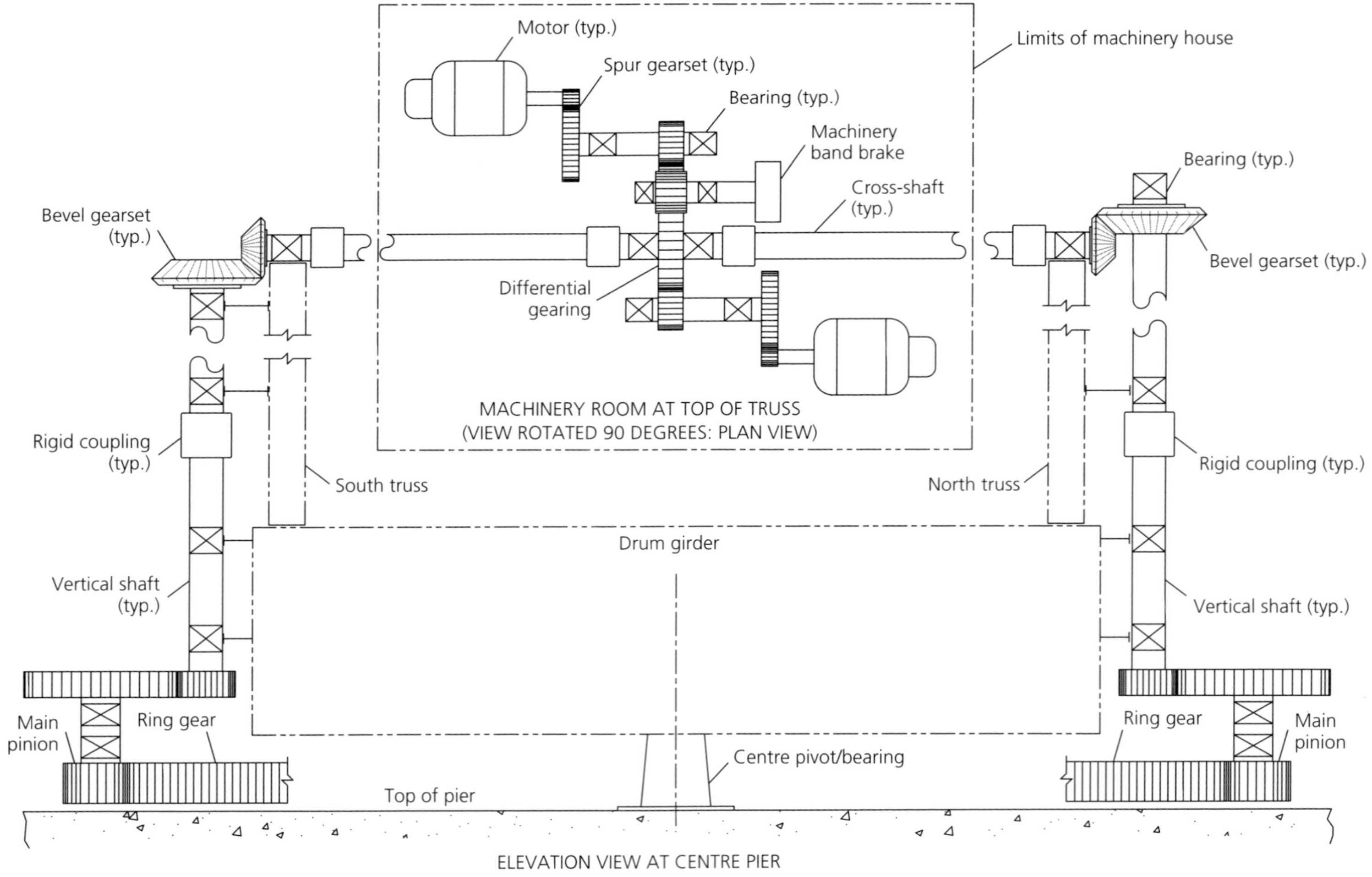

Figure 8.7 Span drive vertical lift bridge machinery layout. All drive machinery is on the lift span

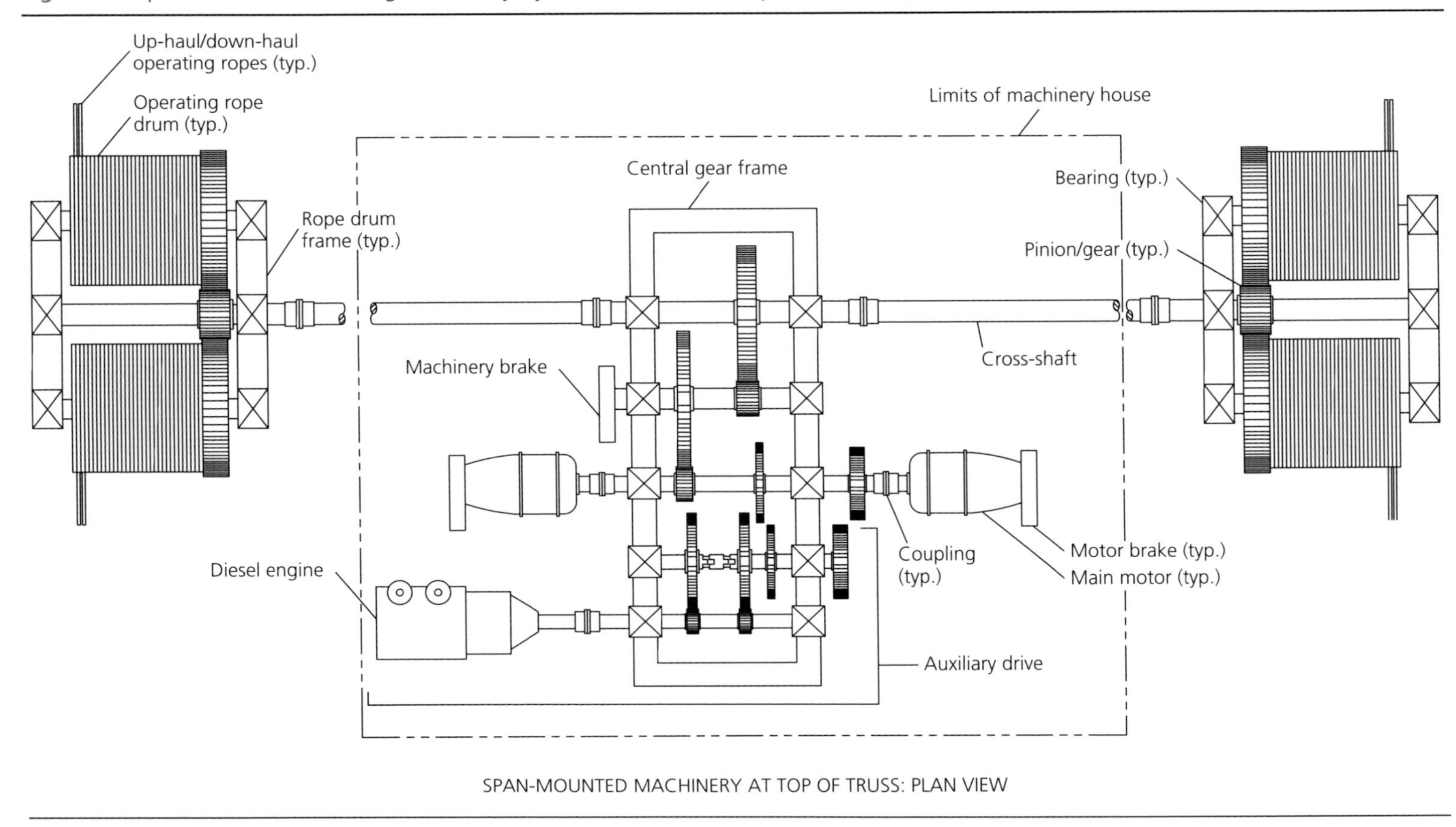

which dates to 1929. The operating machinery for the lift span is mounted on the lift span and is situated in a machinery house above the roadway at mid-span. The machinery is powered by two 150 hp, 600 rpm electric motors. The motors, which are situated on opposite sides of a central gear frame, are coupled to a common pinion shaft while the back end of each motor is equipped with a motor brake. The motor pinion mates with a gear and forms the first of the two open spur gearsets that are housed in the central gear frame. The output gear of the second gearset is mounted on a shaft that couples into and transfers the output torque to cross-shafts on either side of the gear frame. The cross-shafts span between the central gear frame and the machinery support frames for the operating rope drums that are located on either truss. The cross-shafts are coupled to the third and final pinion shaft that mates with the ring gears mounted onto each of the operating rope drums.

One operating rope drum serves each corner of the lift span with two uphaul and two downhaul ropes. The uphaul ropes are terminated at the top of the tower and the downhaul ropes are terminated near the foot of the tower. All ropes run along the tower legs and pass around deflector sheaves at the top chord of the truss, then run through a series of rollers and deflector sheaves back to the operating rope drums. Figure 8.8 illustrates the operating rope arrangement. When energised, the electric motors rotate the operating rope drums, which pay in or pay out the operating ropes to raise or lower the lift span.

Auxiliary operation is provided by way of a diesel engine located inside the machinery room. The diesel engine powers two gearsets at the input end of the central gear frame that are engaged or disengaged by way of a manual clutch mechanism and couple into the primary gear train. A manual brake is provided for span control when using the diesel engine.

8.7. The tower drive vertical lift bridge

An alternate operating arrangement to the span drive vertical lift bridge is the tower drive vertical lift bridge. For the tower drive arrangement, the drive machinery is used to directly turn the counterweight sheaves and the lift span raises or lowers as the ropes pass over the sheave due to friction between the ropes and sheave. Whereas the span drive arrangement provides mechanical skew control, the tower drive arrangement is reliant on electrical skew control.

Figure 8.9 illustrates a typical tower drive machinery configuration. This arrangement is for a through-truss vertical lift bridge with a lift span 540 ft long (164.7 m) and was built *c.* 1936. The operating machinery is located in an enclosed room at the top of each tower. The machinery in each tower is essentially the same. The two drives are mechanically independent. The drives are tied together electrically to ensure that both ends of the span move at the same rate during operation of the bridge in order to prevent a skew condition.

The normal power for each span drive is obtained from two 200 hp, 580 rpm, electric motors and emergency power is obtained from two 50 hp, 575 rpm, electric motors. The normal and emergency drives operate the same machinery with the exception that the emergency power passes through additional gearing, which reduces the speed of the lift span to approximately 25% of normal speed. A clutch is provided to engage and disengage the additional gearing for the emergency drive as necessary. Each of the motors is equipped with a motor brake on the outboard (non-driving) side of the motor.

The power from the motors is transmitted through a primary differential reduction gearset. A clutch is provided to lock the differential during operation, which forces both sides of the span to raise together, thereby preventing transverse skew if there is no slip between ropes and sheaves. The differential is

Figure 8.8 Span drive vertical lift bridge operating rope arrangement. The operating ropes extend from the operating drums at mid-span to terminations on the towers

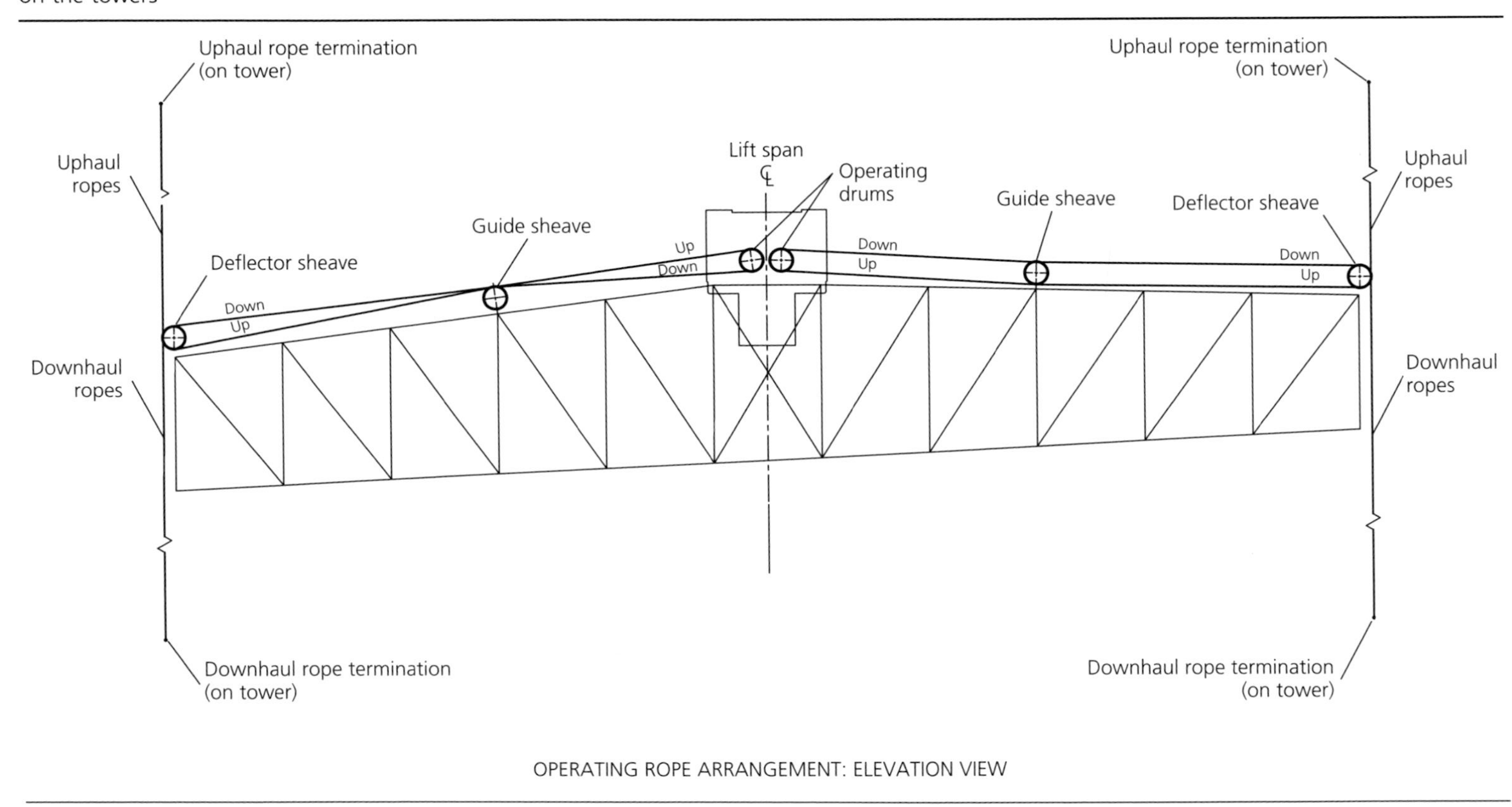

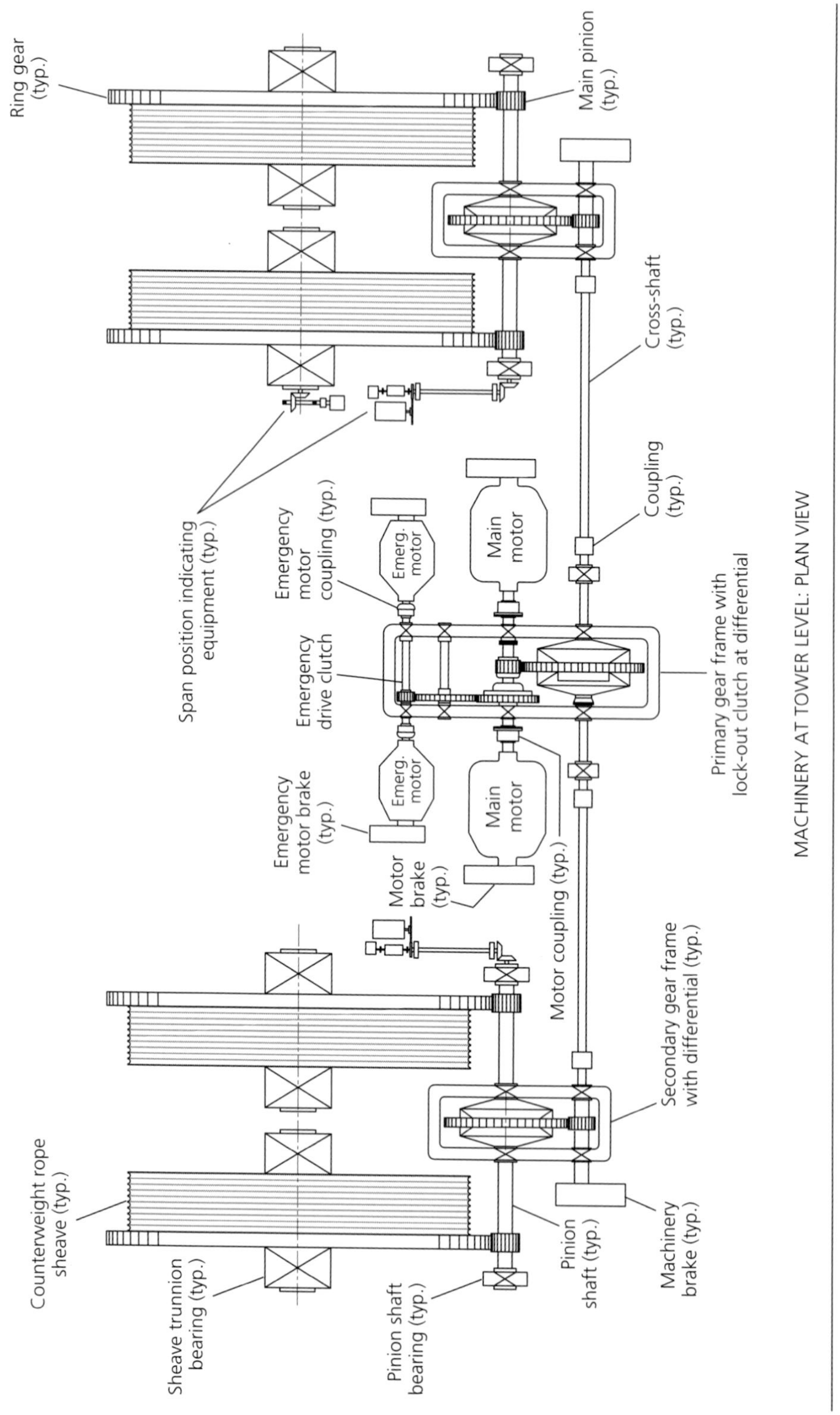

Figure 8.9 Tower drive vertical lift bridge machinery layout. All drive machinery is mounted atop the tower

provided to level the span transversely at initial installation and to adjust the span if necessary throughout the life of the bridge.

Transverse shafts extend from the primary differential east and west to a secondary differential reduction gearset and machinery brake at each side of the tower. The secondary differentials are located between the two sheaves at each side of the tower and ensure that driving force is evenly distributed between the sheaves. Shafts extend from both sides of the differential to drive pinions which mate with large ring gears mounted to each of the sheaves.

During normal operation, the main electric motors power the drive machinery to rotate the counterweight sheaves, which enables the span to be raised and the counterweights lowered as the wire ropes pass over the sheaves. The weight of the span and the counterweights provides sufficient friction between the ropes and the sheaves to prevent rope slippage during operation. In the event of a fault with the main motors, the emergency motors can be engaged by way of an emergency clutch and used to operate the bridge.

The foregoing examples of span drive arrangements for trunnion bascule, rolling bascule, heel trunnion bascule, swing, and span and tower drive vertical lift bridges have been used for nearly a century. Bridges are being designed today with similar machinery arrangements but, of course, using modern components. In the next chapter some of the mechanical components and parts that are used to assemble these drives will be illustrated. Space is insufficient to show all types of parts and details and recourse may be necessary to manufacturers' literature. The writer has inspected many movable bridges but seldom found two that are mechanically identical, even when the units form a pair at a particular crossing.

REFERENCES

Hopkins LO (1936) Hopkins Bascule Bridge. US Patent 2066110A, Dec.

Hovey OE (1926) *Movable Bridges*, vols I and II. Wiley, New York, NY, USA.

Scherzer AH (1908) *Scherzer Rolling Lift Bridges*. Scherzer Rolling Lift Bridge Company, Chicago, IL, USA.

Scherzer W (1893) Lift Bridge. US Patent 511,713, Dec.

Sööt O (1990) The need for single failure proof design for movable structures. *Proceedings of the 3rd Biennial Symposium of Heavy Movable Structures*. St Petersburg, FL, USA, 12–15 Nov.

Strauss Bascule Bridge Company (1920) *Catalog*. Chicago, IL, USA.

Waddell JAL (1916) *Bridge Engineering*, vols 1 and 2. Wiley, New York, NY, USA.

Movable Bridge Design
ISBN 978-0-7277-5804-0

http://dx.doi.org/10.1680/mbd.58040.199

Chapter 9
Machinery components

George A Foerster

9.0. Introduction

Machinery systems for movable bridges are comprised of many different mechanical components. This chapter will focus on the general components used for these systems. Many of the components described can be purchased as pre-engineered products from a variety of manufacturers. It is important to purchase such components from reputable manufacturers that have produced like components of a proven design. The manufacturers of such products provide catalogue cuts and/or certified drawings that show dimensions, tolerances, required assembly details, mounting, ratings, weights and lubrication information.

Generally, if components are not available as pre-engineered products they are referred to as custom (bespoke) machined parts. The details of such components are given on shop drawings which include information on materials, machining dimensions, surface finishes, and tolerances. Integration of custom parts and pre-engineering manufactured products into a machinery system is typically shown on assembly drawings. They show how the parts are to be assembled, and their relationship to the system. Details would include mounting, fasteners, welding, alignment requirements, lubrication, and painting. As with pre-engineered products, it is important when purchasing custom parts to utilise a reputable manufacturer and/or machine shop that has produced components of similar design.

Requirements for manufacturing machinery components are also typically given in design specifications and recommended practices. The majority of main machinery components are made from carbon or alloy steel. An exception to this is copper alloys such as bronze, which are used for bearing components (see Chapter 6). Bridges in the USA are covered under the AASHTO (American Association of State Highway and Transportation Officials) 2007 design standard (AASHTO, 2007) for highway bridges and the AREMA (American Railway Engineering and Maintenance-of-Way Association) recommended practice (AREMA, 2014) for railway bridges. These publications include requirements for materials, fits and finishes for mechanical components.

The components depicted in this chapter are only intended for demonstrative purposes. They are examples of those used for movable bridge machinery. Some components are models. It should be noted that in many cases, for actual components, the moving surfaces would be covered to protect personnel from contact during operation. The paint colour would follow a safety code system. One basic colour code is: orange for movable parts and green for stationary parts. Colour and guard requirements vary from project to project and need to be checked and addressed with the specific requirements for the actual designs.

9.1. Shafts, couplings and universal joints

Shafts are provided as machined parts required by a design unless they are part of a manufactured product. Shafts that are used to transmit torque are typically referred to as transmission shafting.

Figure 9.1 Types of shafts

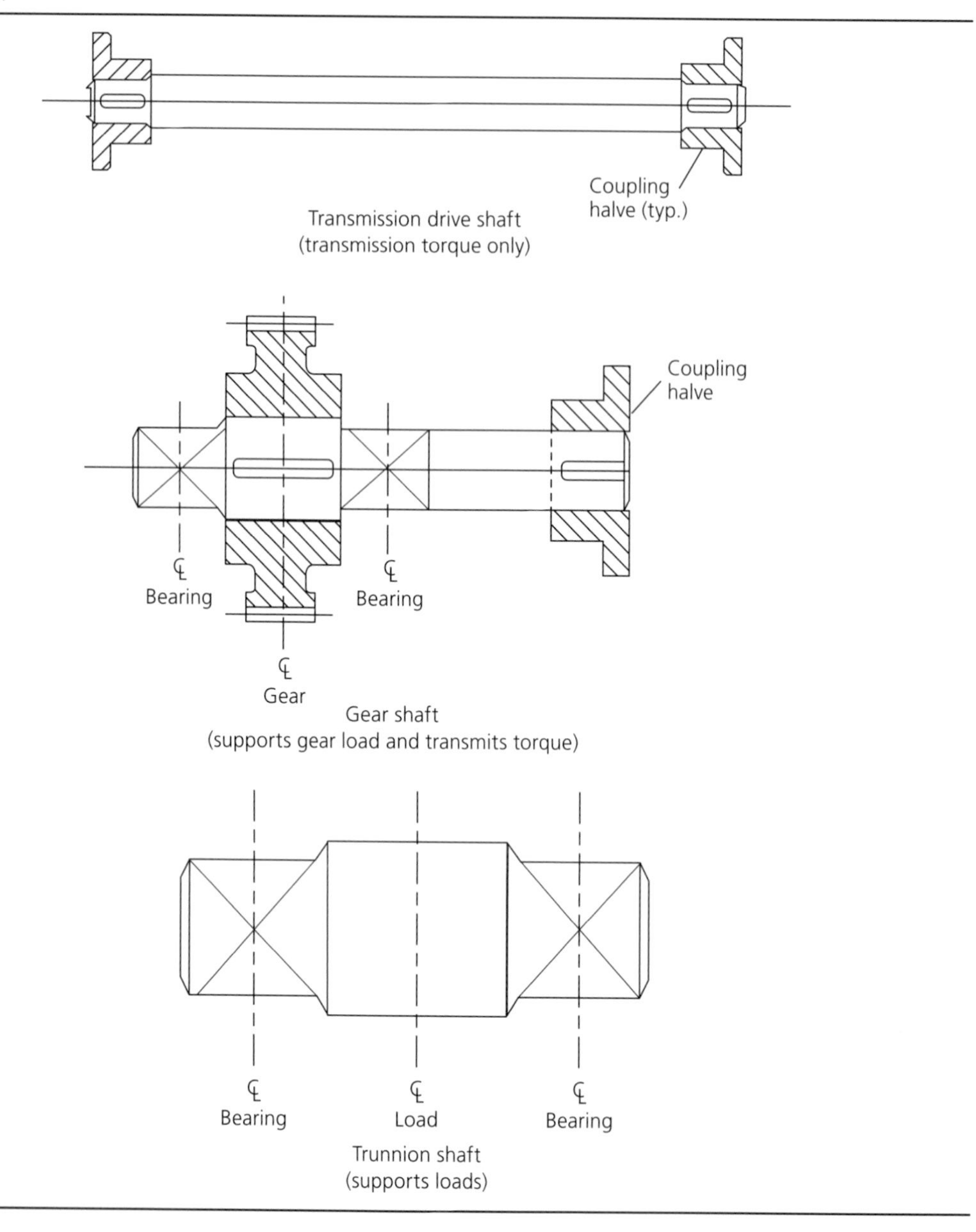

Other types of shafts are used to support loads and transmit torque or just to support loads and are allowed to rotate. These types of shafts include gear, sheave, drum, roller and trunnion shafts. Common shaft types are shown in Figure 9.1.

Couplings are used to join two rotating shafts. They transmit torque from one shaft to another. There are many coupling designs that permit some misalignment of the shafts. They are commonly known as flexible couplings. These types of couplings include double engagement gear, single engagement gear,

Figure 9.2 Gear coupling: flex–flex. (Photograph by George A Foerster)

grid and jaw-type couplings. Non-flexible couplings are sometimes used to connect two shafts. They are known as rigid couplings.

A double engagement coupling has two flex halves (see Figure 9.2). This type is used to connect two fixed shafts (shafts supported by two or more bearings). A single engagement coupling has flex and rigid coupling halves (see Figure 9.3) and is typically used on both ends of a floating shaft. Pre-engineered catalogue standard gear couplings are generally available for movable bridge shafts up to 10 in. (254 mm) in diameter. Heavy-duty models are generally available for shafts from 4 in. (102 mm) to above 20 in. (508 mm) in diameter. A rigid–rigid coupling is just two rigid halves bolted together. This type of coupling does not have any misalignment accommodation capabilities.

Figure 9.3 Gear coupling: flex–rigid, disassembled. (Photograph by George A Foerster)

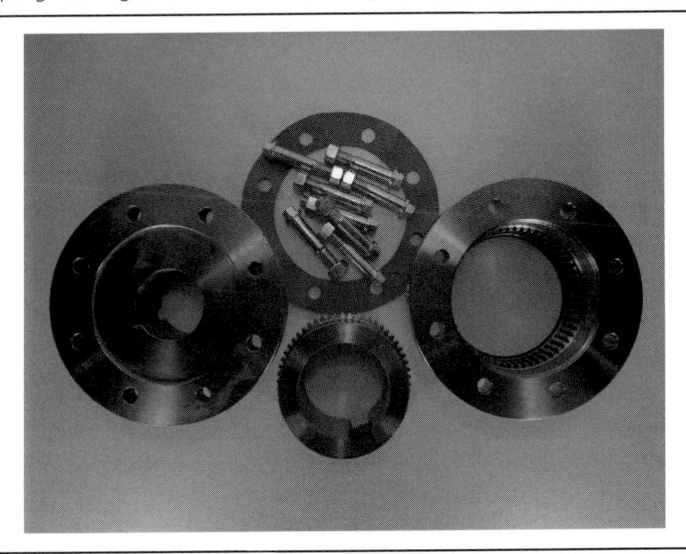

Figure 9.4 Grid coupling: closed cover. (Photograph by George A Foerster)

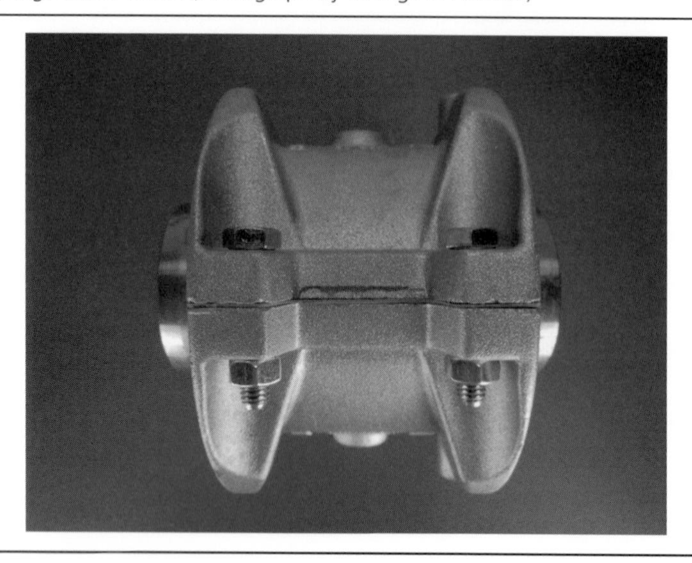

A grid coupling is a flexible coupling that has a steel grid to connect the two coupling halves (see Figures 9.4 and 9.5). The grid is also useful in damping abrupt changes in torque. This coupling is typically used for connecting prime movers to driven equipment. A jaw coupling is a flexible coupling that transmits torque between the jaws on each coupling half. Typically, there is an insert between the jaws to help accommodate misalignment (see Figure 9.6). Couplings are available from many manufacturers. Universal joint couplings act similarly to flex couplings. They transmit torque between two rotating shafts (see Figure 9.7) that are not coaxial. Universal joints can typically tolerate greater misalignment between shafts than couplings. Universal joints transmit constant angular rotation from one half to non-constant angular rotation of the other coupling half as a function of the misalignment between the two.

Figure 9.5 Grid coupling; half of cover removed. (Photograph by George A Foerster)

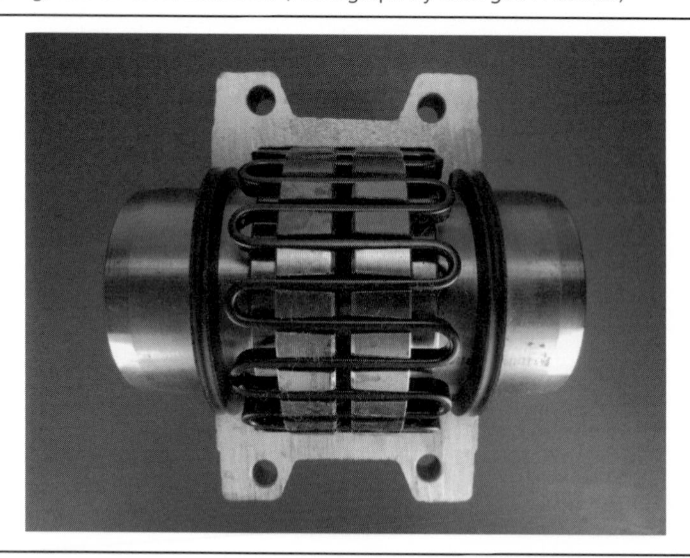

Figure 9.6 Jaw coupling. (Photograph by George A Foerster)

Figure 9.7 Universal joint coupling. (Photograph by George A Foerster)

As with couplings, universal joints are available from many manufacturers. Coupling selection and shaft design is further described in Chapter 12.

9.2. Bearings and bushings

Bearings and bushings are sometimes available as pre-engineered units in pillow-blocks that include housings and seals. They also can be used separately, mounted into custom housings or incorporated with other components such as an enclosed gear box. In such cases, seals are also typically incorporated as part of the custom housing.

Bearings are used in machinery systems to support radial or thrust loads. The main types of bearings are roller bearings (also known as anti-friction bearings) and plain sleeve bearings.

Plain radial sleeve bearings are typically bronze-lined bearings that also have a flange to also take thrust loads (see Figures 9.8 and 9.9). These types of bearings are also available as Babbitt-lined units without

Figure 9.8 Plain bronze bearing pillow block

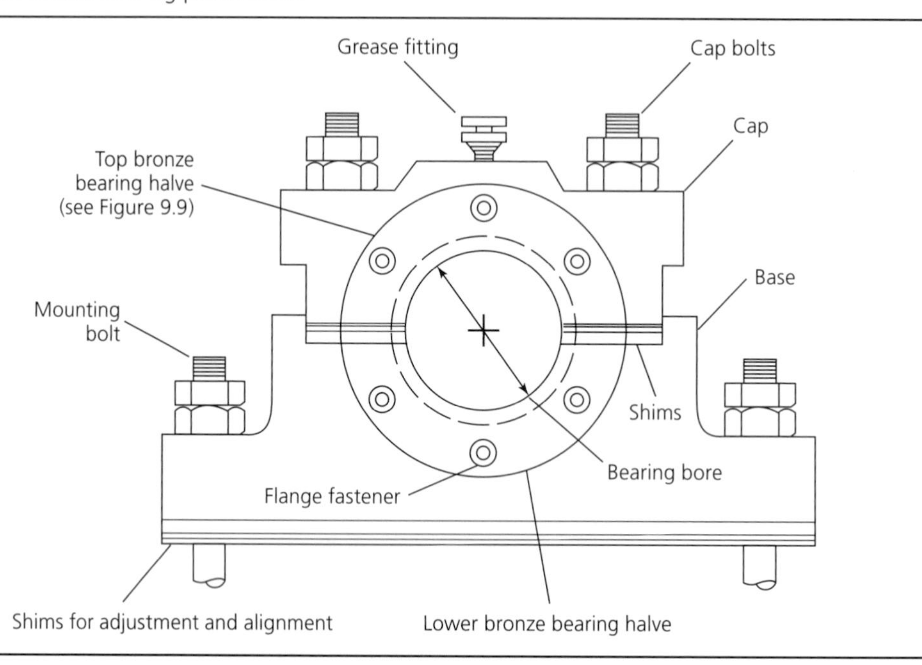

flanges (see Figure 9.10). However, Babbitt bearings are not commonly used in modern machinery installations except where shafts rotate at very high speeds. Larger-sized plain bronze bearings are typically required to be of custom design. Pre-manufactured radial roller bearings are more readily available in a larger range of sizes. One of the more common radial roller bearings used for movable bridges is the spherical roller bearing. This type of roller bearing can self-correct to some degree for relative misalignment of the shaft with respect to the housing (see Figures 9.11 and 9.12). Spherical roller bearings can also generally support some thrust load but the capacity is dependent on the specific type of bearing selected.

Figure 9.9 Half of plain bronze sleeve bearing. (Photograph by George A Foerster)

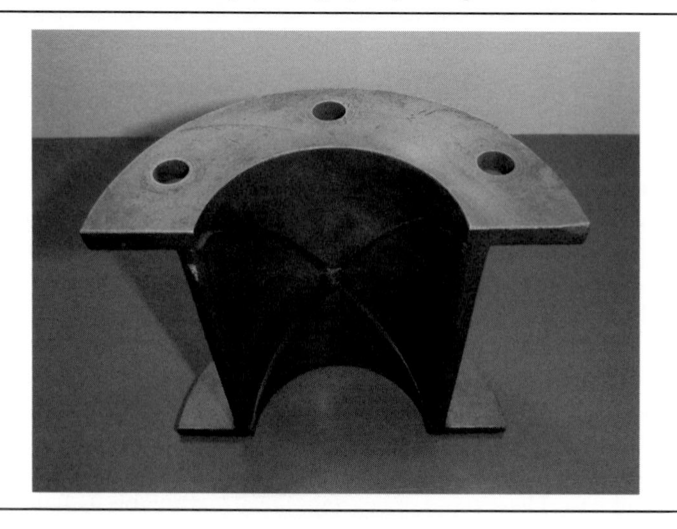

Figure 9.10 Babbit bearing pillow block. (Photograph by George A Foerster)

Figure 9.11 Roller bearing pillow block. (Photograph by George A Foerster)

Figure 9.12 Roller bearing pillow: block-cap removed. (Photograph by George A Foerster)

Plain bearings are also used to take substantial thrust loads. A common application of this type of bearing would be a centre pivot bearing for a swing bridge. Roller bearings can also be used for substantial thrust applications – for example the spherical roller thrust bearing. This type of bearing is similar in some ways to a spherical radial bearing with the exception that the spherical radius is much larger in order to accommodate large thrust loads in comparison with the radial loads. Plain and roller bearings are further described in Chapter 12.

9.3. Drums and sheaves

Drums are used to support wire ropes and they pay out and pull with force as the drum rotates. See Chapters 6 and 12 for further details about wire rope. Drums are typically a custom component fabricated and machined for a specific application and use (see Figure 9.13). Drums for movable bridge machinery have single or often double grooves helically cut on the outer surface or rim to support and guide the rope. It is noted that there are other applications not specific to movable bridges that do not have grooves and the rope is guided by a separate device and the number of wraps sometimes overlap in multiple layers. Grooved drums are far superior because the grooves help maintain the circular cross-section of the rope when it is bent about the drum. This has an important effect on rope life where the rope is under significant tension. For drum construction, the drum rim is typically connected to a hub with webs, spokes and stiffeners. Designs can include hubs that can have bearings mounted in the hub in which case the drum rotates about a static shaft and is driven by a ring gear connected to the side of the drum. The alternate situation is that the drum is fitted with a shaft which is supported in bearings and the drum is rotated by application of torque.

Sheaves are similar in construction to drums. They have a rim, web(s), spokes, stiffeners and hub. Sheaves can have grooves for either single or multiple ropes. The grooves are machined into the

Figure 9.13 General elevation of wire rope drum

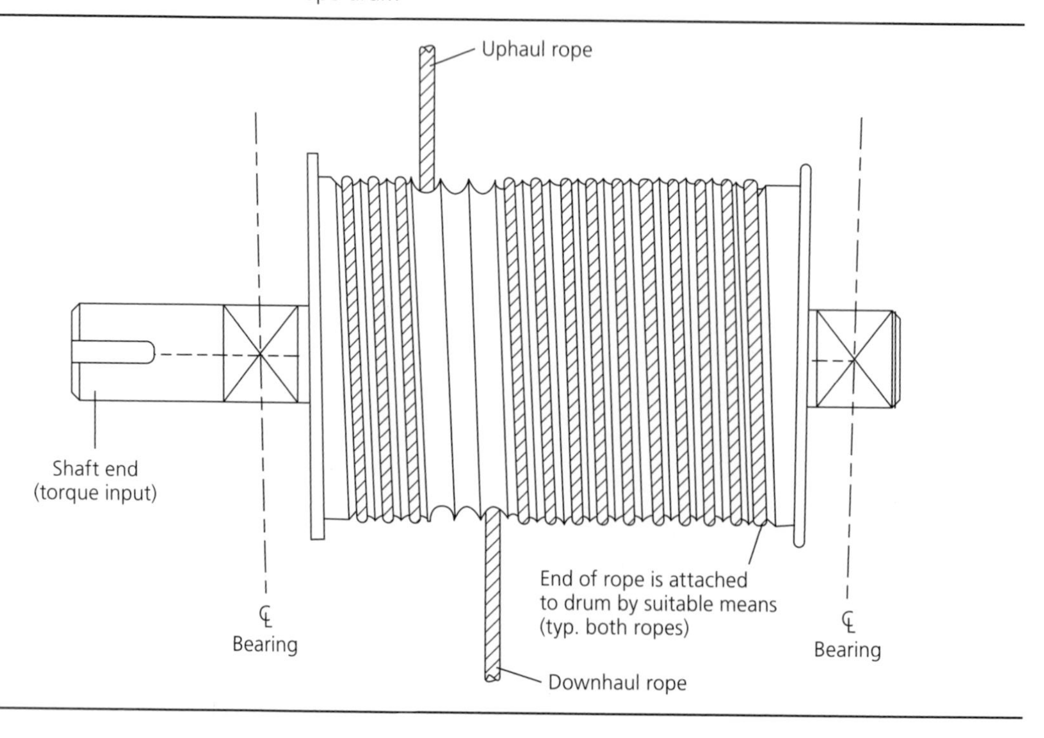

Figure 9.14 Counterweight sheave. (Courtesy of Hardesty and Hanover, LLC)

periphery of the rim. The grooves support the wire ropes as the sheaves are rotated. The wire ropes wrap around the sheave approximately 180°, or less for deflector sheaves. Sheaves are typically fitted with shafts that are supported in bearings (see Figure 9.14). The sheave and shaft rotate in the bearings as the wire rope translates. However, in some cases the sheave hubs are fitted with internal bearings and are supported by a static, fixed, shaft. Large sheaves are sometimes rotated by a ring gear attached to the side of the sheave. Smaller sheaves sometimes can be purchased as a catalogue product. Drum and sheave design and construction are further discussed in Chapter 12.

9.4. Linkages

Linkages include cranks, connecting rods and struts. These components are used to operate limit switches, lock bars, wedges or even movable bridges. With the exception of limit switches, the design of linkages can be a complex and involved process (Barton, 1984). The design of larger structural linkages goes beyond the scope of this book. The design of these components utilises both structural and mechanical principles. Linkage assemblies that drive locks, bars and wedges are discussed in Chapter 12. Linkages that drive limit switches include crank and connecting rod assemblies.

9.5. Power screws, mechanical jacks and pulling devices

Power screws are used in movable bridge design where mechanical jacks or pulling devices are required. One use of mechanical jacks is for end and centre supports for a swing bridge. The jack is a ram which is machined with power screw threads. For a common assembly, the power screw thread is driven by a suitable nut. Power screw devices can be either custom designed or purchased as pre-engineered units. Various Acme and square power screw thread forms are discussed in Oberg *et al.* (2012).

Power screws can also be used as pulling devices. A typical application is for an operating rope take-up assembly. By adjusting the power screw position of the rod, the rope tension can be adjusted. Power screw devices are also used to drive lock bars and wedges as discussed in Chapters 10 and 12.

9.6. Clutches and torque limiters

Clutches and torque-limiting couplings are used to control the torque transmitted between two shafts. A clutch is a component which when released allows two shafts to turn independently. In most cases a clutch is used to allow one shaft to turn and the other to stay stationary when released. When the clutch is engaged the two shafts become connected to transmit torque. The torque is typically transmitted through the clutch by friction between friction plates and a pressure plate assembly or a cone assembly. A cone clutch is typically a custom design for a specific application, whereas friction plate clutches are available as pre-engineered products. A typical application of clutches is for an auxiliary drive. Normally the auxiliary drive machinery would remain stationary with the clutch disengaged and would be engaged when it is needed to drive the bridge with the auxiliary drive.

A friction torque-limiting coupling works on the same principles as a friction plate clutch. However, for this type of coupling the pressure between the friction plates is set to normally transmit a certain maximum torque and then should slip if the torque is exceeded.

9.7. Gears, racks and differentials

When gears, racks and differentials are specified on design drawings, as opposed to being included in an enclosed unit, they are referred to as open gearing (see Figure 9.15). Many older machinery systems were made up of multiple sets of open gears. These gears were custom designed and machined as per specific design requirements. Typically the designs are the basis of the shop drawings, which would generally be approved by the designer for fabrication and machining. The most common open gear tooth form is the spur tooth. See Chapters 12 and 13.

Sets of gears in span drives are utilised to convert high rpms, low torque shafts to low rpm, high torque. The most common set would be a pinion and gear for conversion of torque and rotational speed. In some cases a pinion and rack is used. A straight rack has an infinite radius and, when used with a pinion, will convert torque and rotational speed to linear motion and force. Right-angle bevel gears are sometimes used to permit the rotating shafts to be at a right angle with respect to one another.

Figure 9.15 Internal ring gear and drive machinery of Naestved Swing Bridge. (Danish Road Directorate in Parke and Hewson, 2008, Figure 32, p. 449)

Open differential gear sets were used in older span drive designs. The differential equalises the torque of the two output shafts. A common type is a bevel gear and pinion assembly. This type of system consists of a carrier which is rotated with a certain torque and speed. Bevel pinions, typically two or three of them, are mounted radially with respect to the rotating axis of the carrier. Each side of the carrier with bevel pinions is engaged with a bevel gear which is mounted to an output shaft. If one shaft locks up, the speed will transmit to the opposite shaft. Since the torque must remain equal on both sides, the locked side will have the same torque but with zero speed and zero power. The other side will have all the power and with equal torque the rotational speed will be double. From this it can be seen that the differential will equalise the torque by portioning the power and speed of each of the output shafts. It is noted that there are also spur gear type differentials. However this type is not commonly used for movable bridge applications.

Differentials and open gearing are described further in Chapter 13.

9.8. Parallel shaft, right angle and planetary speed enclosed gear box reducers

Enclosed reducers are gear sets which are mounted inside an enclosed box. The enclosed box is typically cast steel or of welded steel, fabricated from plates and rolled shapes. The gear shafts are supported by bearings which are mounted in precision bored holes in the gear box in order to provide accurate assembly/alignment of the gear sets. The enclosed box can also serve as the reservoir for lubricant.

The most common and straightforward enclosed gear box is the parallel shaft box. As the name implies the shafts are mounted parallel to each other. The gears are typically helical or herringbone. This type of gearing can yield higher relative torque capacity as opposed to spur gears. These types of gears also tend to run quieter and more smoothly than spur-cut gears.

Right angle gear boxes, as the name states, are units in which the input and output shafts are at a right angle. The details of the enclosed box are similar to the parallel unit, except the mounting of the input shafts and output shafts are at right angles (see Figure 9.16). Typically spiral bevel gears are used for this type of gear box. These gear boxes range from small units which are used for limit switch drives to larger sizes used for auxiliary drives and for swing bridge operating machinery.

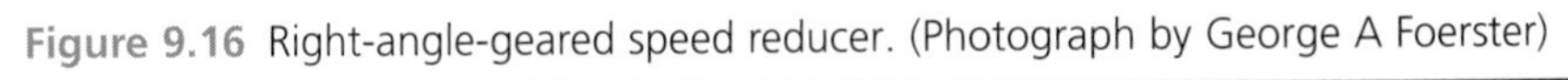

Figure 9.16 Right-angle-geared speed reducer. (Photograph by George A Foerster)

Figure 9.17 Enclosed gear box with differential: cover removed. (Steward Machine Co. in Parke and Hewson, 2008, Figure 17, p. 439)

Combinations of different gear types can be used and accommodated into one enclosed unit (see Figure 9.17). This box includes spur, helical, herringbone and bevel gears.

Another type of gear box is the planetary type. A gear box of this type utilises a central sun and planet gears with a fixed housing internal gear to increase the ratio between the input and output. Figure 9.18 shows a model of a planetary single stage gear. Multiple stages of planet gears with associated sun gear and fixed housing gear can be assembled in line to provide much higher reduction gear ratios, in a smaller envelope than in a parallel shaft gear box. Also, the use of multiple planet gears increases the torque capacity of the planet and sun gear sets over that of a single gear mesh.

Figure 9.18 Plastic model of a one stage planetary gear box. (Photograph by George A Foerster)

9.9. Roller chains and sprockets

Roller chains and sprockets are used to transmit torque and speed between two parallel shafts (see Figure 9.19). These components are arranged to have a driving sprocket on the shaft that delivers the power to a driven sprocket on a shaft that is receiving the power. A roller chain connects the two sprockets. Different numbers of sprocket teeth can be used to achieve different speed ratios between the two shafts. The force in the chain is the torque in the driving shaft divided by the pitch radius of the driving sprocket. The torque in the driven shaft would be the force in the chain times the pitch radius of the driven sprocket. Because there is a positive connection between the chain and sprockets and power is constant, except for minor losses, the ratio of the speed and torque of the sprockets is directly related to the ratio of teeth between the two sprockets. The output sprocket will turn at a rotational speed of the driving shaft (rpm) times the number of teeth of the driving sprocket divided by number of teeth on the driven sprocket. The torque of the output would be equal to the input torque times the number of teeth on the output sprocket divided by the number of teeth on the input sprocket.

Standard chains and sprockets are available from a number of different manufacturers. There are speed and force limitations for the specific type and size of chain. These limitations as well as mounting, alignment, and lubrication requirements are available from the manufacturer.

Figure 9.19 Roller chain and sprockets. (Photograph by George A Foerster)

9.10. Keys, pins, retainer rings, shrink fit devices, collars and threaded fasteners

This section describes components which are typically used to secure and connect machinery parts.

9.10.1 Keys

Keys are used to secure rotating parts to shafts that transmit torque (see Figure 9.20). They are also used to prevent sliding between two surfaces in contact and between thrust-bearing discs to prevent movement of parts that are not intended to move. An example of this is a centre pivot bearing of swing bridges. The top disc is keyed to the top support and the bottom disc is keyed to the bottom support so that all movement occurs between the two surfaces intended to slide.

For keyway assemblies, opposing keyways are cut in the materials and the key is fitted into them so as to act as a shear connection. The key is subject to bearing stress on the faces perpendicular to the shear plane and shear on the plane of the mating surfaces. Keyway and key details are further described in Chapter 12.

Figure 9.20 Keys. (Photograph by George A Foerster)

9.10.2 Pins

Alignment pins are used to fix two components by inserting them into a hole through the pieces being connected. For movable bridge machinery the pins are typically solid straight pins or tapered pins. The pin acts similarly to a key in a keyway. The inside of the bored hole applies the bearing on the pin and the pin acts in shear at the plane of the two connected pieces. Typically, alignment pins are used to secure and hold components in place after final alignment has been achieved. Large pins, also called dowels, are also used as a secondary means to hold or secure large hubs to their shafts. Examples of this are trunnion shafts for bascule bridges and sheave shafts for vertical lift bridges. The dowels are fitted into bored holes through the hub into the shaft, and act as a shear connection between the two parts. Other types of pins include spring pins and cotter pins (see Figure 9.21).

9.10.3 Retainer rings

Retainer rings are used to hold components in a fixed axial position on a round shaft. External rings can be flexed over the shaft and inserted into machined grooves in the shaft. There are both internal and external types of rings (see Figure 9.22). They are typically used to hold bearing races or hubs in place.

Figure 9.21 Pins. (Photograph by George A Foerster)

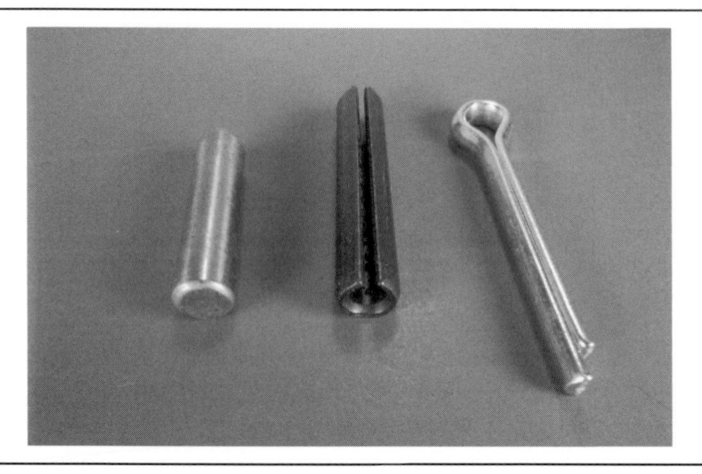

Figure 9.22 Retainer rings. (Photograph by George A Foerster)

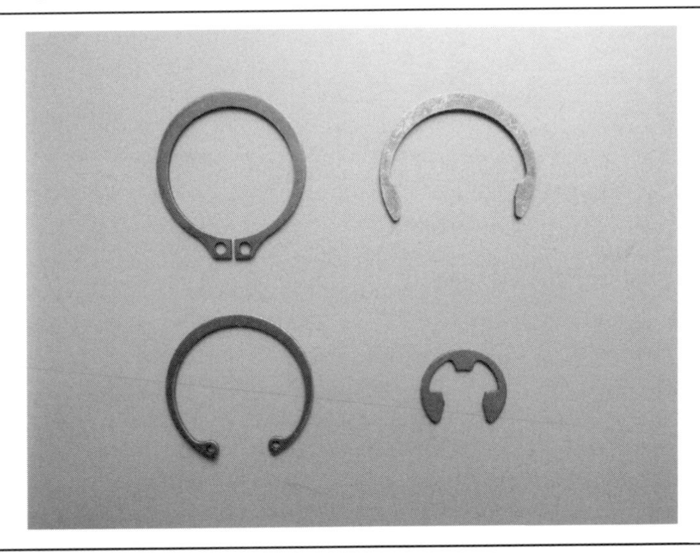

9.10.4 Shrink-fit devices

Shrink-fit devices are used to connect hubs to shafts. They create pressure between inner bore of the hub and outer diameter of a shaft. The pressure is generated by pulling two sloped (wedge) surfaces together using threaded fasteners (see Figure 9.23). These devices are generally used for transmitting torque and force between a hub or component mounted to a shaft. They typically would not be used where large bending moments, such as those due to bridge dead load, need to be supported in addition to torque.

9.10.5 Collars

Shaft collars are used to hold the position of the shaft axially (see Figure 9.24). They can either be a solid circular ring with a certain thickness, and have a set screw to hold them in place, or they can

Figure 9.23 Shrink-fit ring. (Photograph by George A Foerster

Figure 9.24 Collars. (Photograph by George A Foerster)

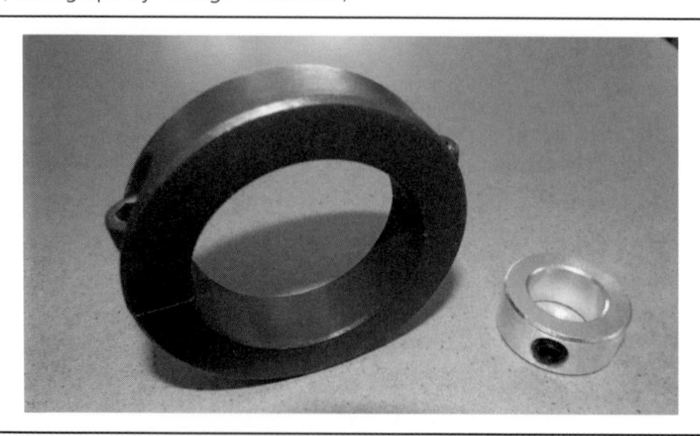

be split halves, so that when they are bolted together they clamp onto the shaft. Collars can either be of custom design or be selected from offerings of many manufacturers that make different sizes.

9.10.6 Threaded fasteners

Threaded fasteners include bolts, nuts and cap screws. The general definition of a bolt is where a nut is used to tension the fastener and a cap screw is where the head of the fastener is used to tension the fastener inserted into a tapped hole. General bolt, cap screw, nut and washer specifications are listed in Table 9.1. For movable bridge design, bolt and cap screws are generally of the heavy hex size. A general sketch of a heavy hexagonal bolt is shown in Figure 9.25. The internal threads for assembly of a cap screw are generally a minimum depth of $1\frac{1}{2}$ times the diameter. However, the depth is dependent on specific details and the materials used. Bolt and cap screw are installed to produce an initial bolt preload. Further details of bolts and preload are described in Chapter 12. Bolts which have been torqued to a tension close to or above the yield strength are typically not reused.

Table 9.1 Common bolting material specifications

Description	Material	Specification
High-strength-finish body bolts, turned bolts and studs	Quenched and tempered steel	ASTM A449-14 (ASTM, 2014a)
Structural bolts	Quenched and tempered steel	ASTM A325/A325M (ASTM, 2014b)
Socket head cap screws	Alloy steel	ASTM A574/A574M (ASTM, 2013a)
Countersunk head cap screws	Alloy steel	ASTM F835/A835M (ASTM, 2013b)
Nuts	Alloy steel	ASTM A563/A563M (ASTM, 2014c)
Washers	Hardened steel	ASTM F436/F436M (ASTM, 2011)

Figure 9.25 Bolt

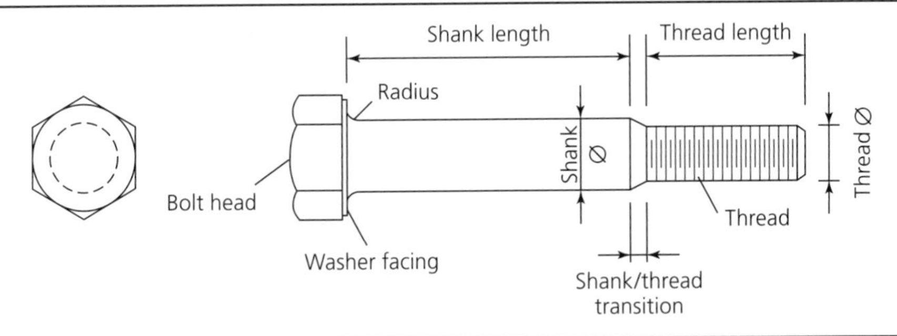

9.11. O-rings, lip seals and packing material

Seals are important components for machinery design. They are typically used to keep lubricants in and contaminants out of bearings. Lubricants include both oil and grease. There are a few different types.

One type is the labyrinth seal. This type of seal is commonly used in pre-engineered radial roller bearing pillow block housings (see Figure 9.26). The labyrinth consists of multiple slots which are enclosed on three sides.

O-rings are elastic rings that usually have a round cross-section; they fit in grooves and create a seal against another surface. Dimensions of the grooves are critical.

Another type of seal is a lip seal (see Figure 9.27). There are many different types of lip seals available from different manufacturers. They are typically fitted into a machined counter bore and are held in place with a cover plate or pressed in when the seal comes in a metal casing. The lip of the seal rides on the rotating surface to create the seal.

Packing material is used in some cases, typically more common in older designs, to create a seal between two parts. This type of material can be used to seal covers as well as shafts. It is commonly

Figure 9.26 Labyrinth seal. (Photograph by George A Foerster)

Figure 9.27 Lip seal. (Photograph by George A Foerster)

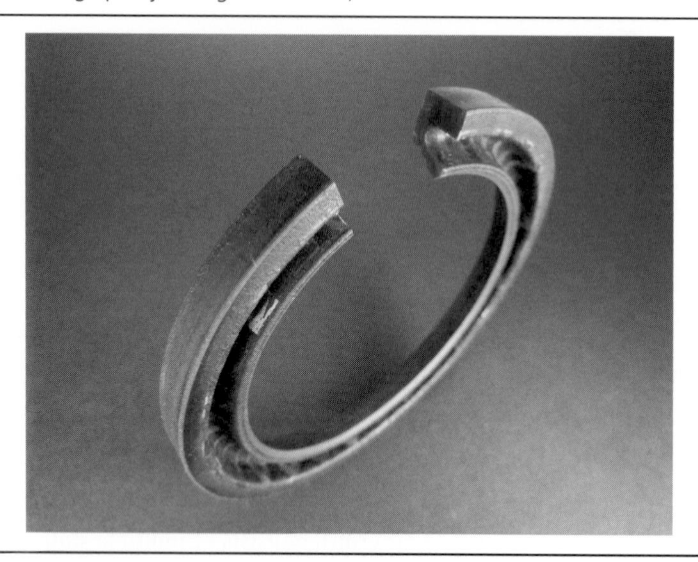

fitted into one part and is pressed into the other part to create a seal. On shafts a packing glad is usually installed so that the packing may be compressed as it wears due to shaft rotation. Friction between the seal and shaft generally creates heat during operation. The proper installation is very important for the seals to function as desired. This includes installation alignment, mounting, machining tolerance, sliding surface hardness and finish.

9.12. Springs

Springs are used in machinery systems to exert force as a function of the relative displacement of the spring ends. Two common types of springs are coil springs (see Figure 9.28) and Belleville springs (see Figure 9.29).

Figure 9.28 Coil springs. (Photograph by George A Foerster)

Figure 9.29 Belleville springs. (Photograph by George A Foerster)

A coil spring has a large displacement with a smaller change in force compared to a Belleville spring. The Belleville spring can be used to achieve a large force with small displacement. The force a coil spring exerts is equal to the spring constant of the spring times the displacement. Coil springs can be used to apply a force from the spring's free length to some desired compression or from a preloaded condition throughout a range of displacement.

A coil spring, as its name suggests, is made from wire of a certain diameter. The common materials used for coil springs are steel, stainless steel, or bronze. The material is wrapped in a coil at a radius with a number of coils as deemed necessary. The ends are typically ground flat for bridge machinery applications. The material, diameter of the wire, and number of coils will determine the spring's performance.

Belleville springs of many different characteristics are available from many manufacturers, as is also the case for coil springs. Belleville springs can be stacked in series, parallel, or parallel-series to attain desired forces and displacements. For specific applications, the forces and displacements for Bellville springs would need to be determined based on the actual springs, characteristics, and arrangements used.

9.13. Closing summary

The foregoing describes some of the common components of mechanical span drives such as those described in Chapter 8. Other parts are necessary for a complete installation. Illustrations of machinery assemblies may be found in Parmley (2005), Winsmith (1980) and Steward Machine Company (2008). Photographs and line art of components are included in Childs (2014), Lynwander (1983), Mancuso (1986), Oberg *et al.* (2012), Shigley and Mischke (2001) and Spotts *et al.* (2004). Those trade and textbooks treat machine design much more thoroughly than is possible in the space available for this overview chapter on mechanical components. Hydraulic components are discussed in Chapter 14.

REFERENCES

AASHTO (American Association of State Highway and Transportation Officials) (2007) Standard specifications for movable highway bridges, 2nd edn. AASHTO, Washington, DC, USA.

AREMA (American Railway Engineering and Maintenance-of-Way Association) (2014) Movable bridges – steel structures. Chapter 15 in *AREMA Manual for Railway Engineering*, AREMA, Lanham, MD, USA.

ASTM (American Society for Testing and Materials) (2011) ASTM F436-11 Standard specification for hardened steel washers. ASTM International, West Conshohocken, PA, USA, http://dx.doi.org/10.1520/F0436-11.

ASTM (2013a) ASTM A574-13 Standard specification for alloy steel socket-head cap screws. ASTM International, West Conshohocken, PA, USA, http://dx.doi.org/10.1520/A0574.

ASTM (2013b) ASTM F835-13 Standard specification for alloy steel socket button and flat countersunk head cap screws. ASTM International, West Conshohocken, PA, USA, http://dx.doi.org/10.1520/F0835.

ASTM (2014a) ASTM A449-14 Standard specification for hex cap screws, bolts and studs, steel, heat treated, 120/105/90 ksi minimum tensile strength, general use. ASTM International, West Conshohocken, PA, USA, http://dx.doi.org/10.1520/A0449-14.

ASTM (2014b) ASTM A325M-14 Standard specification for structural bolts, steel, heat treated 830 mpa minimum tensile strength (metric). ASTM International, West Conshohocken, PA, USA, http://dx.doi.org/10.1520/A0325M-14.

ASTM (2014c) ASTM A563-07a Standard specification for carbon and alloy steel nuts. ASTM International, West Conshohocken, PA, USA, http://dx.doi.org/10.1520/A0563.

Barton LO (1984) *Mechanism analysis*. Marcel Dekker, New York, NY, USA.

Childs PRN (2014) *Mechanical Design Engineering Handbook*. Butterworth-Heinemann, Oxford, UK.

Lynwander P (1983) *Gear Drive Systems: Design and Application*. Marcel Dekker, New York, NY, USA.

Mancuso JR (1986) *Couplings and Joints: Design, Selection, and Application*. Marcel Dekker, New York, NY, USA, http://dx.doi.org/10.1115/1.3187853.

Oberg E, Jones FD, Horton HL and Ryfel HH (2012) *Machinery's Handbook*, 29th ed. Industrial Press, New York, NY, USA.

Parmley RO (2005) *Machine Devices and Components: Illustrated Sourcebook*. McGraw-Hill, New York, NY, USA.

Parke G and Hewson N (eds) (2008) *ICE Manual of Bridge Engineering*, 2nd edn. ICE, London, http://dx.doi.org/10.1680/mobe.34525.

Shigley JE and Mischke CR (2001) *Mechanical Engineering Design*, 6th edn. McGraw-Hill, New York, NY, USA.

Spotts MF, Shoup TE and Hornberger LE (2004) *Design of Machine Elements*, 8th edn. Pearson Prentice-Hall, Upper Saddle River, NJ, USA.

Steward Machine Company (2008) *Earle Speed Reducers*. Steward Machine Company, Birmingham, AL, USA.

Winsmith (1980) *The Speed Reducer Book: A Practical Guide to Enclosed Gear Drives*. Peerless-Winsmith, Springville, New York, NY, USA.

Movable Bridge Design
ISBN 978-0-7277-5804-0

http://dx.doi.org/10.1680/mbd.58040.219

Chapter 10
Stabilisation machinery

George A Foerster

10.0. Introduction

The main supports of movable bridges support most of the dead load, and even live load in some cases, and also allow the bridge to operate. For bascule bridges the main supports can include either a trunnion assembly (main or, sometimes, counterweight), track and tread assembly for a rolling lift, Rall wheel or heel trunnion and linkages. For a swing bridge the main dead load and some live load can be supported by a main centre pivot bearing or a rim bearing assembly. The dead load of a vertical lift span is commonly supported by wire ropes reeved over counterweight sheave assemblies that are connected to a counterweight at each end of the span. In some cases, small and special lift bridges can be supported by hydraulic cylinders. See Chapter 12 for additional details for common main supports for movable bridges.

All these bridge types do require stabilisation machinery in order for them to perform their function. This includes supporting live loads, transferring shear loads, guiding the movable span and locking the movable span in place in the seated position. This chapter will focus on some of the common types of machinery systems used to stabilise movable spans.

The required stabilisation for movable bridges in the USA is also typically covered under the AASHTO (American Association of State Highway and Transportation Officials) design standard (AASHTO, 2007) for highway bridges and AREMA (American Railway Engineering and Maintenance-of-Way Association) recommended practice (AREMA, 2014) for railway bridges. The AASHTO *Movable Bridge Inspection, Evaluation and Maintenance Manual* (AASHTO, 1998) depicts some of the components required for the stabilisation machinery of movable bridges. In this chapter only common arrangements and basic details for stabilisation machinery is presented. Refer to AASHTO (2007), AREMA (2014) or other specific project requirements for further details on stabilisation machinery.

10.1. Bascule bridges

The stabilisation machinery for bascule bridges includes live load bearings, span locks, tail locks, centring devices and air buffers. The details of these components are described below. See Chapter 3 for different variations of deck joint positions, live load locations and locking machinery for double-leaf bascules.

10.1.1 Live load bearings

Live load bearings are structural bearings that provide reaction points for live load on the bridge. These are typically two flat plates or a curved bearing surface on a flat plate that bear on one another at the contact point between the movable spans and the fixed structure. For double-leaf bascules there are two common types: rear and forward live load bearings. Rear live load bearings provide a reaction point behind the trunnion or centre of roll. Forward live load bearings on some double-leaf trunnion bascules

Figure 10.1 Double-leaf bascule. The illustration shown here is intended to show only the general location of some of the stabilisation machinery. Illustration is not to scale. Tail locks are not shown. Tail locks are required when live loads have the tendency to open the leaf. See Figures 10.4 and 10.5 for typical tail lock machinery

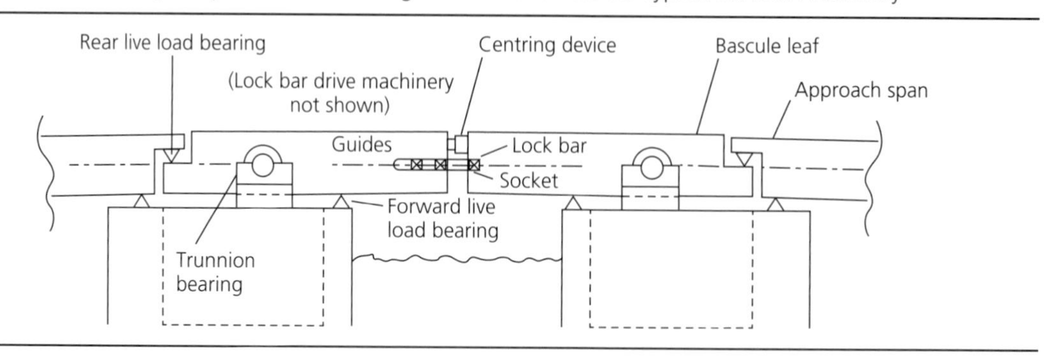

provide a reaction point in front of the trunnion. The forward reaction point reduces the moment arm of the live load traffic on the girder (see Figure 10.1). A single-leaf bascule has a live load bearing at the toe of the leaf (see Figure 10.2).

10.1.2 Span locks

Mid-span lock bars are used to transfer shear on double-leaf bascule bridges while keeping the tips of the mating leaves aligned (see Figure 10.1 and Chapter 3). They are also used to secure single-leaf bascule bridges in the closed position (see Figure 10.2). The basic components of a lock bar assembly are the bar, socket, front guide, rear guide and drive machinery. See Chapter 12 for further details on lock bar and drive machinery design.

The typical analogy of a shear lock system for a double-leaf bascule is a hinged joint. This type of connection experiences leaf-tip rotation at the joint between leaves for a double-leaf bascule. The deflection curves of the two leaves intersect at a single point. This action at the tips of the leaves is prohibitive for double-leaf bascules for supporting heavy rail traffic, due to the abrupt transition of profile. One way to address tip rotation for highway traffic double-leaf bascules is to use cylindrical bearing surfaces for the sockets to accommodate the rotation.

Figure 10.2 Single-leaf bascule. The illustration shown here is intended to show only the general location of some of the stabilisation machinery. Illustration is not to scale. Tail locks are not shown. Tail locks are required when live loads have the tendency to open the leaf. See Figures 10.4 and 10.5 for typical tail lock machinery

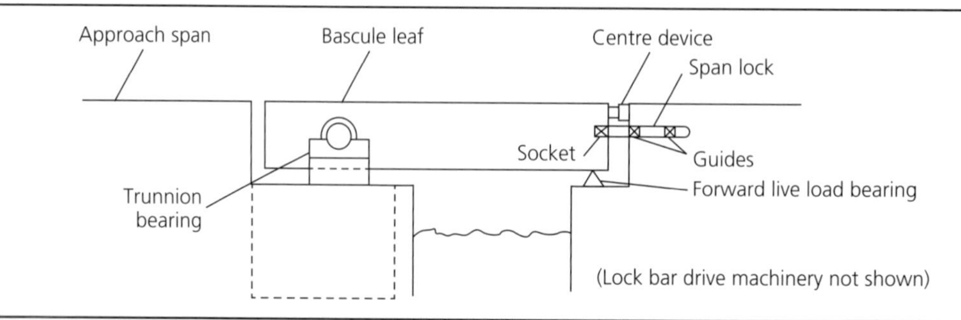

Figure 10.3 Double-leaf rolling leaf. The illustration shown here is intended to show only the general location of some of the stabilisation machinery. Illustration is not to scale. Tail locks are not shown. Tail locks are required when live loads have the tendency to open the leaf. See Figures 10.4 and 10.5 for typical tail lock machinery

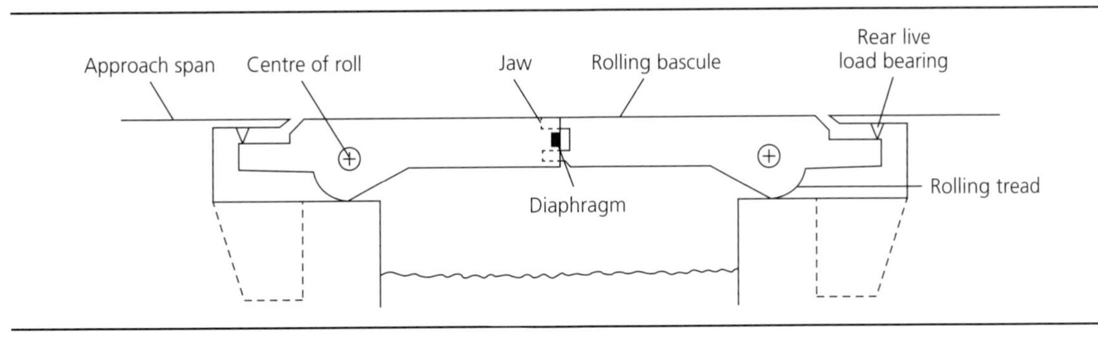

Another type of locking system used on double-leaf bascules are moment locks. This type of system is less commonly used on double-leaf bascules, due to the requirement for double the machinery and larger lock bar cross-sections. Moment lock systems transfer bending moment as well as shear forces from live loads on the bascules' spans. See Chapter 3 for further information and details on moment locks.

A third type of double-leaf span locking system is a jaw and diaphragm. The jaw and diaphragm perform the same function as a span lock bar system to transfer the shear load between mating leaves. This type of system is used on double-leaf rolling bascule bridges where one leaf has a diaphragm that rolls into or out of a jaw on the other leaf during bridge operation (see Figure 10.3).

10.1.3 Tail locks

On double-leaf bascules when live loads cause the tendency to open the leaves, tail locks are used to prevent them from rotating open. Tail locks are used on bascule bridges to secure the counterweight end of the leaf. Common types of tail locks include a sliding bar that engages a socket and a rotating column that locks a point underneath the counterweight. For the sliding bar, typically the tail-lock bar would be supported in guides and driven with machinery mounted on the fixed portion of the bridge. The bar when driven would engage a socket on the rear girder end or the counterweight to prevent the leaf from rotating open (see Figure 10.4). The design of tail lock machinery is similar to that of span

Figure 10.4 Bascule-leaf tail-lock bars – transverse elevation at counterweight (CWT). The illustration shown here is intended to show only the general location of some of the stabilisation machinery. Illustration is not to scale

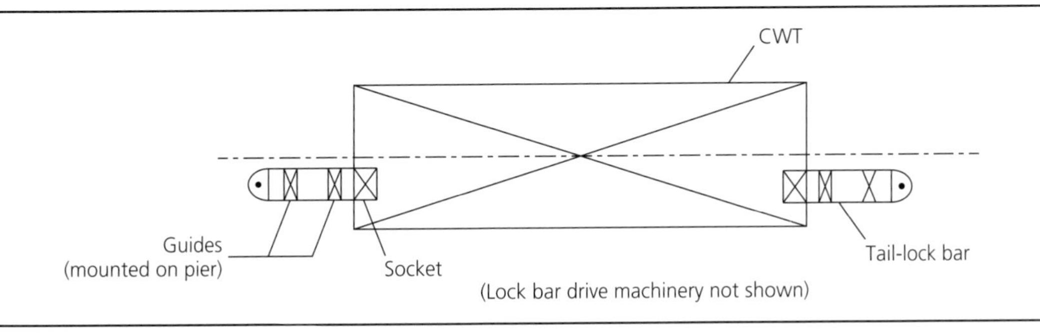

Figure 10.5 Bascule-leaf tail-rotating lock column – transverse elevation at counterweight. The illustration shown here is intended to show only the general location of some of the stabilisation machinery. Illustration is not to scale

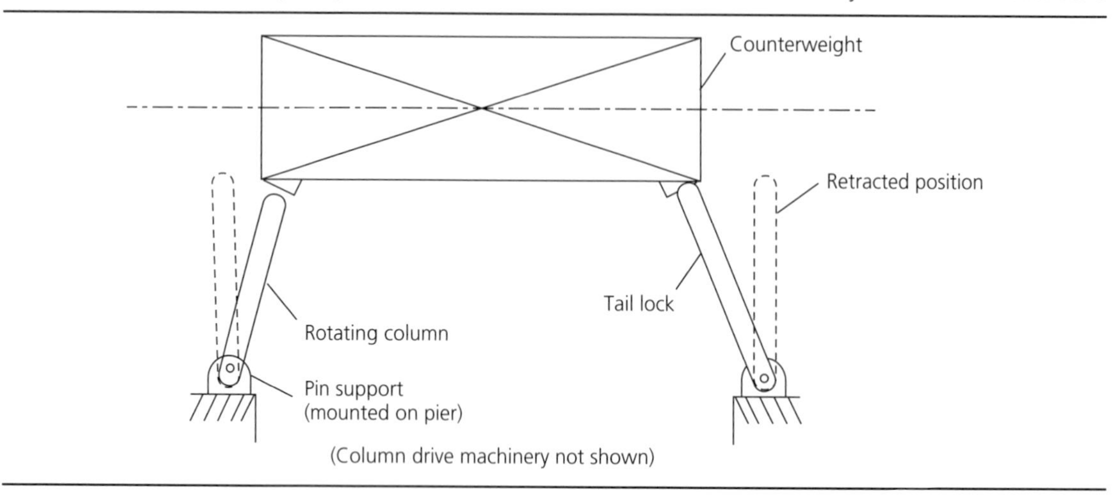

lock machinery. Similarly the rotating column, rear support pivot bearing and drive machinery would be mounted on the fixed portion of the bridge. The column would then be rotated to engage a reaction point on the bottom of the counterweight to prevent the leaf from rotating open (see Figure 10.5).

10.1.4 Centring devices

Centring devices are used on double-leaf and single-leaf bascules to centre or align the bridge in the transverse direction. The most common type is a socket on one leaf or rest pier and a pin or protrusion on the opposing side that engages the socket to align the leaves.

10.1.5 Buffers

In some cases buffers are used at the ends of operational travel to decelerate the movable span when modern electrical controls are not utilised. These can be either custom air buffers, which dissipate energy through compressed air and its regulated release, or modern industrial shock absorbers. Modern industrial shock absorbers are available as pre-engineered manufactured units from multiple manufacturers. In some cases wooden bumper blocks are used for the open position of the bascule.

10.2. Swing bridges

The stabilisation machinery for swing bridges includes end supports, centre supports, balance wheels, centring devices, locking devices, rigid stops and buffers. The details of these components are described below.

10.2.1 End supports

End-lift devices deflect the ends of the swing span to put a preload reaction into the swing span and support additional loads from traffic on the bridge (see Figure 10.6). The preload reaction is intended to avoid uplift at the ends due to a continuous span loaded only on one side (see Figure 10.7). Typically, the minimum reaction is 1.5 times the uplift load, but the specific required load would need to be checked with the governing project design requirements. The end lifts for swing bridges are typically either wedges or an eccentric shaft with a cam. See Chapter 12 for further details on wedge design. The eccentric shaft can also be fitted with a roller to reduce frictional resistance when driving into the

Figure 10.6 Centre bearing swing bridge. The illustration shown here is intended to show only the general location of some of the stabilisation machinery. Illustrations are not to scale

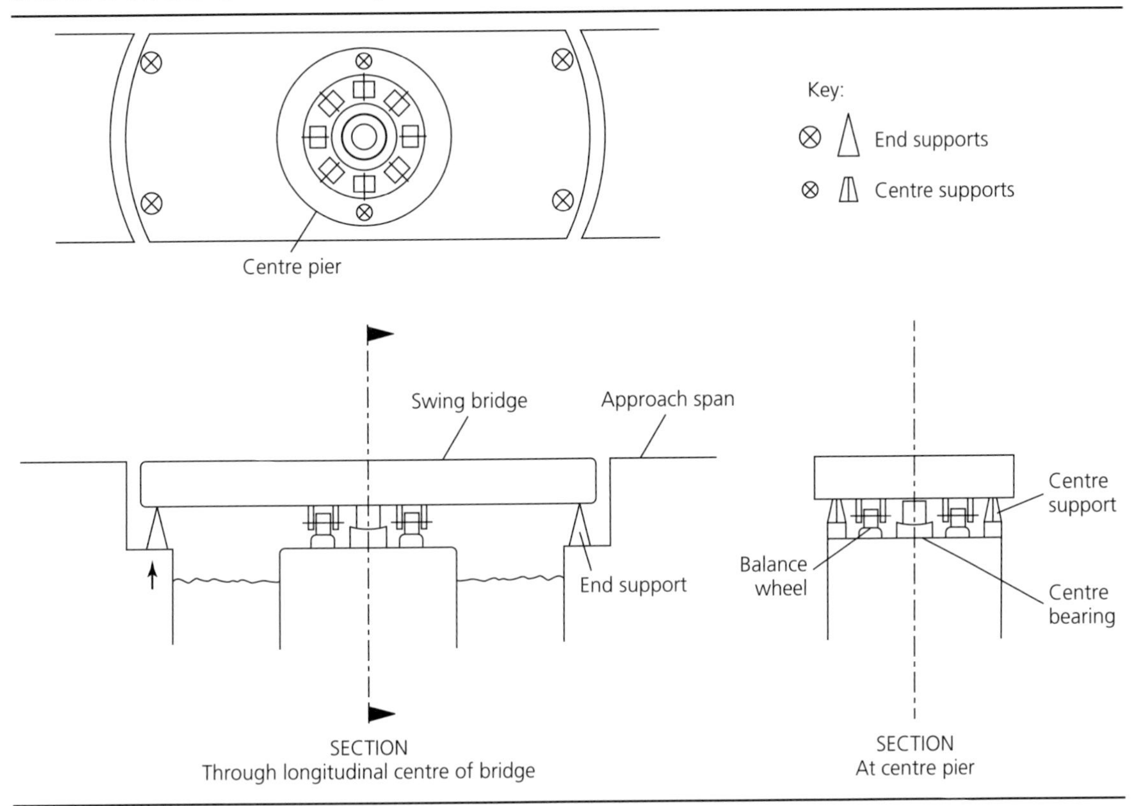

engaged position. The end lifts would be engaged to support live load reactions and then disengaged to allow the swing span to rotate (see Figures 10.8 and 10.9).

10.2.2 Centre supports

On centre pivot swing bridges, centre wedges are used to provide a live load support when the swing span is in the closed position with traffic on the bridge. Because the centre pivot bearing will allow the bridge to pivot side to side under traffic load or wind, the centre wedges are designed to restrict this

Figure 10.7 Swing bridge: exaggerated loading diagram. The illustration shown here is intended to show only the general location of some of the stabilisation machinery. Illustration is not to scale

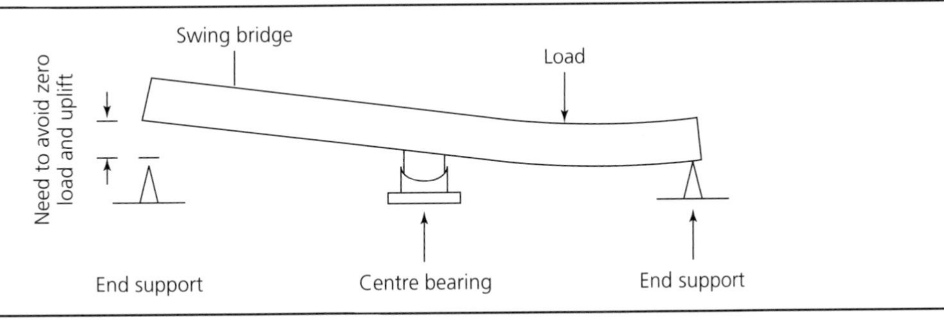

Figure 10.8 End support wedge. The illustrations shown here are only intended to show the general components of an end-support wedge. Illustrations are not to scale

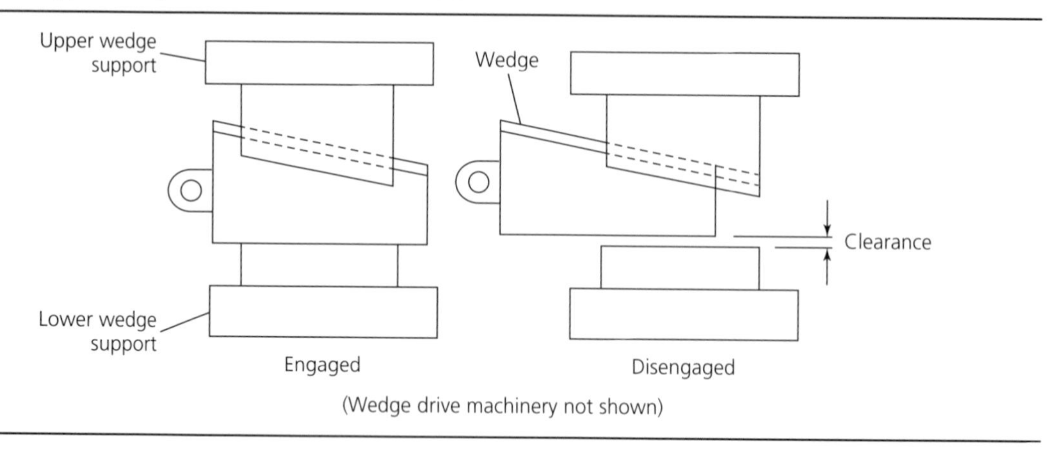

action and provide for live load support (see Figure 10.6). A commonly used assembly is a spring-loaded wedge driven by a crank and connecting rod. The spring will allow the wedge to seek its final position and still provide a minimum and maximum driving load on the wedge when driven to its final position (see Figure 10.10).

10.2.3 Balance wheels

As centre pivot bearing bridges can pivot in a vertical plane at the vertical centre of rotation, balance wheels are utilised to provide support to the swing span when the end and centre support devices are retracted before and as the bridge rotates. The balance wheel assembly consists of a series of wheels mounted to the swing span at a diameter close to the width of the bridge. See Chapter 12 for further details on wheel or roller design. Typically there are a minimum of eight wheel assemblies. The wheels ride on a track that is mounted on the top of the centre pier (see Figure 10.6). When a rack and driving pinion(s) is utilised for the operating machinery, the rack is also, typically, connected to the same track.

Figure 10.9 End-support eccentric shaft and cam. The illustrations shown here are only intended to show the general components of an end-support eccentric shaft and cam. Illustrations are not to scale

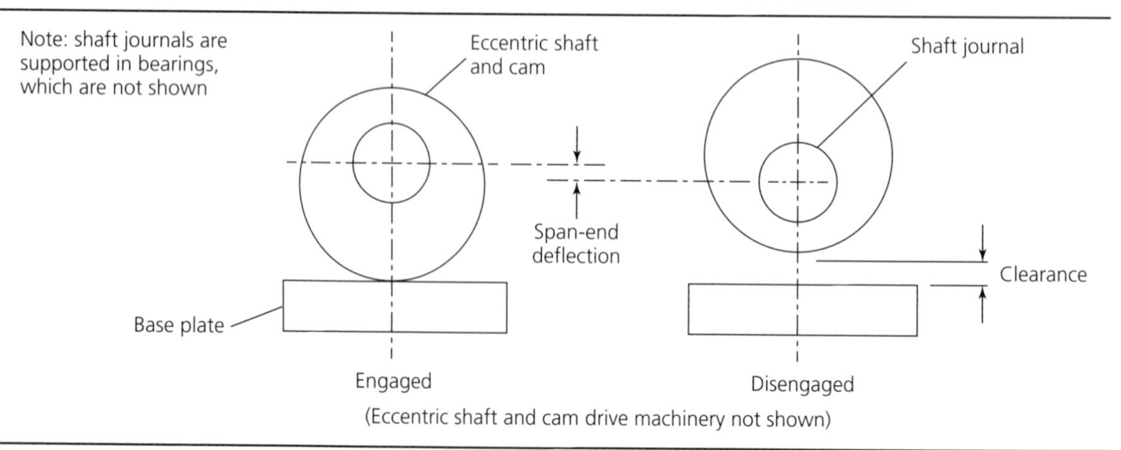

Figure 10.10 Centre support spring-loaded wedge. The illustrations shown here are only intended to show the general components of a spring-loaded wedge. Illustrations are not to scale

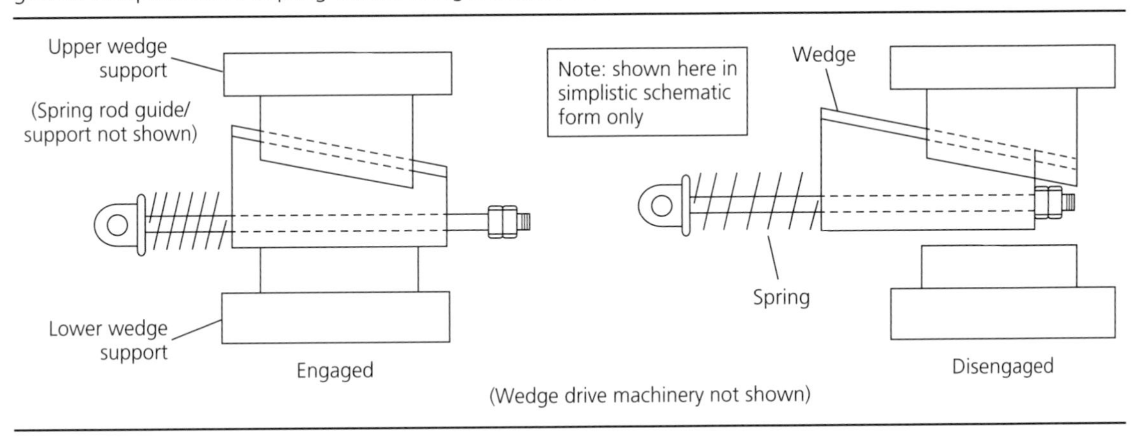

10.2.4 Centring and locking devices

The centring and locking devices are sometimes the same device. In other cases they can be separate. Typically, these devices consist of a sliding bar similar to the span lock bar described for bascule bridges. However, the bar for swing bridges would typically have more taper to allow for a greater range of misalignment. The bar and machinery is commonly mounted on the swing span and engages a socket on the pier. In some cases a vertical bar with a roller is used as a centring and locking device. For this type of system, machinery is used to release the bar to allow the span to open. When the bridge closes, the vertical bar is allowed to slide up and down. The socket has a sloped surface that inclines up to the socket. As the bridge closes, the bar and roller ride up the incline and then drop into the socket at the final centred position of the bridge.

Also for swing bridges with end wedges, the wedges and bases can be used to centre and also to lock the bridge. The base in this kind of bridge has guides on the sides of the wedge that position and lock the wedge (attached to swing span) into place as the wedges are driven.

10.2.5 Rigid stops and buffers

Rigid stops are utilised on swing bridges that do not rotate 360° to bring the span to a set position when closed. Typically the electrical control system will bring the bridge close to its final position. However, rigid stops are used to stop over travel or achieve a more defined position of the span at the end of its travel. They are rigid so that they stop the span without excessive displacement. In some cases wood blocking or a buffer is used. From the closed position after striking the rigid stop, or where stopped by the electrical controls, the span is brought to its final position by a centring device.

10.3. Vertical lift bridges

The stabilisation machinery for vertical lift bridges includes live load bearings, centring devices, span guides, counterweight (CWT) guides, span locks, auxiliary counterweights, balance chains and buffers. The details of these components are described below.

10.3.1 Live load bearings

Live load bearings are used to support the lift span when it is lowered, and the bridge supports live load traffic. Due to the length of the span, one end has fixed bearings and the other end has expansion

Figure 10.11 Vertical lift bridge. The illustration shown here is intended to show only the general location of some of the stabilisation machinery. Illustrations are not to scale

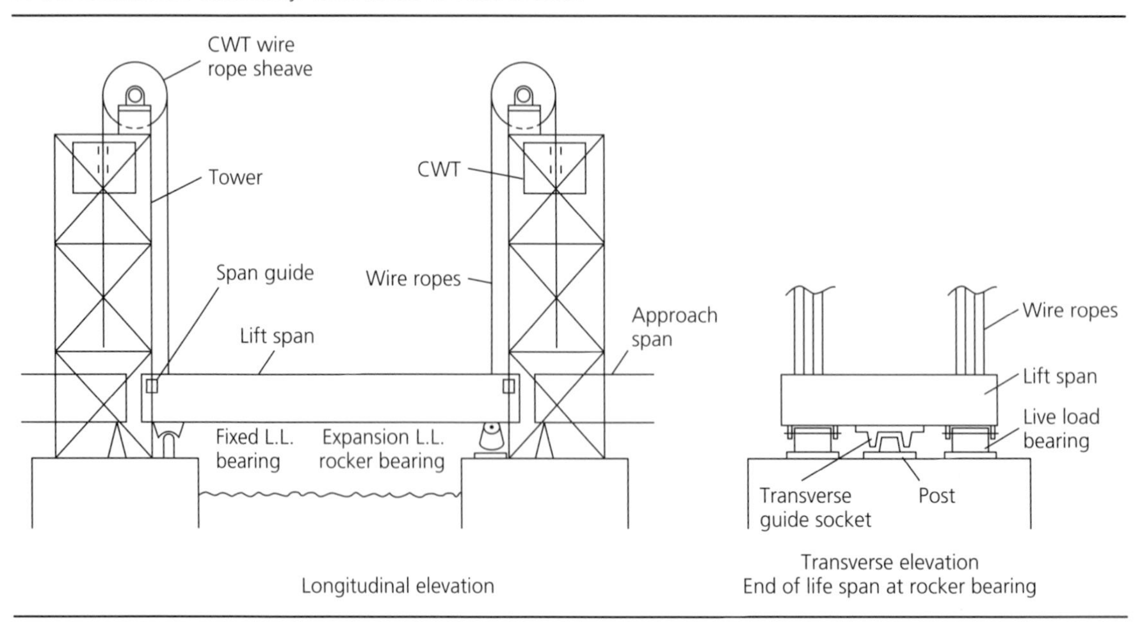

bearings to permit movement due to ambient temperature changes. Typically, the fixed bridge reaction is a combination of a socket and post. As the lift span is lowered, the socket engages a post to fix one end of the lift span longitudinally and provide a support for live load (see Figure 10.11). For expansion either a rocker or a curved surface bears on a flat plate. These types of bearings will provide a support for live load but also allow the end of the lift span on this end to expand and contract.

10.3.2 Centring device

A common centring device for a lift span is a socket on the lift span and a post on the pier. As the bridge seats, the socket engages the post and centres the bridge in the transverse direction (see Figure 10.11). This device is also typically designed to resist wind loads when the lift span is in the seated position.

10.3.3 Span and counterweight guides

The lift span guides keep the bridge in its proper position during vertical motion. This includes guiding the span in the transverse and longitudinal directions. For the longitudinal direction, allowance must be considered for thermal expansion because of the length of the span. This is commonly taken into account by making the corners on one end fixed and permitting expansion at the other end of the span. The fixed side also needs to be coordinated with the fixed live load bearings. The transverse guides are typically at each corner of the span (see Figure 10.12). Because of the relatively short width of the span, temperature change deformation for the transverse direction can be typically accommodated in the clearances between the guides and rails.

The guides are either of the sliding or of the roller type. Sliding guides are much simpler; however, roller guides have less friction under load. The guides bear against rails mounted onto the tower columns. Some lift bridges have upper and lower guides at each corner of the lift span.

Figure 10.12 Vertical lift bridge: plan view span and CWT guides. The illustrations shown here are intended to show only the general location of some of the stabilisation machinery. Illustrations are not to scale

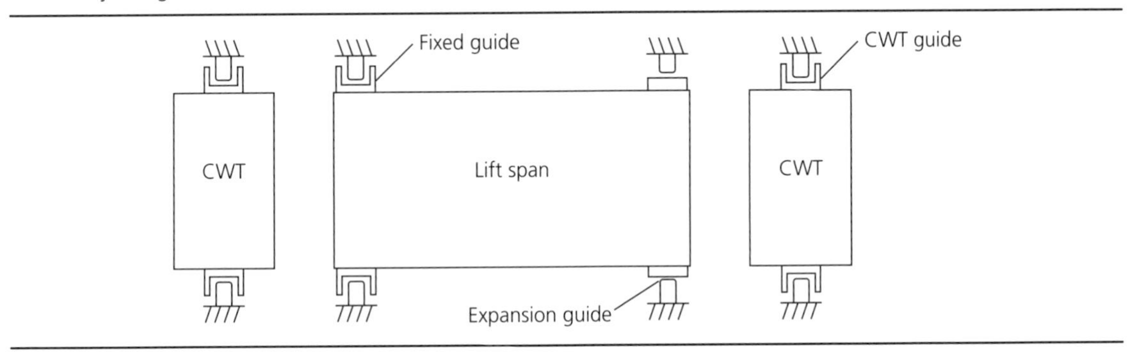

Counterweights are typically guided at the two ends transverse to the bridge. Due to the shorter dimensions relative to the span, each end typically positions the counterweight during operation in both the transverse and longitudinal directions (see Figure 10.12). Typically, there are upper and lower guides. Like span guides, they are either of the sliding or of the roller type.

10.3.4 Span locks

The span locks for a lift bridge are similar to those for a bascule bridge. They consist of a sliding bar mounted in guides that engage a socket. The socket is either on the lift span or on the pier with the bar, guides and the driving machinery mounted on the pier or on the lift span respectively. The bar retracts to let the bridge operate and engages the socket to lock the bridge in a fixed position. The lock bar is typically designed to resist a factored stall torque of the operating machinery in case of accidental operation.

10.3.5 Auxiliary counterweights and balance chains

When a vertical lift operates, as the lift span raises a given amount, the same linear amount of main counterweight cable weight is transferred from the span side to the counterweight side of balance about the vertical centreline of the sheave. Therefore, as the lift span raises the lift span side, with respect to the sheave's vertical centreline, becomes lighter and the counterweight side becomes heavier.

Auxiliary counterweights and balance chains are used to balance the lift span to within an acceptable range of criteria developed specifically for a project to account for the transfer of weight of the main counterweight ropes from the span side to the counterweight side of the sheaves' vertical centreline during operation. This becomes more of a factor for larger bridges, bridges with heavier/larger diameter wire ropes, higher lift heights, and where it is desired not to utilise larger machinery to account for the greater variation of imbalance during operation.

A common auxiliary counterweight system for a bridge has an auxiliary counterweight assembly located at each corner, with four assemblies in total required. For each corner, the auxiliary counterweight assembly would consist of an auxiliary counterweight, a sheave connected to the tower and a pair of wire ropes that connect the lift span to the auxiliary counterweight. The auxiliary counterweight would also be guided for vertical travel on the tower. The angle of the rope, with respect to the fixed position of the sheave, changes as the lift span raises and lowers (see Figure 10.13). This variation of angle provides a reaction force on the lift that changes and can be used to account for the balance changes caused by the transfer of main counterweight wire rope from the span side to the counterweight side of balance.

Figure 10.13 Vertical lift bridge: auxiliary CWT system. The illustrations shown here are intended to show only the general location of some of the stabilisation machinery. Illustrations are not to scale

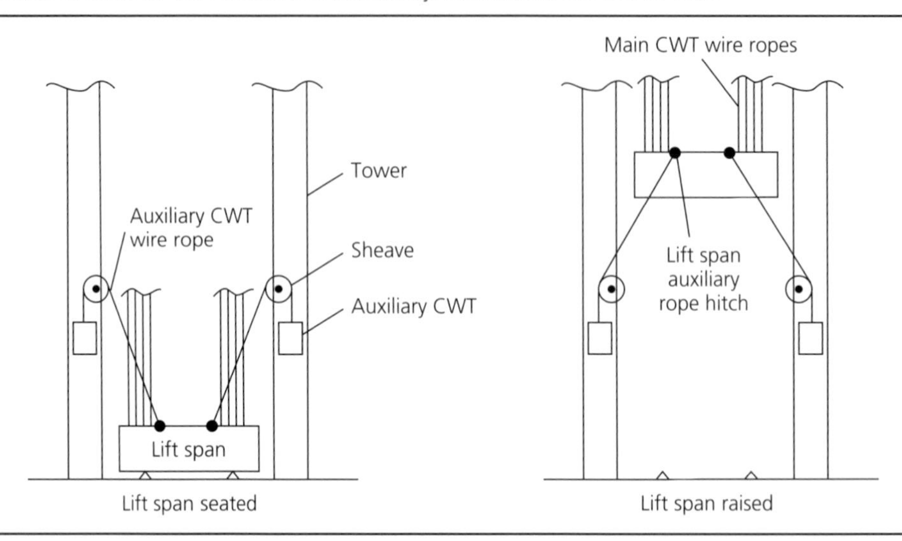

Another type of system that is used to account for main wire rope balance changes during operation is the system of balance chains. The chains are connected to the bottom of the counterweight and to a fixed point on the tower around mid-height. Balance chain assemblies are typically connected in a multiple number of pairs attached to the bottom of each counterweight. When the bridge is seated the chain provides a maximum reaction on the bottom of the counterweight. As the bridge lifts the reaction reduces and is at its minimum when the bridge is fully raised and counterweight lowered (see Figure 10.14). Like the auxiliary counterweight system, the varying chain reaction can be used to account for the transfer of the main counterweight wire rope weight during lift span operation.

Figure 10.14 Vertical lift bridge: CWT balance chain system. The illustrations shown here are intended to show only the general location of some of the stabilisation machinery. Illustrations are not to scale

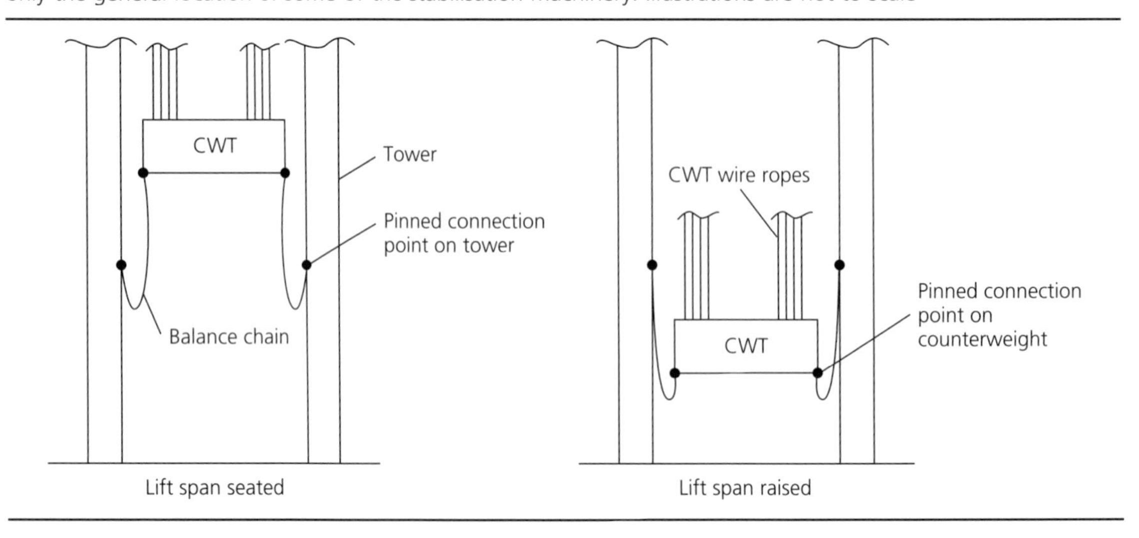

10.3.6 Buffers

In some cases, buffers are used at the ends of lift span travel to decelerate the movable lift span when modern electrical controls are not utilised. These can be either custom air buffers which dissipates energy by compressing air and regulating its release, or by using modern industrial hydraulic shock absorbers. Modern industrial shock absorbers are available as pre-engineered units from multiple manufacturers.

REFERENCES

AASHTO (American Association of State Highway and Transportation Officials) (1998) *Movable Bridge Inspection, Evaluation and Maintenance Manual*. AASHTO, Washington, DC, USA.

AASHTO (2007) *Standard Specifications for Movable Highway Bridges*, 2nd edn. AASHTO, Washington, DC, USA.

AREMA (American Railway Engineering and Maintenance-of-Way Association) (2014) Movable bridges – steel structures. Ch. 15 in *AREMA Manual for Railway Engineering*. AREMA, Lanham, MD, USA.

Movable Bridge Design
ISBN 978-0-7277-5804-0

http://dx.doi.org/10.1680/mbd.58040.231

Chapter 11
Superstructure structural design

Charles Birnstiel

11.0. Introduction

Movable bridge superstructures of many configurations are shown in Chapters 1–4. The structural design of those members does not differ significantly from that for corresponding members of fixed bridges, except for additional loads. For more than 150 years the most commonly used structural materials were wrought iron and steel. During the past 50 years a few major movable swing bridges of reinforced and prestressed concrete have been built. Space restrictions preclude treating the subject of structural design in detail herein. Besides, most readers probably have taken engineering courses on the subjects of determinate and indeterminate structural analysis and also structural steel design, or at least studied statics and strength of materials.

Design procedures for fixed metallic bridges in common use prior to about 1970 have been presented in many books such as Kunz (1915), Waddell (1916), Shedd (1934), Young and Morrison (1949) and Williams and Harris (1957). In North America, railway bridge design follows the historic allowable stress design approach (AREMA, 2014; Unsworth, 2010). Structural analysis in the pre-computer age was treated thoroughly in the classic series *Modern Framed Structures* by Johnson, Bryan and Turneaure which appeared in nine editions between 1893 and 1916 (Johnson *et al.*, 1916). Important later books in English on elementary structural analysis are Parcel and Maney (1936), Shedd and Vawter (1941), Wilbur and Norris (1948) and Michalos and Wilson (1965).

The advent of large-storage capacity digital computers and the load and resistance factor structural design philosophy in the 1970s changed bridge design practices. Design of fixed metallic bridges based on the new approach was treated by Merritt (1972), Salmon and Johnson (1996), Chen and Duan (1999), Parke and Harding (2008) and Wright (2012). Structural analysis resources that discuss post-1970 approaches include Weaver and Johnston (1984), Megson (1996), Ghali and Neville (1997), McGuire *et al.* (2000), Hibbeler (2002) and Shanmugam and Narayanan (2008). Of practical use to the engineer is documentation available from finite element software houses such as LUSAS, 2014) and RISA (2014).

The resources for guidance on structural design and analyses cited above were selected because of their relevance to bridges and because they are, in the author's opinion, written in an elementary straightforward manner. This is not to minimise the contributions to engineering of others, such as Rankine. His works on mechanics for civil and mechanical engineers were of enormous significance. According to Addis, 'Rankine was the first person who gave serious and clear attention to the process of creating a mathematical model of an actual structure' (Addis, 1990). On the Continent, some who contributed to steel bridge engineering in the 20th century were Friedrich Hartmann (Melan, 1951), Hawranek (1958) and Stüssi (1971).

Design of a movable bridge superstructure generally follows the concepts and procedures described for fixed bridges. However, for the movable bridge there are additional actions that need to be considered: inertia forces of the moving mass due to acceleration or deceleration, impact from the moving mass in case of defective electro-mechanical control, and frictional resistance of machinery. In addition, there are some factors relating to the interaction of the structure with the stabilising machinery that require attention. They will be discussed subsequently under separate headings for the major types of bridges.

11.1. Common movable bridge superstructure forms

Movable bridge superstructures built during the past two centuries have been described from the engineering perspective (Waddell, 1916; Hovey, 1926; Hawranek, 1936; Hool and Kinne, 1943; Birnstiel, 2000, 2008a, 2008b; Koglin, 2003). Hool has an excellent treatment of a double-leaf simple trunnion highway bascule in straddle bearings with complete design computations, including those for span balancing. He also presented computations for the bar forces (stresses) in swing bridge trusses in another chapter. Hovey describes swing bridges thoroughly, especially centre bearing swings, and explains the precautions necessary for transferring superstructure loads to the rim bearings of rim-bearing swing bridges. The vertical lift bridge is discussed in detail by Waddell (1916) and Howard (1921). These authors treat mostly American bridges. Hawranek shows an international array of movable bridges and is strong on articulated bascules. From today's perspective, most of the examples in these books could be classed as practical, straightforward, industrial architecture: the kinds of bridges public agencies needed to enable tens of thousands of motor vehicles and hundreds of railway trains to cross a navigable waterway daily and to operate reliably. As an aside, the average daily traffic (ADT) on the Woodrow Wilson Bridge when it was replaced was about 200 000 vehicles. The replacement bridge was designed for 300 000 vehicles per day.

More modern forms, in the sense that they are considered dynamic and 'playful', have been built or proposed and are shown in Bennet (1997) and Wilkinson and Eyre (2001). But these were mostly built for the use of pedestrians and bicyclists. In the 19th and 20th centuries the design of movable bridges was dominated by engineers whose objectives were to build sound structures (really, machines) that were economical with respect to first and operating costs. Aesthetics was a secondary consideration. These priorities led to movable bridge designs, which were straightforward industrial architecture which a segment of the public no longer appreciates. The public agitates for 'better-looking' bridges; indeed, they should preferably be 'iconic'. Hence, architects are playing an increasing role on bridge design teams.

In the 1950s the architectural philosophy of Ludwig Mies van der Rohe that 'less is more' became influential. Svensson's guidelines for bridge aesthetics (Svensson, 2000) are generally consistent with that approach. His important factors are 'clear structural statements, good proportions, order, compatibility with the surroundings including colouring and above all, simplicity'. However, society's concept of 'better looking' changes with time, and now the dictum of Robert Venturi that 'less is a bore' (Venturi, 2002) is accepted by many. Hence, architects are being included on movable bridge design teams – to propose something different and dynamic to the public. As part of this process, movable bridge concepts that engineers once considered obsolete for practical reasons are being reinvented by architects.

The 'obsolete' movable bridge types are architecturally interesting but the reality of providing for many thousand ADT and reliable operation at reasonable cost practically limits the choice of movable bridge type for highway and railway crossings. Most new movable bridges are bascules, swings, or vertical lifts. This chapter will focus on those three types.

11.2. Bridge decks

Decks have historically been major components of the movable bridge mass. The swing bridge at Newcastle upon Tyne, completed in 1875, originally had a deck comprised of wrought-iron buckle plates riveted to the tops of the stringers on which was placed at least a $2\frac{1}{2}$ in. (64 mm) thickness of asphalt plus two thicknesses of $1\frac{1}{4}$ in. net (32 mm) Greenhart timber planking laid diagonally to the roadway and on top of that $4\frac{1}{2}$ in. (114 mm) high Memel (a region in the Baltic States) wood blocks, with the grain vertical, embedded in asphalt. The deck weighed over 100 psf (488 kg/m^2) (Homfray, 1878). Waddell's Halstead Street vertical lift bridge opened to traffic in 1894 and had a deck made of timber plus sand plus wood blocks that weighed about 70 psf (340 kg/m^2). Contrast those deck weights with that of the recent re-decking of the Broadway Bridge in Portland, OR, which weighs about 20 psf (100 kg/m^2) – a marked reduction in weight.

Reducing the self-weight of existing movable bridges is an important objective because it makes available more supporting capacity for live and seismic loads. Until recently, few movable bridges in the USA were designed to resist seismic events. However, seismic design is now required for new movable bridges and many of the older bridges are being retrofitted so as to resist possible earthquakes.

Movable bridge decks are now designed using many materials. The choice depends on the owning agencies' historical experience maintaining bridge decks in their environment, anticipated ADT and average daily truck traffic (ADTT), in-place costs, and the effect on cost of structural framing. Deck materials in use include

- timber
- steel reinforced regular and lightweight cast-in-place concrete
- pre- and post-tensioned regular and lightweight concrete panels
- steel grids
- extruded aluminium panels
- fibre-reinforced polymers (plastics)
- orthotropic steel deck.

It is important to evaluate them for the local conditions as replacement is costly and a great inconvenience to the travelling public. A few comments on the available alternative materials follow.

11.2.1 Timber

Timber was typically used in the USA for roadway decks until about 1940, especially for replacements, and it is still used for railway bridges. Preservative coal-tar creosotes and preservatives in petroleum solutions usually have little effect on the mechanical properties of wood, but waterborne solutions have a more significant effect (USDA, 2007). The American Wood-Preservers' Association (AWPA) Standard has recommendations. Highway bridge decks were usually nominally 2×6 in. (50×150 mm) sticks set on edge and nail laminated which supported wearing surfaces of end-grain wood blocks or asphaltic concrete blocks. A typical railway deck of timber ties on steel stringers is shown in Figure 11.1.

11.2.2 Reinforced concrete

Cast-in-place regular or lightweight concrete slabs reinforced with steel bars were and are used for movable bridge decks. In order to minimise future cracking at the Woodrow Wilson Bridge, stainless steel slab reinforcement was used in the cast-in-place concrete decks of the bascule leaves which act compositely with the structural steel (Arzoumanidis and Bluni, 2007).

Figure 11.1 Typical timber railway bridge deck. (Photograph by Charles Birnstiel)

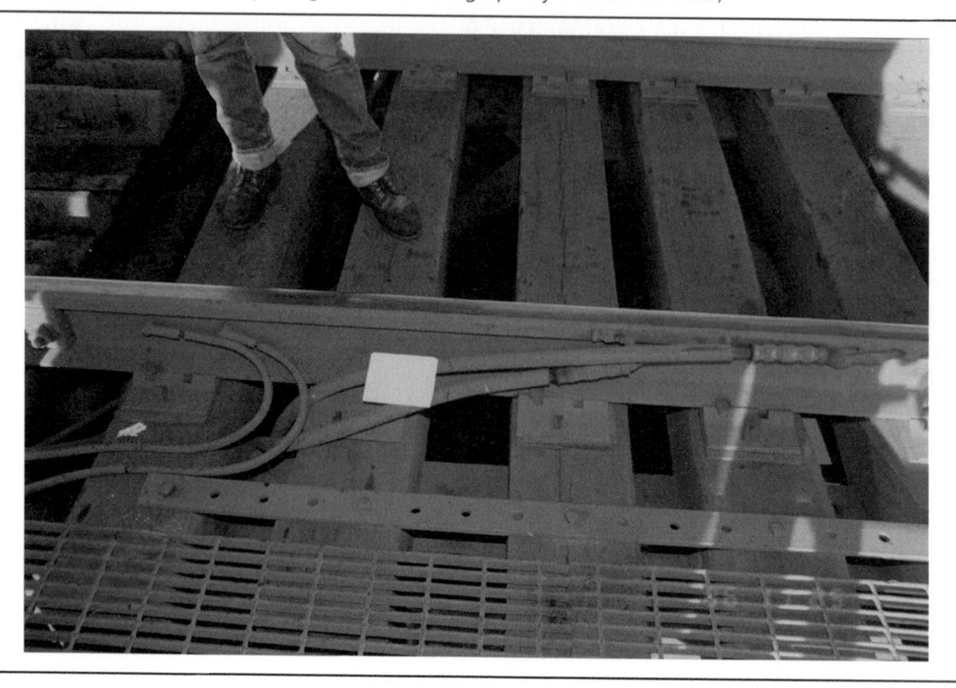

11.2.3 Pre- and post-tensioned concrete panels

Instead of, or in addition to, regular steel reinforcement, concrete slabs are pre- and/or post-tensioned in order to produce and maintain compressive stress near the surface thereby minimizing cracking. It is usual to pretension slabs transversely to the bridge axis and post-tension them longitudinally (Versace and Ramirez, 2004).

11.2.4 Steel grids

The most common types of grids, shown in Figure 11.2, are

(*a*) open grid deck
(*b*) full-depth concrete filled grid deck
(*c*) half-depth concrete filled grid deck with 1.5 in. (38 mm) overfill
(*d*) exodermic deck.

The FHWA/IN study (Versace and Ramirez, 2004) compared various steel grid decks. It has been noted that

- The open grid type is the lightest, but is fatigue prone, has unpleasant ride quality, is noisy and roadway debris falls through the grids. It is suggested that open grids be used only where minimum self-weight is a priority or there is a need to match existing construction. Open grid panels are field welded to supporting steel stringers and fatigue cracking has been observed. The AASHTO (American Association of State Highway and Transportation Officials) Standard has provisions for open grating design, including fatigue limitations. Selecting grids based on load tables is considered insufficient, a fatigue analysis should be made. Figure 11.3 shows the underside of a bascule leaf having an open grid deck in the fully open position.

Figure 11.2 Common forms of steel grid decks. (a) Rectangular open grid; (b) full-depth concrete fill with overfill; (c) half-depth concrete fill with overfill; (d) exodermic deck

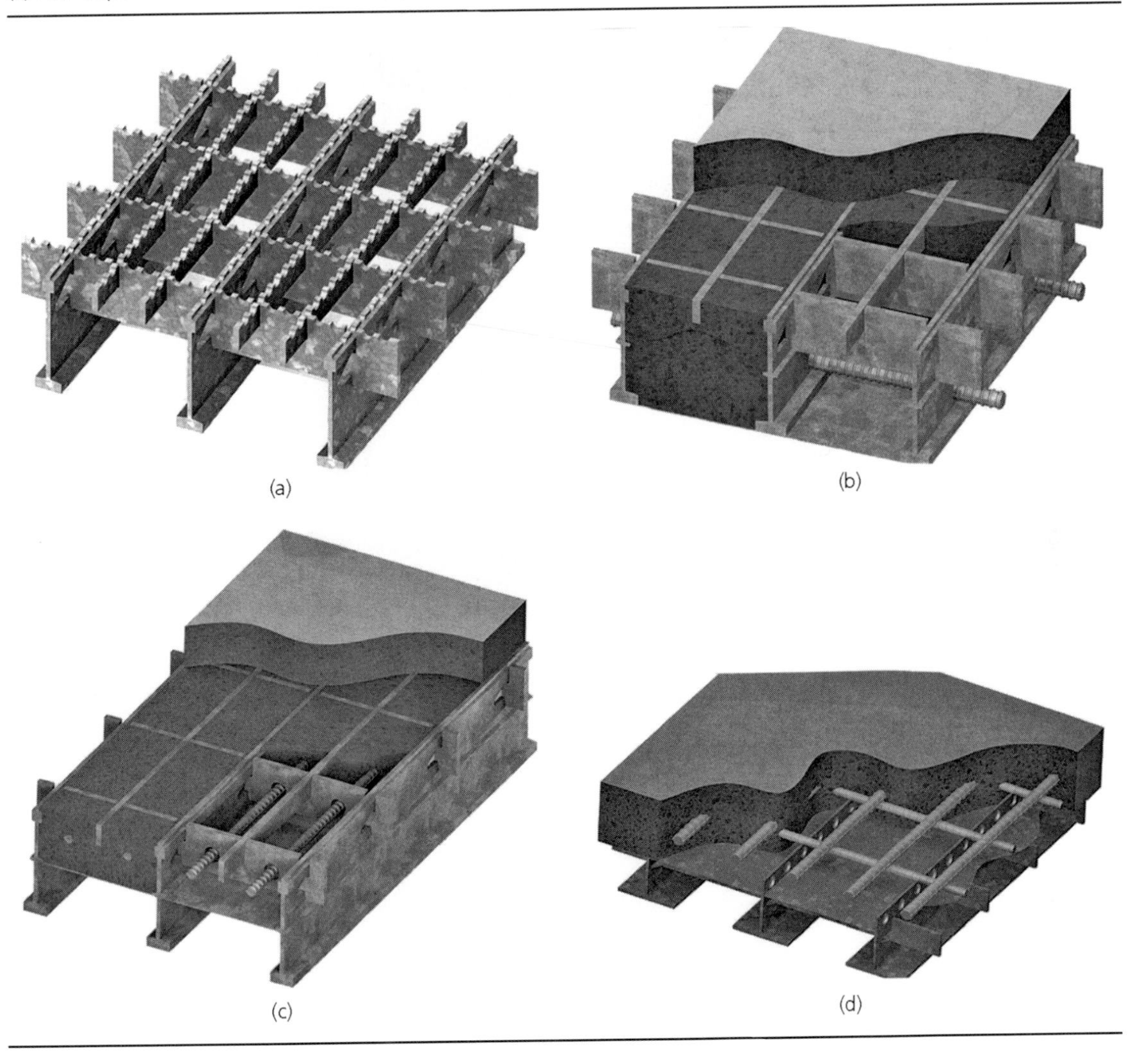

- Exodermic decks and concrete-filled (full or partial depth) grid do not require welding to supports except for shear studs and can span farther than other grid systems. Figure 11.4 depicts the underside of a bascule leaf with an Exodermic deck in the open position. Note the clean appearance.

Some engineers prefer half-filled grating with $1\frac{1}{2}$ in. (38 mm) nominal overfill. One advantage of this system is that the cast-in-place overfill can be screeded to match profile requirements, providing that the overfill thickness is not reduced below 1 in. (25 mm). There have been no reported fatigue problems with any concrete-filled grid deck.

11.2.5 Aluminium deck

Aluminum extrusions have been used for bridge decking in the USA, to date only in small amounts. Figure 11.5 shows an extrusion suggested for movable bridges. They are shop welded together to form

Figure 11.3 Underside of bascule with open grid deck in open position. (Photograph by Charles Birnstiel)

Figure 11.4 Underside of bascule with exodermic deck in open position. (Photograph courtesy of URS Corporation)

Figure 11.5 AlumaDeck extruded shape cross-section

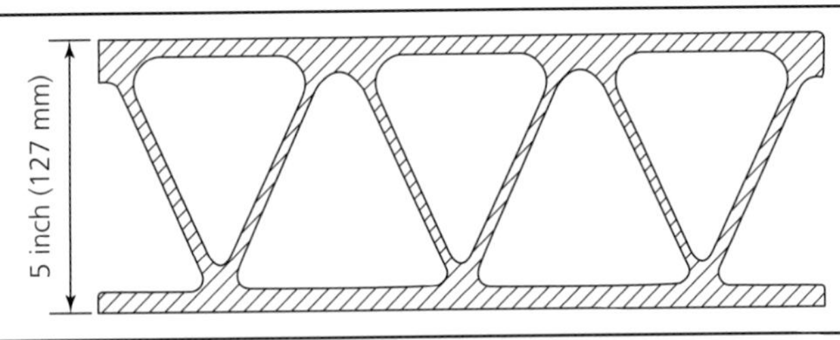

panels, which are then shipped to the site. At the site, mechanical joints connected by blind fasteners hold the panel edges together. A skid resistant epoxy and aggregate wearing surface is bonded to the deck.

AlumaBridge decking was one of the four systems considered as an alternate to open steel grid decking in a study sponsored by the Florida Department of Transportation (URS, 2010). The evaluation findings in the report include the statement on page 11 that 'A friction-stir welded 5-inch deep Aluminum Orthotropic Deck, similar to the 8-inch deep Sapa R-Section Deck, but fabricated specifically to replace 5-inch deep Steel Open Grid Deck, offers a number of advantages and is recommended over the other evaluated alternative deck systems to replace Steel Open Grid Deck on typical Florida bascule bridges.'

11.2.6 Fibre-reinforced polymer composite decking

During the past decade, increased interest in the application of fibre-reinforced polymer composites for construction has been evidenced, including bridge decks. Fibre-reinforced polymer/plastic (FRP) decks are being actively considered to replace deteriorating highway bridge decks of various materials. The lower weight, about 20% of that of concrete decks, and their natural resistance to salt solutions and other negative environmental effects, has spurred interest in the this topic. Copious funding from the US Department of Transportation has assisted the process because of the effort by the US Federal Government to promote military technology transfer of FRP following the end of the Cold War (Telang *et al.*, 2006). State-of-the-art reports have been prepared by Tang and Podolny (1998), Bakis *et al.* (2002) and Keller (2003). Hollaway has discussed the mechanical properties of FRP as well as negative aspects such as thermal effects, behaviour in fire and creep (Hollaway, 2008).

There are many FRP deck systems on the market and the vendors change over the years. Some join the group, others leave. Space is not available for descriptions of all current systems. There are two basic methods of manufacturing FRP decks: sandwich construction and assembling pultruded components. Experience with FRP decks and wearing surfaces in Oregon was reported by Bottenberg (2010).

- Sandwich panel construction comprises strong stiff face sheets which provide the bending resistance to the panel and a foam core that holds the sheets apart. There are edge and intermediate vertical webs of stiff sheets intended to transfer vertical shear. One advantage of foam core sandwich construction is the absence of large voids in which water can accumulate.
- FRP panels are also produced by bonding pultruded cells together using adhesives and adding top and bottom sheets, adhesively bonding them to the prior cell assembly. The pultruded panels are delivered to the field and set on the supporting stringers. Fastening to the

structural steel is either by clips on the underside of the deck or by shear studs welded to the top of the beam flange. In the latter case the access holes in the deck are filled with a suitable cementitious grout.

- Pultruded shapes may be mechanically fastened together. One system utilises a bottom plate with integral upstanding T-shapes that transfer vertical shear. This bottom shape is fastened to the supporting stringer from top-side. After the bottom pultrude has been fastened, the top plate pultrude (upon which the vehicular traffic rolls) is mechanically fastened atop the upstanding T-webs.

Although the materials and shapes of the deck are most important for selecting a deck type, equally important are the financial and technical support considerations when awarding a construction contract. As per URS (2012)

> It is preferred that the lightweight solid deck system be a non-proprietary product, with the opportunity for competitive bidding, and without royalty payments. Where the lightweight solid deck system is proprietary, without similar competitive products, the deck supplier should have sufficient financial support to ensure that the product is available in the future and that there is corresponding technical support.

The matter of the manufacturer-supplier's technical support is of utmost importance. Beside chemical and industrial process support there should be engineering support for preparing the shop fabrication and erection drawings needed to guide the construction contractor in making a proper installation. A thorough report on the field inspection of FRP bridge decks with information on various deck types and composite materials has been presented (Telang *et al.*, 2006). Some suggestions for the bridge designer may be found in O'Connor (2014).

11.2.7 Steel orthotropic decks

Modern orthotropic steel deck (OSD) systems were developed in Germany in the 1950s. A motive was to economically reconstruct many of the bridges destroyed during World War II. In the USA, orthotropic deck bridges first received attention in California because lighter structures may be designed for lower seismic forces. OSDs are well suited for movable bridges because they are lightweight, and beneficially interact with the remainder of the bridge structure. According to Connor (2012) about 50 European bascule bridges have OSDs, as do the recently constructed large French vertical lift bridges described in Chapter 2.

Design of othrotropic steel decks is beyond the scope of this book. The topic is treated in Mangus (1999), Parke and Harding (2008) and Connor (2012). The Mangus presentation is well-coordinated with the AASHTO requirements.

11.3. Bascule bridges

As is evident from Chapters 1–4, many subtypes of bascule bridges have been patented and built. This section will be limited to the simple trunnion bascule and the rolling bascule (Scherzer rolling lift).

11.3.1 Simple trunnion bascule

Simple trunnion bascules rotating about straddle bearings are essentially parallel cantilever girders supporting a floor deck system as depicted in Chapters 1–3. More girders were used per leaf in the early bridges than is the usual practice now (seldom more than two now). Scott Street in Hull (built 1902) has eight girders per leaf connected to a continuous trunnion passing through seven bearings. Each leaf of Tower Bridge has four bascule deck trusses (Birnstiel, 2000). For extremely wide roadways, such as at Woodrow Wilson, the bridge is divided into multiple parallel units each having two girders (Arzoumanidis and Bluni, 2007).

With regard to structural action, the girder of a single-leaf bascule is a cantilever for dead load and a simple span for live load. For double-leaf bascules the girder is a cantilever for dead load. The design criterion for behaviour under live load differs among engineers. Some rely on a functioning mid-span shear lock, in which case the girder is an elastically propped cantilever (Hool and Kinne, 1943). A mid-span lock designed to transfer live load bending moment, as well as shear, was installed on the double-leaf bascule bridge over the Eider River at Friedrichstadt, Germany, *c.* 1919 (Hawranek, 1936) and the type has been used elsewhere in Europe since then. Theoretically, such locks make the bascule girders continuous over four supports for resisting live load; however, deformations in the lock-bars and mountings reduce their effectiveness somewhat. Moment locks are often installed on large double-leaf bascules designed for railway traffic, for example at Galata (Saul and Zellner, 1991), Valencia (Ibarguen *et al.*, 2002) and on the Woodrow Wilson Bridge (Foerster, 2006). Some degree of structural continuity at the mid-span joint reduces the cusp-like depression in the girder deflection curve there due to rolling live load.

A typical steel framing layout for a bascule leaf is visible in Figure 11.3. The opening grid floor spans transversely between longitudinal stringers (usually WF shapes), the stringers are framed into (or bear on) floorbeams and the floorbeams are framed into the bascule girders. The diagonal bracing is necessary to support the floorbeams in bending about their weak axes when the leaf is open. It is also necessary to maintain the leaf geometry. When Exodermic decking is installed (with its concrete fill) it is sometimes considered sufficiently rigid in its plane so that diagonal bracing may be omitted.

A thoroughly worked example of a bascule bridge design may be found in Hool and Kinne (1943). Although the graphical method of analysis therein is now considered obsolete, as are many of the specification requirements in that chapter, the design procedure is valid for any balanced bascule bridge. The sample balance and clearance computations are important.

The application of orthotropic steel deck design to simple trunnion bascule bridges seems appropriate, especially where seismic design requirements are severe. Participation between the deck and the main structure is beneficial. The availability of large finite element computer programs such as LUSAS makes possible the analysis of complete leaves composite with the remainder of the framing.

An important consideration in layout of the leaf framing members are the joints between the movable leaf and the fixed structure (especially the clearances for all positions of the leaf) and a means of deck drainage to avoid dousing the trunnion bearings and other machinery with rainwater.

11.3.2 Rolling bascule

Design considerations for rolling bascules are similar to those for the simple trunnion bascules except that early double-leaf rolling bascules (Scherzer) were constructed so as to act as three-hinged arches in resisting live load located forward of the centre of roll. This action is illustrated in Figure 11.6. The concentrated live load W is equilibrated by the pressure lines passing through the three hinges – a mid-span hinge and the two hinges formed at the bascule piers by the front teeth of the track and the corresponding sockets in the treads.

Figure 11.6 Three-hinged arch action of double-leaf rolling bascule under live load. (Based on Parke and Hewson, 2008)

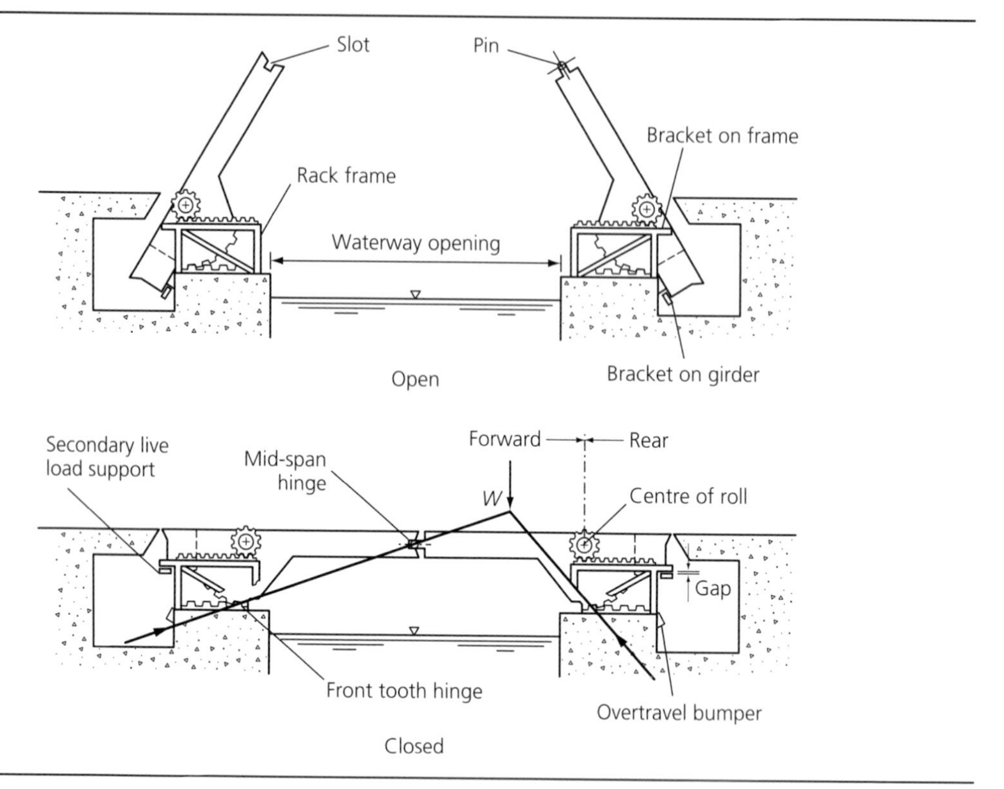

Later double-leaf deck Scherzers designed for arch action under live load were also equipped with supplementary live load reactions either at the rear of the leaf or at the front wall. Normally they are not active; they only become active when a leaf is lowered too far, as can happen when only one leaf is being closed while the other leaf is open. Live load rearward of the centre of roll produces moment, which tends to open the bridge. In the early Scherzer bridges this action is resisted by the machinery, which is 'wound up' when the mid-span hinge is seated and then the brake applied with the motor torqued. This causes fluctuating stresses in the span drive machinery as live load passes over the bridge. Depending on machinery to keep the leaf in equilibrium in the closed position is now a controversial design issue.

For double-leaf rolling bascules where the live loads are not to be equilibrated by arch action (e.g. the Scherzers in Chicago) uplift anchorages are provided at the tails of the leaves (see Chapter 3).

In Europe, a few double-leaf rolling bascule bridges were constructed with three-hinged arch action for live load and for a portion of the dead load. In these bridges, part of the counterweight is lifted by machinery at the rear wall of the bascule pit as the bridge is closed, thereby shifting the centre of gravity of the whole leaf forward towards the channel. This compresses the arch springings and the mid-span hinges prior to the addition of live load. These bridges are very rigid in the closed position. Because the leaves are very span-heavy after the centre of gravity is shifted, live loads on the rear of the leaf do not lift the mid-span hinges. Examples of bascules with this feature are the Langebro and Knippelsbro Bridges in Copenhagen, Denmark.

During rehabilitation design projects, some engineers have altered double-leaf rolling bascule (Scherzer) bridges originally designed for three-hinged arch action from that behaviour to double cantilever action in order to simplify electrical control redesign. However, such an alteration may reduce the rigidity of the closed bridge, unless some remedial measures are included in the rehab to compensate for the loss of the arch action.

The 'modern' rolling bascule (Scherzer) has been a popular type of bridge since the early 1900s. Types of Scherzer framing are shown in Chapter 2. They were originally of riveted steel construction, built up of rolled plates, channels and angles as were the fixed metal bridges at the time. In the past 30 years the welding process has advanced so that the use of welded box sections is common and even field splices are welded in Europe. The rolling bascule in Halsskov, Denmark, that is depicted in Figure 2.19 is an example of the possibilities of clean form that can be developed with welded construction. A closer view of the rolling end of the leaf is depicted in Figure 11.7. Figure 11.8 shows the top of the leaf underside. The orthotropic steel deck is built compositely with the bottom chord of the bascule trusses. The two lines of wider and deeper ribs support the railway track rails embedded in the deck, which are not visible from the underside. The rolling bascule can be an attractive movable bridge type.

Figure 11.7 Rolling end of Halsskov Bridge, Denmark. (Photograph by Charles Birnstiel)

Figure 11.8 Underside of bascule leaf at Halsskov, Denmark. (Photograph by Charles Birnstiel)

11.4. Swing bridges

The main structural members of double-arm swing bridges are trusses or girders. In the open position they are double-armed cantilevers balanced on the pivot pier, and in the closed position they are supported on the pivot and the rest piers. The centre bearing swing truss is supported at three points and the rim-bearing swing truss usually at four points. The elevations of the rest pier supports are adjustable so that the reaction at the end of the draw is always upward for any condition of live load and temperature. Figure 11.9 illustrates designs for rim-bearing bridges and Figure 11.10 shows the centre bearing arrangement. Maintaining upward reactions at A and D is especially important for railway bridges in order to avoid slapping of the draw at the rail joints because of the heavy moving live load.

During the era of major railway expansion, 1890–1910, at least 400 swing bridges were constructed in the USA. Some had draws more than 500 ft (150 m) long and many of these bridges are still operational. Waddell designed a Phoenix rim bearing swing bridge with a 520 ft (158.6 m) draw for a railway across the Missouri River at East Omaha, NB, that was built in 1894. After the Missouri River changed its channel, it was necessary, in 1905, to add another draw of the same length in series with the first. A photograph of the two bridges appears as Figure 11.11. The 1905 bridge was demolished *c.* 2010.

Swing bridge structural forms varied mainly in order to obtain required stiffness and strength and also to reduce the computational effort required for numerical structural analyses of externally and internally indeterminate bridges in the era before mechanical calculators or electronic digital computers. For rim bearing and combined bearing swing bridges, there was the additional consideration that live load should not cause uplift at the girder or truss reactions at the rim bearing drum girder.

Suppose a continuous girder of constant cross-section serves as a three-span bridge between A and D in Figure 11.9(a). According to the conventional beam theory (bending strains only), there would be uplift at support B due to a live load placed on span CD. Because span CD is so much longer than BC the uplift at B could, theoretically, be quite large. Of course, the uplift at B would be counteracted by the downward dead load. However, for open-deck railway bridges, the ratio of live to dead load is often

Figure 11.9 Rim or combined bearing swing bridge. (a) Continuous girder; (b) pratt; (c) warren with verticals; (d) lattice truss; (e) modified warren; (f) Phoenix

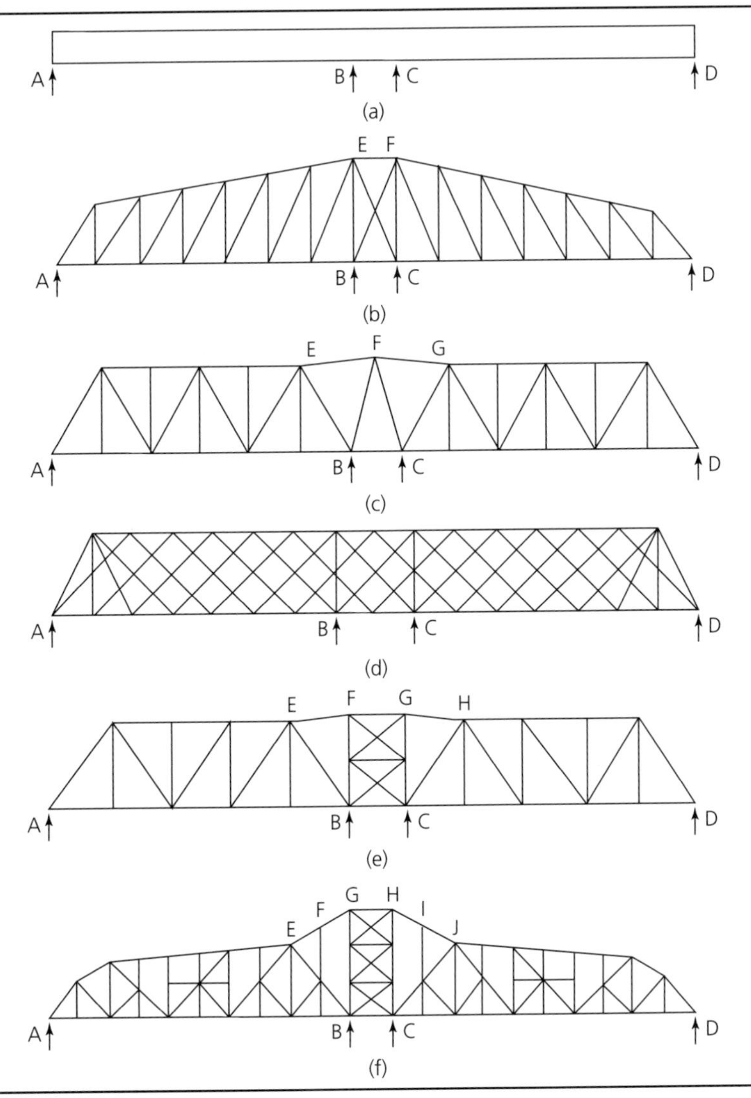

large and, according to conventional beam theory, the net reaction at B could be uplift. Uplift at B would be impractical to accommodate in the rim bearing. However, in the USA, trusses were usually used as spanning members (mainly for economy), instead of solid web girders and their shear deformations are more significant than shear deformations in solid web girders, with the result that the uplift reaction at B for a truss would be smaller than for a solid-web beam. Nevertheless, designers wanted to minimise the possibility of uplift at B because they wanted the drum girder to be as uniformly loaded by the superstructure as possible. So the truss diagonals in panel B–C were omitted, or configured so that they were ineffective in transmitting much vertical shear across the panel B–C. Configuring the truss so that panel B–C could not transmit much shear also had the advantage that it reduced the degree of structural indeterminacy by a factor of 1. This was important at the end of the 19th century because

Figure 11.10 Centre bearing swing bridge. (a) Continuous girder; (b) continuous truss

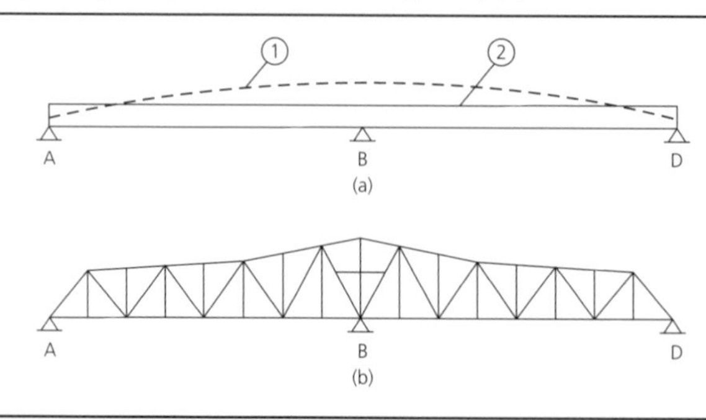

it significantly reduced the complexity of structural analysis. This topic was discussed by Johnson *et al.* (1916) and by Parcel and Maney (1936).

Some popular forms of swing span trusses used for rim-bearing and combined-bearing swing bridges in the USA are shown in the bridge-closed position in Figure 11.9 (b–f). In each diagram, A and D are the ends of trusses at which end lifting machinery is located. B and C are supports for the trusses at the rim bearing. For each truss, the equations of condition and the structural analysis for the case of gravity loading will be described. Because it is stipulated that there are no horizontal loads for these examples, the static requirement that the sum of the horizontal forces equal zero is, by definition, satisfied.

(*a*) Continuous plate girder from A to D. Because horizontal equilibrium is identically satisfied. two equations of equilibrium remain to satisfy the four reactions. Hence, the girder is externally indeterminate to the second degree.

Figure 11.11 Swing spans of the Missouri River Bridge at East Omaha, NE. (Old postcard)

(*b*) Pratt truss having 'weak' diagonals BF and CE. With BF and CE inactive (or virtually so) no vertical shear can be transmitted across the centre panel. This is an equation of condition. The truss is externally indeterminate to the first degree.

(*c*) Warren truss with verticals. Members EF and FG are eyebar tension members. When ends A and D are lifted, these eyebars are compressed and buckle because they have minimal compressive capacity and are thereby effectively removed from the system. No live load shear transfers across panel BC because of the buckled bars in the top chord. The truss sections A–B and C–D are effectively simply supported. At least one engineer designed slotted holes in eyebars EF and FG to ensure that these members were inactive when the whole truss was raised by end machinery at A and D. Truss is externally determinate.

(*d*) Lattice truss. No equation of condition. Externally indeterminate to second degree. Used for British swing spans and also by railways in midwest USA. Uplift of truss at B or C is possible, depending on the live to dead load ratio. Rim bearing cannot develop downward reaction.

(*e*) Modified Warren truss. Bars EF and GH are tension-only eyebars giving two equations of condition. No live load vertical shear transfers across panel BC because of buckled top chord. Truss is externally determinate. (Similar to (*c*) above.)

(*f*) Phoenix draw truss. Members EF, FG, HI, and IJ are eyebars. As ends A and D are lifted, these eyebars may buckle. Hence EG and HJ are equations of condition. No live load shear transfers across panel BC because of buckled bars in the top chord. Externally determinate. A combined bearing Phoenix draw 521 ft long was built over Williamette River in Oregon in 1907. It operated until 1989 under very heavy rail and marine traffic and was replaced by a span drive vertical lift bridge in 1989.

The above examples of engineering approaches for simplifying swing bridge truss models by considering them partially continuous, or simply supported, in the closed position are seldom necessary today. The digital computer now makes it possible to analyse almost any elastic structure using the displacement method in matrix form. Nevertheless, the simple engineering approach is beneficial for understanding and preliminary analysis.

In order to distribute the reactions of the trusses at B and C onto the drum girder of rim bearing, and to the centre bearing in the case of a combined bearing swing bridges, various arrangements of grillages have been utilised. Examples of such framing are shown in Johnson *et al.* (1916), Hovey (1926) and Hool and Kinne (1943).

During the first decade of the 1900s, while an engineer with the Pencoyd Iron Works, which was merged into the American Bridge Company (ABC), and later while he was a vice president of ABC, Charles C Schneider strongly advocated the centre-bearing swing bridge in preference to the rim-bearing or combined-bearing types which had heretofore been the most common swing bridges. In the 'Specification' which accompanied his 1908 paper Schneider recommended the centre bearing support for single-track railway bridges up to 500 ft (150 m) draw length (Schneider, 1908). Recently, some very heavy centre-bearing draws were built for highways, utilising spherical roller thrust bearings as centre bearings – for example, the Third Avenue and Willis Avenue bridges in New York City.

Various styles of swing span trusses built in the late 19th and early 20th centuries are described by Waddell (1916), Hovey (1926), Hawranek (1936) and Hool and Kinne (1943). Structural analysis of some styles of swing span trusses and girders is discussed by Johnson *et al.* (1916) and Parcel and Maney (1936).

11.4.1 Secondary motions of swing bridges

11.4.1.1 Double-arm swing bridges

The structural behaviour of a truss or girder draw differs in the closed and open positions. In the closed position the stress due to dead load in the bottom chords (or flanges of girders) is tension near the middle of the arm. In the open position, the stress is compression. Hence the deflected shapes due to self-weight are different and it is necessary to account for this situation in designing the stabilising machinery, as well as the superstructure.

For symmetrical swing bridges, the ends of the draws are lifted after the draw is rotated to the closed position. Tandem draws are an exception. The distance that the ends are raised is chosen such that the ends will not rise under any combination of unsymmetrical live load and temperature change. The intent is that the end reactions at the rest piers will always be directed upward, to avoid 'slapping' on the end bearings.

11.4.1.2 Tilting swing bridges

Some bobtailed centre bearing swing spans are supported on a spherical pivot bearing (called a throne) and are tilted on opening and back after the span is returned to the closed position. They are sometimes called tilting swing bridges. If the draw is counterweight-heavy, jacks at the tail end lift the tail and thereby force the toe downward onto the forward rest pier. A description of the reconstruction of a counterweight heavy tilting bobtail swing bridge with excellent graphics of the new machinery is due to Perrier (1994). A few bobtailed swing bridges are span-heavy and, at closing, the toe is lifted by jacks, and bearing blocks are inserted to support the toe in a raised position.

11.4.1.3 Raise and turn swing bridges

The secondary motions for the 'lift or lower then turn' swing spans are illustrated in a description of the secondary motions for the Selby Swing Bridge, included in Chapter 2.

11.5. Vertical lift bridges

Description of vertical lift bridges built during the 20th century may be found in Waddell (1916), Howard (1921), Hawranek (1936), Hool and Kinne (1943) and Koglin (2003). Since Waddell's design and patent of the first practical vertical lift bridge, many such bridges have been built.

In Europe a larger portion of vertical lift bridges are constructed with reinforced concrete towers. Independent towers along the sides of the roadway are popular, especially if the ratio of roadway width to lift span length is large. Independent towers usually require independent main counterweights. Some interesting bridges of this type have been built in the UK, Belgium and, recently, in France.

The Katy Railroad Bridge over the Missouri River at Boonville, MO, has been selected to illustrate structural features of a tower drive vertical lift bridge. Figure 2.32 shows the lift span in its raised position, where it has been located most of the time since the last freight train crossed the bridge in 1986. Ownership of the bridge was transferred from the Union Pacific Railroad to the City of Boonville as a tourist attraction in 2013. It is to be restored to operation in order to close a gap in the Katy Trail.

The bridge was opened to railway traffic on 1 February 1932, replacing a swing bridge on a nearby alignment (Anon., 1932, 1932a). The American Bridge Company erection drawings were approved 16 March 1931 by RM Stubbs, the Bridge Engineer, and F Ringer, the Chief Engineer of the Missouri–Kansas–Texas Railroad Co. With a lift-span length between end bearings of 408.33 ft

Figure 11.12 North Tower span superstructure framing of Katy Bridge

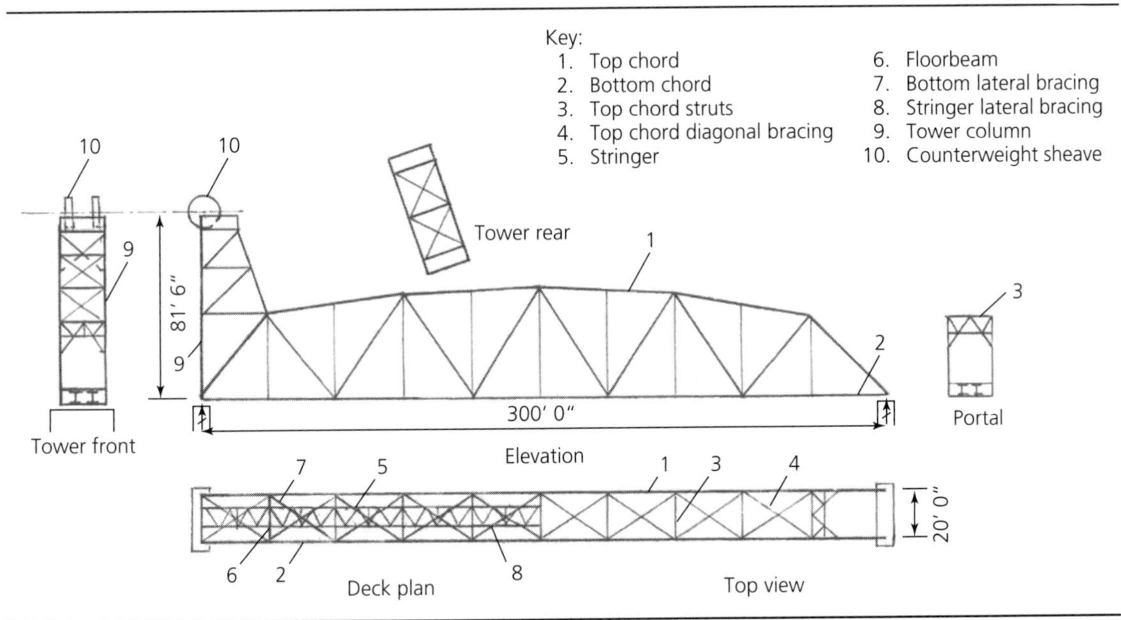

(124.46 m) it was the longest railway lift span in the nation. It was the second tower drive vertical lift, and the first tower lift span drive in the USA that incorporated the electrical synchronous tie levelling feature in order to maintain levelness of the lift span during raising and lowering (see Chapter 2). The bridge was the forerunner of many tower drive vertical lift bridges in the USA.

Most members of the north tower flanking span are shown in Figure 11.12. Much of the structural steel was rolled from silicon steel (a high-strength alloy that had been developed a few decades earlier). The tower span length is 300 ft (91.44 m) between working points. The height of the tower is 81.50 ft (24.84 m) from the working point at the lower chord to the counterweight trunnion axis. The truss depth at mid-span is 48.00 ft (14.63 m). One half of the floor (deck) plan is shown. It shows the stringers, the floorbeams, and the bottom lateral diagonal bracing for the truss and the lateral bracing for the stringers. A half top view of the framing is shown; the top lateral diagonal bracing for the truss and the horizontal struts between the top chord panel points. A lower chord panel point connection is duplicated in Figure 11.13.

Some aspects typical of vertical lift bridge structures that should be considered during design are

- Guidance of the lift span during vertical movement when subjected to wind, also considering temperature change of the structure.
- Tower columns should be plumb when loaded by lift span and counterweight.
- Provision should be made for the span drive machinery to remain aligned after the counterweight sheaves are loaded. May need to erect machinery pads high temporarily (until sheave is loaded).
- Vertical motion of counterweights should be guided.
- Provide detail for hanging counterweight directly from tower in order to facilitate the changing of counterweight ropes.

Figure 11.13 Typical connection at lower chord panel point of Katy Bridge. (Photograph by Charles Birnstiel)

A description of a modern highway tower drive vertical lift bridge of medium span is presented in Capers *et al.* (2005).

11.6. Movable bridge balancing

Most movable bridges in North America are nearly balanced, in order to minimise the power required to operate them and for safety in case of span drive failure. The greater the imbalance the greater the potential impact if a span drive malfunctions.

Much has to do with the history of transportation development in the USA and the competitive economic environment produced by the movable bridge patentees. The early large movable bridges were necessitated by the railways and they operated as private businesses (although they accepted enormous financial grants from all levels of government) and they sought the least expensive movable bridges. Hence, the bridge manufacturing companies competed to produce the cheapest bridges and because mechanical operating machinery was a significant portion of the cost they wanted to minimise that cost by minimising the required operating power. The bridge owners also wanted to minimise operating power and this could only be accomplished by balancing the movable span weight by means of counterweights. It may be difficult to appreciate now the importance of power costs then. Now the average annual number of bridge openings may number a few thousand. In the first part of the 20th century, annual openings of a bridge exceeding 25 000 were not unusual in American ports. Bridge imbalance is, of course, a more important issue for certain movable bridge types such as bascules and vertical lift bridges, where a portion of the bridge moves vertically, than for swing bridges that rotate in a horizontal plane.

11.6.1 Bascule bridges

For balance computations, it is convenient to consider the moving leaf as two individual items

(*a*) all the moving mass except the counterweight
(*b*) the counterweight.

For a bridge with a counterweight rigidity attached to the moving leaf (simple trunnion bascule or Scherzer-type rolling bascule) the centre of gravity of the leaf (A), the trunnion centre (B) and the centre of gravity of the counterweight (C) must be collinear in order to obtain perfect balance neglecting trunnion friction (see Figure 2.1). Furthermore, the distances A–B and B–C must be inversely proportional to the ratio of leaf weight at A and the counterweight at C.

For Strauss bascules with an underdeck counterweight (Figure 2.12) the centres of gravity of the leaf, the trunnion, and the counterweight trunnion must be collinear for perfect balance. The situation is similar for the Strauss bascule with the vertical overhead counterweight (Figure 2.10) because the Strauss parallelogram forces the counterweight to remain plumb while moving horizontally during opening.

It can be shown that for a Strauss heel trunnion bascule (Figure 2.13) balance requires that the line between the leaf centre of gravity and the heel trunnion be parallel to the line between the counterweight trunnion and the centre of gravity of the counterweight. However, the effect of the moving operating strut must be considered when determining system imbalance.

11.6.1.1 Imbalance of bascule bridges

Opinions differ about the desirable amount of imbalance for movable leaves of bascule bridges. There is the European view that the imbalance should be substantial so that the forward reaction (at the toe) of a single-leaf bascule due to dead load imbalance will be directed upward by a significant amount because the dynamic behaviour of such a span under roadway traffic is superior to that for the opposite situation when the leaf is nearly balanced and the leaf 'bounces'. (The leaf should press down on the rest pier.) In fact, some short-span trunnion bascules have been built without any counterweight (Anderson and Jensen, 2011). Of course the span drive machinery for a bridge without a counterweight must be more substantial than for a nearly balanced bridge. Imbalanced bascules are usually operated by hydraulic cylinders because they provide the most economical power to capacity ratio of any machinery and hence are the logical choice for counterweight-less systems because of the need to provide greater operating forces.

In North American practice, for many years, the policy was to closely balance the bascule leaf so as to minimise the power to operate the leaf. Short- and medium-span single-leaf bascules with two bascule girders were designed for girder dead load toe reactions of as little as 1 kip (454 kg) directed upwards. In structural analysis terminology, the force exerted by the support onto the structure is called a 'reaction'. When the leaf presses downwards onto the support, the reaction is upwards. But the movements of these spans in the closed position under vehicular traffic has resulted in high maintenance costs for the stabilising machinery. A current view is that the imbalance reaction at the toe of the span should be at least 2 kips (908 kg) directed upward per girder – that is, the leaf should really be in the span-heavy condition when closed.

Opinions also differ regarding the balance condition in the leaf open position. Some owners want the leaf to be span-heavy for the full opening angle. Others maintain that the leaf shall be counterweight-heavy (tail-heavy) in the fully open position so that it will not start to rotate closed without the application of power.

It may be convenient for describing imbalance in the design and the field measurement stages to plot the centre of gravity (g) of the leaf plus the attached counterweight (W) on a diagram as shown in Figure 11.14. The diagram is directly valid for simple trunnion and rolling bascules. It may be adapted to Strauss bascules as shown in Hool and Kinne (1943). The leaf opening angle is denoted by θ, the

Figure 11.14 Leaf imbalance relationship

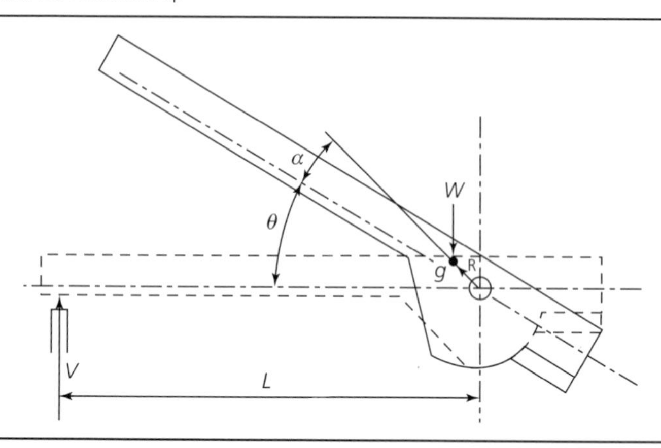

maximum opening angle is θ_m, the angle between the horizontal axis of the leaf and the centre of gravity, g, of the moving mass is α, and R is the radius from the centre of rotation to g.

For example, consider simple trunnion tower bascules with their centre of gravity of the total moving mass (including counterweight), W, located at, g, and with R as a constant greater than zero, and with a maximum opening angle $\theta_m = 75°$. What is the balance condition at the closed and fully open positions for $\alpha = 20$, 100, 200 and 280°? From Figure 11.14 there are

$\alpha°$	Closed $\theta = 0°$	Fully open $\theta_m = 75°$
20°	span-heavy	counterweight-heavy
100°	counterweight-heavy	counterweight-heavy
200°	counterweight-heavy	span-heavy
280°	span-heavy	span-heavy

From Figure 11.14 it is obvious that the moment necessary to move the leaf solely to overcome imbalance, M_b, is given by

$$M_b = WR \text{ cosine } (\theta + \alpha)$$

A positive value of M_b indicates moment necessary to open the leaf: a span-heavy condition. That means that the toe of the leaf presses downwards on the pier due to the dead load imbalance.

11.6.1.2 Balance computations

In order to size the counterweight it is first necessary to compute the weight of the moving structure and locate its centre of gravity. An example of such computations is shown in Hool and Kinne (1943). These days they are typically done in spreadsheet format. The work needs to be done shortly before the design is finalised so as to include all structural steel members, welds, bolts, decking, railings, and coatings such as paint and hot-dipped galvanising – a tedious, but necessary process. Movable bridge balance and counterweight design considerations are discussed in Giernacky and Tosolt (2010).

Most North American movable bridges are still being designed with concrete counterweights. Void pockets need to be provided in the counterweights to permit 'final' and future weight adjustment to account for changes in weight of the movable structure in case of future application of surface treatment to the roadway deck, painting, etc., for highway bridges and loss of deck weight of railroad bridges due to evaporation of preservative in timber ties or loss of material due to fungal action. The location of the void pockets should be considered in the design so as to permit both high and low adjustment of the centre of gravity. Adjustments are made by adding or removing steel plates or cast iron or precast concrete blocks. Guidelines for counterweight void design may be found in AASHTO (2007).

11.6.1.3 Clearance verification

Prior to finalising the design, especially of bascule bridges, span operational clearances should be confirmed numerically. Besides clearances at floor and sidewalk breaks and the counterweight, the clearances at ends of roadway barriers and railings need to be investigated. The topic is discussed in detail in Hool and Kinne (1943).

REFERENCES

AASHTO (American Association of State Highway and Transportation Officials) (2007) *Standard Specifications for Movable Highway Bridges*, 2nd edn. AASHTO, Washington, DC, USA.

Addis W (1990) *Structural Engineering: The Nature of Theory and Design*. Ellis Horwood, London.

Anderson IB and Jensen MB (2011) Bridge crossing of the River Nidelva, Trondheim, Norway. *Bridge Engineering ICE Proceedings* **164(BE3)**: 157–165, Sept.

Anon. 1 (1932) M-K-T constructs longest lift span. *Railway Age* **92(18)**: 729–733, 18 Apr.

Anon. 2 (1932) Longest railway lift bridge is built by M K T Railroad. *Engineering News-Record*, pp. 755–756, 26 May.

AREMA (American Railway Engineering and Maintenance-of-Way Association) (2014) Movable bridges – steel structures. Chapter 15 in *AREMA Manual for Railway Engineering*. AREMA, Lanham, MD, USA.

Arzoumanidis S G and Bluni SA (2007) Replacement of the Woodrow Wilson Memorial Bridge bascule span. *Symposium Report CD-ROM, IABSE Report* **93**: Weimar, Germany, 19–21 Sept.

Bakis CE Bank L, Brown V *et al.* (2002) Fiber-reinforced polymer composites for construction – state-of-the-art review. *Journal of Composites for Construction*, ASCE, May, http://dx.doi.org/10.1061/(ASCE)1090-0268(2002)6:2(73).

Bennet D (1997) *The Architecture of Bridge Design*. Thomas Telford, London.

Birnstiel C (2000) Movable bridges. Chapter 12 in *The Manual of Bridge Engineering* (Ryall MJ, Parke GAR and Harding JE (eds)). Thomas Telford, London.

Birnstiel C (2008a) The Mississippi river railway crossing at Clinton, Iowa. In *Historic Bridges: Evaluation, Preservation, and Management*, CRC Taylor & Francis, Boca Raton, FL, USA, http://dx.doi.org/10.1201/9781420079968.pt1.

Birnstiel C (2008b) Movable bridges. In *ICE Manual of Bridge Engineering* (Parke G and Hewson N (eds)), 2nd edn. ICE, London, http://dx.doi.org/10.1680/mobe.34525.

Bottenberg RD (2010) Fiber-reinforced polymer deck for movable bridges. *Structural Engineering International* **20(4)**: 418–422, http://dx.doi.org/10.2749/101686610793557816.

Capers Jr HA, Schetelich GE, Coates AC and Hann CD (2005) The Route 7 Bridge over the Passaic River. *Structure*, Oct.

Chen WF and Duan L (1999) *Bridge Engineering Handbook*. CRC Press, Boca Raton, FL, USA.

Connor R *et al.* (2012) *Manual for Design, Construction, and Maintenance of Orthotropic Steel Deck Bridges*, Publication FHW-IF-12-027. Federal Highway Administration, US Department of Transportation, Washington, DC, USA.

Foerster G (2006) The support and stabilization machinery for the Woodrow Wilson Bridge bascule spans. *Proceedings of the 11th Biennial Symposium of Heavy Movable Structures*, *Orlando*, FL, USA, 6–9 November.

Ghali A and Neville AM (1997) *Structural Analysis*. E & FN Spon, London.

Gierncky RG and Tosolt RJ (2010) Movable bridge balance and counterweight design considerations for designers and constructors. *13th Biennial Symposium of Heavy Movable Structures, Inc.*, 25–28 October, Orlando, FL, USA.

Hawranek A (1936) *Begwegliche Brücken* (in German). Julius Springer, Berlin, Germany.

Hawranek A (1958) *Theorie und Berechnung der Stahlbrücken* (in German). Springer, Berlin, Germany.

Hibbeler RC (2002) *Structural Analysis*, 5th edn. Prentice Hall, Upper Saddle River, NJ, USA.

Hollaway LC (2008) Advanced fibre polymer composite structural systems used in bridge engineering. In *ICE Manual of Bridge Engineering* (Parke G and Hewson N (eds)), 2nd edn. Thomas Telford, London, http://dx.doi.org/10.1680/mobe.34525.

Homfray S G (1878) *The Tyne Bridge*, presented before the Tyne Engineering Society on February 14, 1878, South Shields, Tyne and Wear: Port of Tyne Authority, UK.

Hool GA and Kinne WS (1943) *Movable and Long-Span Bridges*, 2nd edn. McGraw-Hill, New York, NY, USA.

Hovey OE (1926) *Movable Bridges*, vols I and II. Wiley, New York, NY, USA.

Howard EE (1921) Vertical lift bridges. *Transactions*, ASCE **84**: Paper No. 1478.

Ibarguen PH, Arias AO and Alfonso FT (2002) Construcción del Puente Mouil en el Puerto de Valencia (in Spanish). *Proceedings II Congreso de Ache de Puente y Estructuras*. Madrid, Spain.

Johnson JB, Bryan CW and Turneaure FE (1916) *Modern Framed Structures*, 9th edn. Wiley, New York, NY, USA.

Keller T (2003) Use of fiber reinforced polymers in bridge construction. *Structural Engineering Documents #7*. IABSE-AZPC-IVBH, Zurich, Switzerland.

Koglin TL (2003) *Movable Bridge Engineering*. Wiley, Hoboken, NJ, USA, http://dx.doi.org/10.1002/9780470172902.

Kunz FC (1915) *Design of Steel Bridges: Theory and Practice for the Use of Civil Engineers and Students*. McGraw-Hill, New York, NY, USA.

LUSAS (London University Stress Analysis System) (2014) *Application and Reference Manuals*. LUSAS, Kingston upon Thames, UK.

Mangus AR (1999) Innovative movable bridges with welded orthotropic steel decks. *Proceedings of the 16th Annual International Bridge Conference*, 4–6 June, Pittsburgh, PA, USA.

McGuire W, Gallagher R and Ziemian R (2000) *Matrix Structural Analysis*. Wiley, New York, NY, USA, http://dx.doi.org/10.1115/1.3256379.

Megson THG (1996) *Structural and Stress Analysis*. Arnold, London.

Melan E (1951) *Der Brücken Bau* (in German), vol. 3: *Stahlbrücken* (Hartmann F, pub. posthumously). Ernst Melan, Franz Deuticke, Vienna.

Merritt FS (1972) *Structural Steel Designers' Handbook*. McGraw-Hill, New York, NY, USA.

Michalos J and Wilson EN (1965) *Structural Mechanics and Analysis*. Macmillan, New York, NY, USA.

O'Connor J (2014) *Current Practices in FRP Composite Technology*. See http://ww.fhwa.dot.gov/bridge/frp/frppract.cfm for further details (accessed 21/12/2014).

Parcel JI and Maney GA (1936) *Statically Indeterminate Stresses*, 2nd edn. Wiley, New York, NY, USA.

Parke GAR and Harding JE (2008) Design of steel bridges. In *ICE Manual of Bridge Engineering*, 2nd edn, pp. 235–281. Thomas Telford, London, http://dx.doi.org/10.1680/mobe.34525.

Parke G and Hewson N (eds) (2008) *ICE Manual of Bridge Engineering*, 2nd edn. Institution of Civil Engineers, London, http://dx.doi.org/10.1680/mobe.34525.

Perrier M (1994) La reconstruction due pont mobile sure le pertuis de Southampton au Havre (in

French). *Bulletin Ponts Metalliques* **17**. OTUA Office Technique pour l'utilisation de l'Acier, Paris, France.
RISA (2014) *RISA-3D General Reference Manual and User Guide*. RISA Technologies, Foothill Ranch, CA, USA.
Salmon CG and Johnson JE (1996) *Steel Structures: Design and Behavior*, 4th edn. Harper Collins, New York, NY, USA.
Saul R and Zellner W (1991) The Galata bascule. *Report of the IABSE Symposium Leningrad*, IABSE, Zurich, Switzerland, **64**: 557–562.
Schneider CC (1908) Movable bridges. *Transactions, ASCE*, LX, Paper No. 1071.
Shanmugam NE and Narayanan R (2008) Structural analysis. In *ICE Manual of Bridge Engineering* (Parke G and Hewson N (eds)), 2nd edn. Thomas Telford, London.
Shedd TC (1934) *Structural Design in Steel*. Wiley, New York, NY, USA.
Shedd TC and Vawter J (1941) *Theory of Simple Structures*. Wiley, New York, NY, USA.
Stüssi F (1971) *Vorlesungen über Baustatik*, in two vols (in German). Birkhäuser Verlag, Basel.
Svensson H (2000) Bridge aesthetics – guildelines for the new millennium. *Proceedings of the 5th International Engineering Conference of the Transportation Research Board, Tampa, FL*, 3–5 April, http://dx.doi.org/10.3141/1696-01.
Tang B and Podolny Jr W (1998) A successful beginning for FRP composite materials in bridge applications. *FHWA Proceedings, International Conference on Corrosion and Rehabilitation of Reinforced Concrete Structures*, 7–11 Dec., Orlando, FL, USA.
Telang NM, Dumlao C, Mehrabi AB, Ciolko AT and Gutierrez J (2006) Field inspection of in-service FRP bridge decks. *NCHRP Report 564*, Transportation Research Board, Washington, DC, USA.
URS (2012) *Deck Alternative Screening Report, Bascule Bridge Lightweight Solid Deck Retrofit Project, FPID 419497–1-B2–01*. Prepared for Florida Department of Transportation, Structures Design Office, URS Corporation, Tampa, FL, USA.
Unsworth JF (2010) *Design of Modern Steel Railway Bridges*. CRC Press, Boca Raton, FL, USA, http://dx.doi.org/10.1201/9781420082180.
USDA (US Department of Agriculture) (2007) *The Encyclopedia of Wood*. US Dept of Agriculture, Forest Products Laboratory, Madison WI, USA. Skyhorse Publishing, New York, NY, USA.
Venturi R (2002) *Complexity and Contradiction in Architecture*. The Museum of Modern Art, New York, NY, USA.
Versace JD and Ramirez JA (2004) *Implementation of Full-Width Deck Panels: A Synthesis Study*, Publication FHWA/IN/JTRP-2003/24. Joint Transportation Research Program, Indiana Department of Transportation and Purdue University, West Lafayette, IN, USA.
Waddell JAL (1916) *Bridge Engineering*, vols I and II. Wiley, New York, NY, USA.
Weaver W Jr and Johnston PR (1984) *Finite Elements for Structural Analysis*. Prentice Hall, Englewood Cliffs, NJ, USA, http://dx.doi.org/10.1115/1.3167704.
Wilbur JB and Norris CH (1948) *Elementary Structural Analysis*. McGraw-Hill, New York, NY, USA.
Wilkinson C and Eyre J (2001) *Bridging Art and Science*. Booth-Clibbon, London.
Williams CD and Harris EC (1957) *Structural Design in Metals*, 2nd edn. Ronald Press Company, New York, NY, USA.
Wright WJ (2012) *Steel Bridge Design Handbook*. Publication FHWA-IF-12–052, in 14 vols. US Department of Transportation, Federal Highway Administration, US Department of Transportation, Washington, DC, USA.
Young CR and Morrison CF (1949) *Structural Problems in Steel and Timber*, 3rd edn, Wiley, New York, NY, USA.

Movable Bridge Design
ISBN 978-0-7277-5804-0

http://dx.doi.org/10.1680/mbd.58040.255

Chapter 12
Mechanical design

George A Foerster

12.0. Introduction

Machinery design is a broad topic and its entirety goes beyond the scope of this book. This chapter will focus on some matters specific to movable bridges. The information presented herein is not intended to be all-encompassing. In addition to basic mechanical engineering principles only portions of AASHTO (American Association of State Highway and Transportation Officials, 2007) and AREMA (American Railway Engineering and Maintenance-of-Way Association, 2014) are discussed in this chapter. In some cases details, equations and allowables are also only given from one of these references. This was not done to exclude details, specifications or recommended practice requirements, but out of the necessity to cover a broader scope of machinery design in one chapter. For a specific movable bridge project the applicable specifications and practices would need to be thoroughly reviewed in their entirety by the engineer. An experienced engineer is required to develop and incorporate all aspects and details of specific specifications or recommended practice requirements into an actual final design.

More general, and in many cases more detailed, design procedures for mechanical components may be found in machinery design textbooks such as Childs (2014), Spotts *et al.* (2004), Shigley and Mischke (2001), Ugural (2004) and Deutschman *et al.* (1975). For American nomenclature and machinery design and manufacturing practice see Oberg *et al.* (2012). References for machinery component designs of older bridges are Hool and Kinne (1943), Hovey (1926), Waddell (1916) and Rankine (1877). Recent books specific to movable bridge design include *ICE Manual of Bridge Engineering* (Parke and Hewson 2008) and *Movable Bridge Engineering* (Koglin, 2003).

12.0.1 Machinery design and the process

The design process involves conceptually assembling parts and components into a functioning system to perform a task efficiently, with benefit to mankind and to our environment. This requires putting ideas, materials, knowledge of current manufacturing capabilities and available monetary resources together to achieve an acceptable functioning design. For a successful movable bridge, the designer should also consider past and present machine design practices as well as field experience from past designs still in service. (Note that figures shown in this chapter are typically only in a basic general form.)

When designing movable bridge machinery the following key points and questions should be considered.

- Safety – in design, safety is one of the most important factors. This includes protection of both life and property. Considerations include: who will operate the machinery and where; and who will be present around the machinery or structure during operation.
- Design life – how long does the machinery system need to function before replacement?

- Ease of operation – is the mechanism easy to use? Does it require multiple controls, multiple operator or actions to function?
- Material selection – what materials are available? Can they be used in an efficient manner in manufacture?
- Fabrication and machining – can the parts be fabricated and machined to meet the requirements of the design?
- Installation – can the parts be assembled and installed easily?
- Lubrication – what type of lubricants are required?
- Ease of maintenance – what type of maintenance is required and how frequently?
- Future corrosion – as machinery systems for movable bridges age, deterioration due to corrosion is common. This is especially the case for components exposed directly to the elements, or, possibly, to flooding due to high water levels, or to salt from roadway maintenance. Because of such concerns special seals are typically used for some machinery to protect internal components and minimum thickness of machinery elements are chosen so as to tolerate some loss of cross-section over time.
- Future additional loads – for some new movable bridge projects, additional future loading is considered at the design stage. An example of this is a movable highway bridge where the potential for light rail to be added to the structure is known during the design. For such a case the possible future addition should be considered in the design to avoid the need for future machinery redesign and any complicated resizing of machinery components when the future loading is added.
- Cost – can the designed machinery be fabricated, machined, assembled and installed at a cost acceptable to the owner or client?

12.0.2 Machinery design basics

Machinery for movable bridges should be of a substantial and proven design. The arrangement of machinery components should be such that it allows routine regular maintenance as well as scheduled condition inspections. Maintenance and lubrication should be standardised as much as practical.

Machinery design should also consider and include a means for installing and positioning components for an acceptable final alignment. Shims (shim packs) are typically used under the mounting surfaces which are bolted to the structure. Shims are a multiple of varying thicknesses of metal plates. The varying plate thicknesses in a shim pack are selected to allow components to be adjusted to a fine position to achieve an acceptable alignment, fit or location by the addition or subtraction of shims. AREMA (2014) specifies full-length shims for assembly and alignment of components.

Safety guards and covers should be installed on machinery in order to protect the public, bridge operators and maintainers as applicable. Requirements for safety shielding are typically prescribed in design specifications or by client and governmental requirements. These requirements are beyond the scope of this book, but should be addressed for the specific movable bridge project.

12.0.3 Main mechanical components that support the structure

Before the design of operating machinery for a movable bridge commences, its main supports need to be selected. These components include the mechanical assemblies which will support the dead and live loads and allow the structure to move. In order to design the support components, the characteristics of the structure must be known. After the support components have been designed, the frictional resistance of the main support system is estimated and those quantities are used, in conjunction with the other required design loads, to design the operating machinery system.

12.0.4 Operating machinery design

Operating machinery design begins with calculation of the forces that the structure needs to resist or overcome in order to move. This includes external live loads such as wind and ice plus internal forces such as friction, inertia, machinery system inefficiencies and moving structure gravity imbalance. In order to calculate inertia the designer must determine the operational time-line for opening and closing the movable bridge. This includes determining times and characteristics for acceleration, constant speed running and periods of deceleration. Note some movable bridges use multiple steps of deceleration to stop the bridge. The total time to open or close the bridge is typically less than two minutes but would be dependent on specific project requirements. If the various load cases, constant operational speed, overall machinery ratio and the assumed machinery system efficiency are known, the prime mover size may be selected. The overall machinery ratio for simple bridge types is typically the motor speed divided by the bridge or counterweight sheave rotational speed. The prime mover can be an electric motor, hydraulic motor, direct-drive internal combustion engine, or manual power. For the purposes of this chapter, examples will be based on an electric motor prime mover with reduction gearing.

The electrical motor and drive can not only produce the torque needed to move the structure but it can produce excess torque up to, in some cases, the stall torque of the motor. Note that motor performance characteristics, loading, operating times and frame selection need to be coordinated with the electrical design. After the size of the prime mover is selected, the machinery is designed based on a percentage of the full load motor torque (FLT). In summary, the generalised steps for operating machinery design are

(*a*) Calculate the operating forces/loads for the various load cases and the operating time characteristics.
(*b*) Estimate the machinery system efficiency numerically.
(*c*) Size the prime mover.
(*d*) Size, design and select the operating machinery components.
(*e*) Review efficiency based on actual components designed or selected.

Generally, design is an iterative process and requires multiple revisions. Therefore, the steps above may need to be repeated several times until an acceptable design is achieved.

12.1. Design methods, loadings and load factors

Two basic methods are used to design and size machinery for movable bridges in the USA. The first is working stress design (WSD), also known as allowable stress design (ASD), and the second is load and resistance factor design (LRFD). It is noted that there are many similarities as well as variations of these two methods used in the USA for movable bridge machinery design. However, for this text and for simplicity, only the basics of these two methods will be discussed. The first design method is the historical, classic, process that has been the basis of past AASHTO movable bridge design specifications (AASHTO, 1988) and in past and current AREMA recommended practice (AREMA, 2014). The second method is now the basis of the current AASHTO specification (AASHTO, 2007).

Machinery loading requirements are also specified by AASHTO design specifications and AREMA recommended practice. These loads include self-weight or dead loads, live loads, wind, ice, imbalance, friction, wire rope bending resistance, inertia, motor stall torque, braking, holding and seismic.

During an opening and closing cycle, the required design operating load for a movable bridge may be categorised in three phases. They are: starting, accelerating and running loads. The starting loads are taken as those associated with static friction for the calculation of frictional resistances. Next, the accelerating load is considered as the dynamic frictional loads with the inertial loads. Inertial loads are included with dynamic frictional loads because movement will not occur until static friction is 'broken' – that is, overcome. However, in some cases static friction is used with inertial loads. These are special conservative cases and they would need to be checked with specific project requirements. Finally, running loads essentially use the accelerating load without inertia. AREMA permits inertia to be neglected for swing bridges in certain cases.

Example simple bridge type torque summary tables based on AASHTO (2007) for each of the load phases for operation are given in Tables 12.1, 12.2 and 12.3 for the three major movable bridge types. In regions with ice it is conservatively recommended to include ice loads in all load phases for operation. Note, AREMA defines three load conditions which are noted as A, B and C. AREMA Condition C is the most stringent load case. See AREMA (2014) for further details on the three load conditions.

12.1.1 Bascule bridge

Table 12.1 Bascule bridge operating torque summary (torques are about bridge centre of rotation)

	Starting	Accelerating	Running (constant velocity)
Static friction	T_{SF}	–	–
Dynamic friction	–	T_{DF}	T_{DF}
Inertia	–	$T_{INERTIA}$	–
Imbalance	$T_{IMBALANCE(\varnothing)}$	$T_{IMBALANCE(\varnothing)}$	$T_{IMBALANCE(\varnothing)}$
Wind	$T_{WIND(\varnothing)}$	$T_{WIND(\varnothing)}$	$T_{WIND(\varnothing)}$
Ice	$T_{ICE(\varnothing)}$	$T_{ICE(\varnothing)}$*	$T_{ICE(\varnothing)}$*
Totals (maximum, see notes)	$T_{STARTING}$	$T_{ACCELERATING}$	$T_{RUNNING}$

Summary based on loads listed in AASHTO (2007). See AASHTO (2007) or governing specifications for specifics about loads.

Key:

Symbol	Description
$T_{ACCELERATING}$	Total of accelerating torques
T_{DF}	Bascule bridge dynamic friction torque
$T_{ICE(\varnothing)}$	Bridge ice torque (as a function of opening angle)
$T_{IMBALANCE(\varnothing)}$	Imbalance torque (as a function of opening angle)
$T_{INERTIA}$	Bridge inertial torque
$T_{RUNNING}$	Total of running torques
T_{SF}	Bascule bridge static friction torque
$T_{STARTING}$	Total of starting torques
$T_{WIND(\varnothing)}$	Bridge wind torque (as a function of opening angle)

Notes:

The designer should also consider and review operating machinery inertial loading for both operating and braking loads

Wind load conditions vary for starting, accelerating and running

*Ice load for accelerating and running is shown here to be conservative. However, AASHTO does not list this as a required load for accelerating and running

The total load for each phase (starting, accelerating and running) needs to be investigated through the complete range of motion to determine maximum totals

12.1.2 Swing bridge

Table 12.2 Swing bridge operating torque summary (torques are about vertical centreline of pivot/rotation)

	Starting	Accelerating	Running (constant velocity)
Static friction	T_{SF}	–	–
Dynamic friction	–	T_{DF}	T_{DF}
Inertia	–	$T_{INERTIA}$	–
Wind	T_{WIND}	T_{WIND}	T_{WIND}
Ice	T_{ICE}	T_{ICE}^{*}	T_{ICE}^{*}
Totals (maximum, see notes)	$T_{STARTING}$	$T_{ACCELERATING}$	$T_{RUNNING}$

Summary based on loads listed in AASHTO (2007). See AASHTO (2007) or governing specifications for specifics about loads.

Key:

Symbol	Description
$T_{ACCELERATING}$	Total of accelerating torques
T_{DF}	Swing bridge dynamic friction torque
T_{ICE}	Bridge ice torque (added friction from weight)
$T_{INERTIA}$	Bridge inertial torque
$T_{RUNNING}$	Total of running torques
T_{SF}	Swing bridge static friction torque
$T_{STARTING}$	Total of starting torques
T_{WIND}	Bridge wind torque

Notes:

The designer should also consider and review operating machinery inertial loading for both operating and braking loads

Wind load conditions vary for starting, accelerating and running. The designer needs to investigate wind load conditions to determine maximum

*Ice loads for accelerating and running are included here to be conservative. However, AASHTO does not list these as required loads for accelerating and running

The total load for each (starting, accelerating and running) needs to be investigated through the complete range of motion to determine maximum totals

12.1.3 Vertical lift bridge

Table 12.3 Vertical lift bridge operating torque summary (torques are about sheave shaft)

	Starting	Accelerating	Running (constant velocity)
Static friction	T_{SF}	–	–
Dynamic friction	–	T_{DF}	T_{DF}
Rope bending	$T_{ROPE\text{-}BENDING}$	$T_{ROPE\text{-}BENDING}$	$T_{ROPE\text{-}BENDING}$
Inertia	–	$T_{INERTIA}$	–
Imbalance	$T_{IMBALANCE(y)}$	$T_{IMBALANCE(y)}$	$T_{IMBALANCE(y)}$
Wind	T_{WIND}	T_{WIND}	T_{WIND}
Ice	T_{ICE}	T_{ICE}^{*}	T_{ICE}^{*}
Totals (maximum, see notes)	$T_{STARTING}$	$T_{ACCELERATING}$	$T_{RUNNING}$

Summary based on loads listed in AASHTO (2007). See AASHTO (2007) or governing specifications for specifics about loads.

Key:

Symbol	Description
$T_{ACCELERATING}$	Total of accelerating torques
T_{DF}	Torque from vertical lift bridge trunnion dynamic friction
T_{ICE}	Torque from ice load
$T_{IMBALANCE(y)}$	Torque from imbalance (as a function of span lift height)
$T_{INERTIA}$	Torque from lift span and counterweight inertia
$T_{ROPE-BENDING}$	Torque from wire rope bending resistance
$T_{RUNNING}$	Total of running torques
T_{SF}	Torque from vertical lift bridge trunnion static friction
$T_{STARTING}$	Total of starting torques
T_{WIND}	Torque from wind load

Notes:

The designer should also consider and review operating and sheave machinery inertial loading for both operating and braking loads

*Ice load for accelerating and running is shown here to be conservative. However, AASHTO does not list this as a required load for accelerating and running

The total load for each phase (starting, accelerating and running) needs to be investigated through complete range of motion to determine maximum totals

Once the loads are calculated, it is desirable to estimate the efficiency of a mechanical system to drive the bridge. If the details of machinery are not initially known, an 80% machinery system efficiency may sometimes be used as a first approximation for preliminary design of a modern movable bridge. As the design progresses, the assumed efficiency needs to be refined based on actual efficiencies of components selected or designed. Machinery efficiencies for various types of machinery components are given in AASHTO specifications and in AREMA guidelines for recommended practice.

Having the machinery system efficiency and the required starting, accelerating and running loads, and with the constant operational speed known, the size of the prime mover can be selected. Allowable motor overloads are given in both the AASHTO design specification and the AREMA recommended practices, which are used to size the motor. The AASHTO (2007) allowable electric motor prime mover overloads are: 1.25 FLT for starting, 1.50 FLT for accelerating and 1.00 FLT for running (constant velocity).

Once a motor size has been selected, the operating machinery can be designed. Operating machinery design is typically based on percentages of full load motor torque.

These percentages are given in AASHTO (2007) and AREMA (2014). For example, the AASHTO service limit states that the electric motor minimum design loading for machinery is 1.50 times full load torque (FLT). This is similar to machinery design using AREMA (2014) with allowable stresses at a motor load of 150% FLT. For additional limit states and conditions which need to be designed for and checked, including overload, stall, braking and holding, refer to AASHTO or AREMA or specific project requirements.

12.2. Machinery component design

The first design method (WSD) requires that under the working load the resulting stresses are to be equal to or less than established maximum allowable stresses. The allowable stresses are given or calculated based on the minimum yield or ultimate tensile strength of the material divided by an appropriate safety factor. The second design method (LRFD) requires that the factored load be less than the factored resistance. The load cases for LRFD include service limit state, fatigue limit state, overload limit state, and extreme limit state. The basic equations for these two methods are presented below:

WSD (general form)

$$\sigma \leq \sigma_{\text{allowable}} \tag{12.1}$$

$$\tau \leq \tau_{\text{allowable}} \tag{12.2}$$

LRFD

$$\sum n_i Y_i Q_i \leq \varnothing R_{\text{n}} = R_{\text{r}} \quad \text{(AASHTO, 2007)} \tag{12.3}$$

in which

n_i	load modifier
Q_i	force effect
R_{n}	nominal resistance
R_{r}	factored resistance
Y_i	load factor
σ	maximum computed combined, bending or bearing stress
$\sigma_{\text{allowable}}$	allowable combined, bending or bearing stress
τ	maximum computed shear stress
$\tau_{\text{allowable}}$	allowable shear stress
$\varnothing$	resistance factor

For movable bridge machinery load and resistance factors see the AASHTO (2007) specification or specific project requirements. See AREMA (2014), AASHTO (2007) or governing specifications as applicable for additional information on materials, allowable stresses and resistance. Movable bridge design requirements are discussed additionally in Chapter 5.

12.3. Main machinery support components for movable bridges

The main dead load support components for movable bridges that are classified as machinery include

- trunnions for bascule bridges (see Figure 12.1)

Figure 12.1 Bascule bridge – trunnion assembly (cross-section)

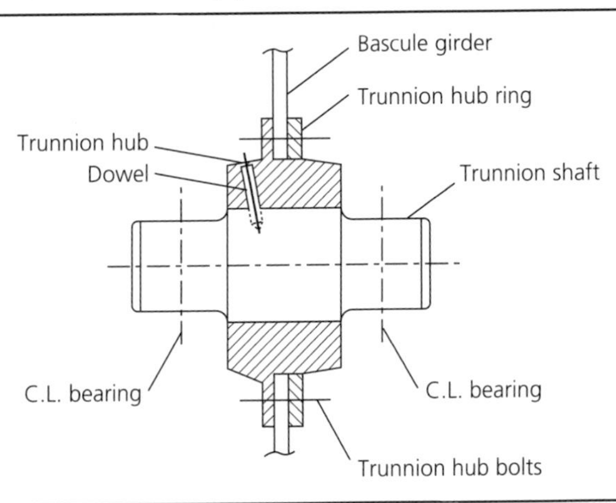

Figure 12.2 Rolling bascule bridge – curved tread and track assembly

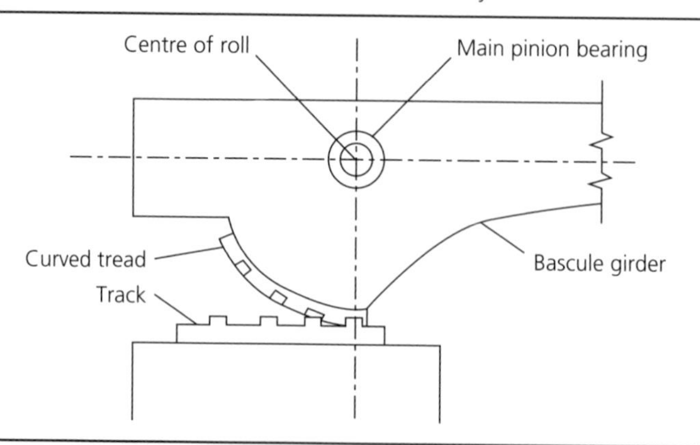

- curved tread and track for rolling bascule bridges (see Figure 12.2)
- rim bearing for swing bridges (see Figure 12.3)
- centre pivot bearing for swing bridges (see Figure 12.4)
- sheave/trunnion and wire rope for vertical lift bridges (see Figure 12.5).

These machinery components support heavy loads, primarily dead load of the movable span. However, in many cases, some of these components will also see live loads, which must be considered. The design of these components is a special case of machinery design – high load and slow speed – which requires

Figure 12.3 Swing bridge – rim bearing assembly

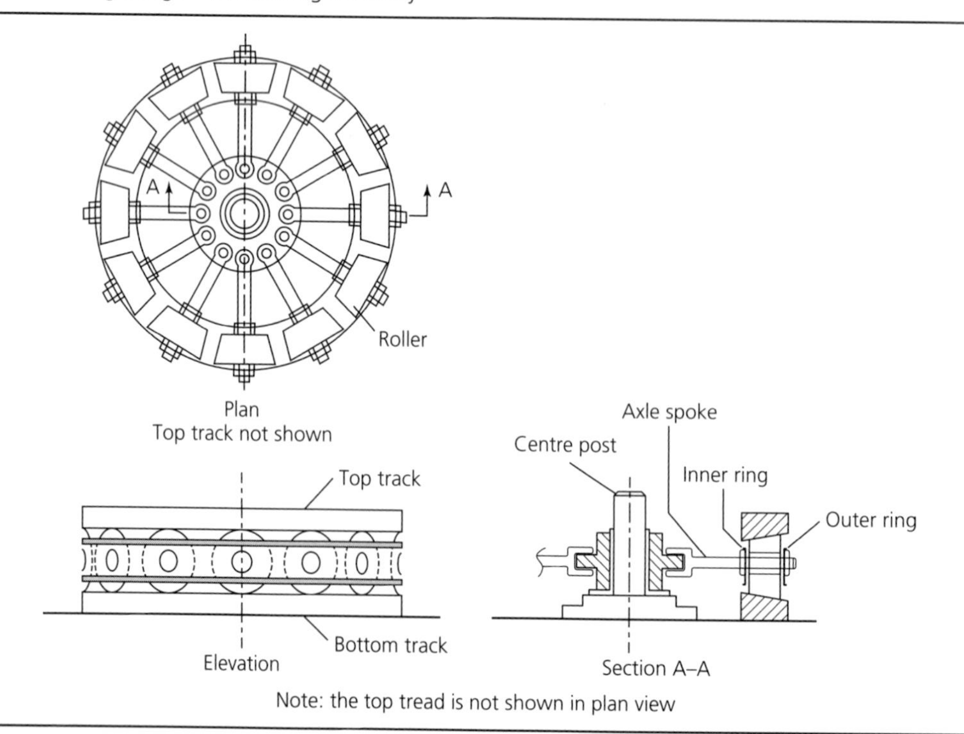

Figure 12.4 Swing bridge – centre pivot bearing assembly (cross-section)

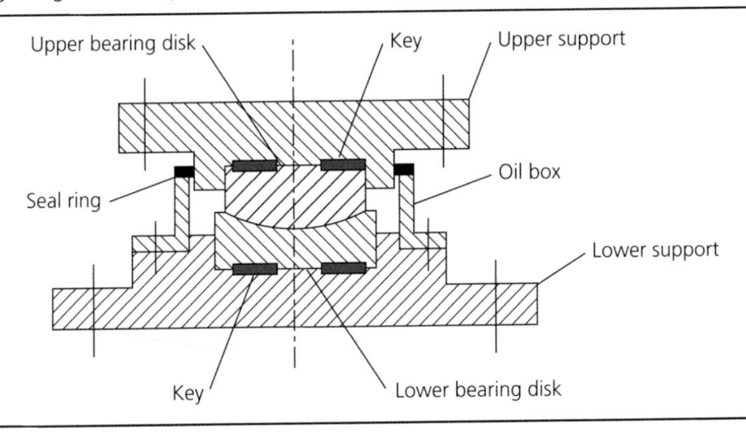

Figure 12.5 Vertical lift bridge – sheave and trunnion assembly (cross-section)

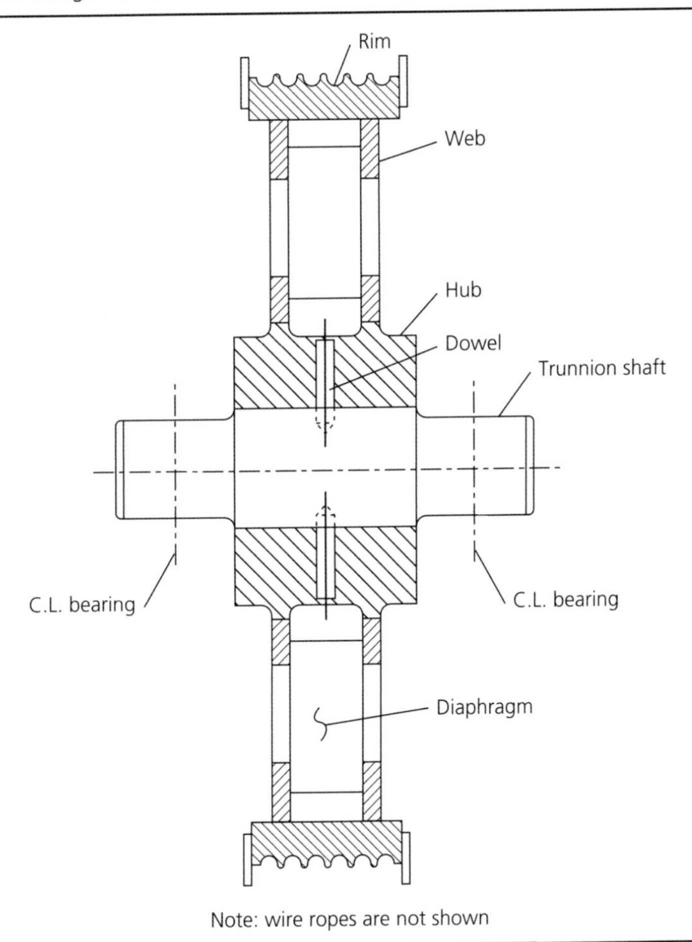

Note: wire ropes are not shown

specific considerations for design. The basic principles and equations for these components may be found in subsequent sections herein. However, the designer should keep in mind the unique and critical function of these components. Generally, these components are much more difficult to replace than other machinery components. Also, the maintenance and future inspection of these components require special considerations because disassembly is more involved due to their nature and function.

12.4. Design of shafts – general

Two types of rotating shafts are common in machinery design. The first are shafts that are loaded in the radial direction, including their self-weight and that can also transmit torque. This type of shaft includes shafts that support dead load of structures or machinery components, shafts in speed reducers or sets of open gearing and shafts that support pulleys or rollers. A fixed shaft is a special case where the shaft does not rotate with respect to its supports. The second type are shafts that primarily transmit torsion and support their self-weight. The latter is also commonly known as transmission drive shafting. Examples of transmission drive shafting include shafts supported by two bearings or by two couplings (see Figure 12.6). Shafts for movable bridge machinery are typically carbon or alloy steel forgings.

One important consideration for shaft design should be how the shafts will be loaded and supported. Achievable alignment should be considered when designing the shafts and selecting the components which support the shaft. An example of this would be a gear reducer shaft. Ideally the designer would like full contact across the face width of the gearset and full support of the bearings. However, in reality the shafts deflect under load and machining tolerances and practical installation errors limit perfect gear contact. The designer should select suitable bearings and consider how they will be aligned, installed, and behave under operating conditions. In some cases spherical plain or rolling element bearings are selected in order to accommodate these factors. Also, as another example, gear reducer housing bearing support bores should be bored in-line and parallel in one set-up of a suitable milling machine to ensure accurate installation of the bearings and shafts.

Additionally, when designing shafts, proper detailing is required to avoid fatigue-sensitive details. When a shaft transitions between different diameters, the steps need to be provided with suitable radii (see Figure 12.7). As typical with all machinery components, sharp edges and corners should be avoided in order to reduce stress concentrations. For static components made from low- or medium-carbon steel, stress concentration may be less of a concern because the material will usually yield at the

Figure 12.6 Examples of transmission drive shafting

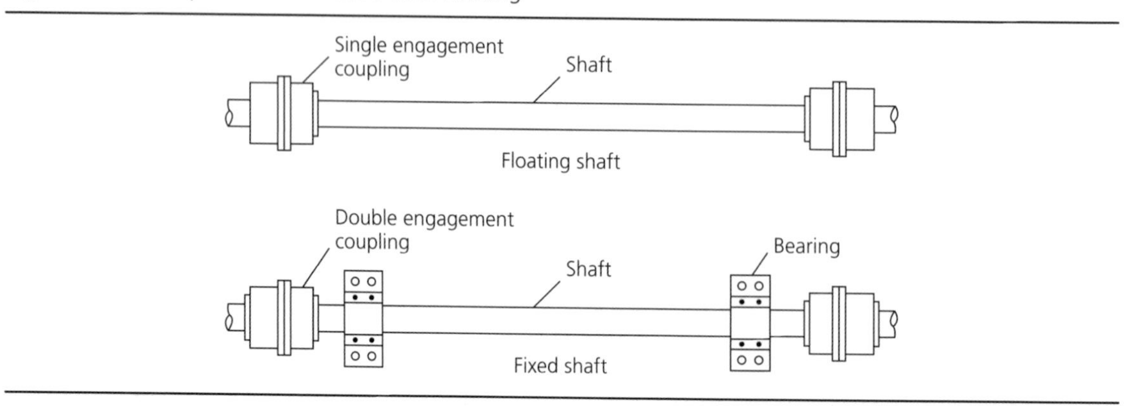

Figure 12.7 Shaft shoulder radii

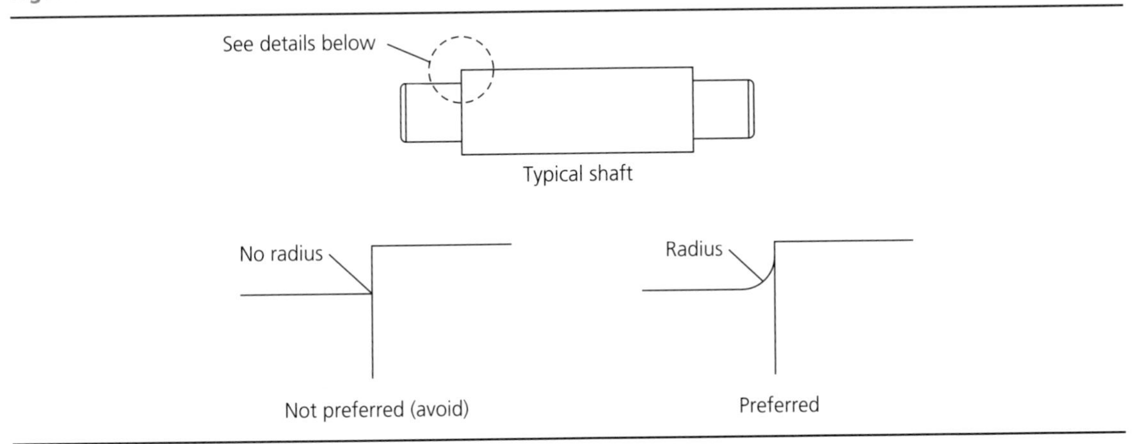

point of stress concentration and partially relieve the stress. However, with a rotating shaft, fatigue can occur because the material goes through stress reversal during every cycle of shaft rotation.

Some example stress concentration factors for a circular shaft are given below (see Figure 12.8).

When designing shafts, keyways and keys are commonly used for securing gears, pulleys, sheaves, drums, sprockets and coupling hubs. Keyways also add stress concentration due to the sharp interruption of the stress field through the shaft. Typical key and keyway sizes are given by applicable specifications.

The hubs of components mounted on shafts are typically detailed with a width equal to or greater than the shaft diameter and the hub radial thickness is usually at least 40% of the gross shaft diameter. (See AASHTO (2007) and AREMA (2014) for further details.) Threads are sometimes also used on shafts to secure roller bearings or hubs. They also introduce stress concentrations due to the sharp peaks and valleys of thread geometry on the shafts.

Other examples of stress concentrations include grease grooves, slots for lock washers or snap rings, and drilled radial holes for grease passages. It is important for the designer to consider these as well as considering any other stress concentrations that might overlap. The combined effect of these is often difficult to calculate and the physical situation should generally be avoided. An example of this would be avoiding having a keyway pass through a fillet. However, sometimes overlap of stress concentrations cannot be avoided and the condition should be studied to estimate the stress-rising effect. Such study might include the review of past data on similar applications or detailed analysis such as computer finite element analysis (FEA).

Figure 12.8 Examples of stress concentration factors

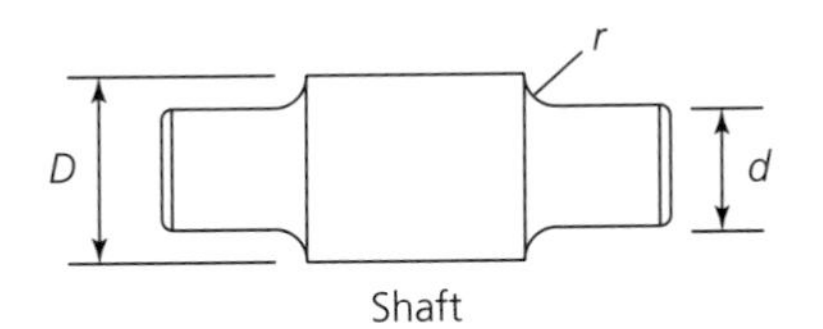

D/d	r/d	K_t^* bending	K_{ts}^* torsion
1.10	0.05	1.94	1.35
1.20	0.10	1.61	1.38

Note: *Approximate values, estimated from (Aashto, 2007)

Although the discussion here is focused on shaft details, the principles of stress concentration can be applied to other mechanical components. Examples of this would be bolts in tension and bending, connecting rods, linkages, hangers, roots of gear teeth or any other component with geometry changes. Additional information on stress concentration and factors can be found in AASHTO (2007) and Pilkey (1997).

12.4.1 Design of shafts for bending and torsion

The design of shafts that support radial loads are primarily based on bending stresses. Rotating shafts experience either partial or full reversals of bending stress. For this reason, current AASHTO bridge design specifications now require that these elements be designed using fatigue failure theory. It is also noted that some specifications and recommended practices still use working stress design with consideration for fatigue having been made when establishing the allowable stresses.

The working stress design method is adequate when care is taken to minimise fatigue-sensitive details. Design based on the AREMA movable bridge recommended practices results in good designs for machinery shafts and trunnions of vertical lift and trunnion bascule bridges when properly applied, using suitable details to avoid high stress concentrations. This method includes estimating the loads, calculating bending stresses and torsional stresses, determining the allowable geometry of the trunnion or shaft and bearings assemblies, and selecting the material for the shaft.

The AREMA equations for computing bending and shear stresses are as follows:

$$f = \frac{16K}{\pi d^3}\left(M + \sqrt{M^2 + T^2}\right) \tag{12.4}$$

$$S = \frac{16K}{\pi d^3}\left(\sqrt{M^2 + T^2}\right) \tag{12.5}$$

in which:

- d diameter
- f AREMA shaft extreme fibre stress, tension or compression
- K impact factor
- M bending moment
- S AREMA shaft shear stress
- T torsional moment

The allowable bending and shear stresses are typically based on the minimum strength of the material, with an appropriate safety factor applied. Allowable bending and shear stresses for basic shaft materials are given in AREMA (2014). The stress in the shaft needs to be checked at locations of maximum bending and torsional moments, stress concentrations and at cross-section changes.

When using ASD or using AREMA recommended practice, the designer should judge if a fatigue analysis needs to be performed in cases where the number of cycles is high, full stress reversal occurs and fatigue-sensitive details are present.

12.4.2 Shaft design for fatigue

The three fatigue analysis methods most often used in machinery design are the stress-life approach, the strain-life approach and the fracture mechanics approach (Bannantine *et al.*, 1990). The stress-life approach was the first fatigue analysis method developed. It is mainly used for long-life applications during which stress and strain remain elastic.

The strain-life method is used when the component becomes inelastic. The inelastic behaviour results in shorter fatigue lives.

The fracture mechanics method utilises linear fracture mechanics principles adapted for cyclic loading. It is used to predict the propagation of an initial defect (crack).

AASHTO included fatigue design provisions in the current movable bridge design specification. They are based on the stress-life method.

For stress-life methods the amplitude and mean stresses are computed and compared to fatigue failure criteria. The computed amplitude stress is compared to the endurance stress, and the mean stress is compared to either the yield strength or the tensile strength of the material. A few different stress-life fatigue design criteria are commonly used in general machine design. The Soderberg, Goodman and Gerber lines for normal (unfactored) stress fatigue design criteria equations are listed below (Shigley and Mischke, 2001). Plotted stresses are to be equal to or lower than the fatigue criteria line. Also, a factor of safety is typically applied to the line in design.

Soderberg Line:

$$\frac{\sigma_a}{\sigma_e} + \frac{\sigma_m}{\sigma_{yt}} = 1 \tag{12.6}$$

Goodman Line:

$$\frac{\sigma_a}{\sigma_e} + \frac{\sigma_m}{\sigma_{ut}} = 1 \tag{12.7}$$

Gerber Line:

$$\frac{\sigma_a}{\sigma_e} + \left(\frac{\sigma_m}{\sigma_{ut}}\right)^2 = 1 \tag{12.8}$$

in which

σ_a amplitude normal stress
σ_e endurance limit
σ_m mean normal stress
σ_{ut} ultimate strength
σ_{yt} yield strength

Formulae for fatigue-resistant design for movable bridge components are given in AASHTO (2007). The formula is based on the Soderberg fatigue design criteria with a factor of safety for shafts that go through full reversals of bending stress. This method requires the calculation of factored von Mises stresses and modified endurance limit.

The AASHTO (2007) equations for calculating the amplitude stresses, mean stresses, modified endurance limit and the fatigue criteria, are given below.

$$\sigma_a = \frac{\sigma_{max} - \sigma_{min}}{2} \tag{12.9}$$

$$\tau_a = \frac{\tau_{max} - \tau_{min}}{2} \tag{12.10}$$

$$\sigma_m = \frac{\sigma_{max} + \sigma_{min}}{2} \tag{12.11}$$

$$\tau_m = \frac{\tau_{max} + \tau_{min}}{2} \tag{12.12}$$

$$\sigma_e = \alpha \sigma_{ut} C_D C_S C_R C_T C_M \tag{12.13}$$

$$\frac{\sigma_a'}{\sigma_e} + \frac{\sigma_m'}{\sigma_{yt}} \leq 0.80 \quad \text{(trunnion shafts greater than 90° rotation, see discussion below)} \tag{12.14}$$

$$\frac{\sigma_a'}{\sigma_e} + \frac{\sigma_m'}{\sigma_{yt}} \leq 1.00 \quad \text{(trunnion shafts less than 90° rotation, see discussion below)} \tag{12.15}$$

where

$$\sigma_a' = \sqrt{(K_F \sigma_a)^2 + 3(K_{FS} \tau_a)^2} \tag{12.16}$$

$$\sigma_m' = \sqrt{(K_F \sigma_m)^2 + 3(K_{FS} \tau_m)^2} \tag{12.17}$$

$$K_F = 1 + q(K_t - 1) \tag{12.18}$$

$$K_{FS} = 1 + q(K_{ts} - 1) \tag{12.19}$$

in which

C_D	size factor
C_M	miscellaneous factor
C_R	reliability factor
C_S	surface roughness factor
C_T	temperature factor
K_F	fatigue stress concentration for normal stress
K_{FS}	fatigue stress concentration for shear stress
K_t	theoretical stress concentration factor for normal stress
K_{ts}	theoretical stress concentration factor for shear stress
q	fatigue design notch sensitivity factor
α	endurance limit factor
σ_a	amplitude normal stress
σ_a'	von Mises applied amplitude stress
σ_e	endurance limit
σ_m	mean normal stress
σ_m'	von Mises applied mean stress
σ_{max}	maximum applied normal stress
σ_{min}	minimum applied normal stress
σ_{ut}	ultimate strength
σ_{yt}	yield strength
τ_a	amplitude shear stress
τ_m	mean shear stress
τ_{max}	maximum applied shear stress
τ_{min}	minimum applied shear stress

Equations 12.14 and 12.15 are in a form using von Mises amplitude and mean stresses for calculations. This form was shown here because of its versatility in different loading conditions and for analysis of existing machinery components. An inspection of the equations shown specifically for shafts and trunnions listed in AASHTO (2007), reveals that they are based on checking each section with a bending moment for amplitude normal stress, zero amplitude shear stress, zero mean normal stress and a one-direction torsional moment for mean shear stress. A conservative approach would dictate that the greatest magnitude of bending moment and torsional moment be calculated for each particular section of the shaft checked, and used for the general application of the shaft and trunnion equations. (See AASHTO 2007 for further forms of the equations, details, and assumptions made.) The AASHTO equations are based on the Soderberg criteria using these stresses with a factor of safety of 1.25 for shafts and 1.00 for trunnion shafts with less than 90° of rotation. The assumptions used, factors of safety, loading and design criteria need to be investigated by the engineer for proper application of the fatigue analysis.

Refer to AASHTO (2007) for additional information on fatigue criteria and factors. The fatigue criteria would need to be checked at areas with maximum bending and torsional moments, stress concentrations, and at cross-section changes.

12.4.3 Other considerations in shaft design

Operating machinery shafting also need to be checked for angular deflection (twist), allowable length and maximum speed as per specific design specification or recommended practice requirements.

The AASHTO (2007) equations for twist, allowable length, and critical speed are given below.

$$\text{Angular shaft twist (degrees/unit of shaft length)} \leq 0.05/D \quad \text{(typical shafts)} \tag{12.20}$$

$$\text{Angular shaft twist (degrees/in. of shaft length)} \leq 0.00667 \quad \text{(when less shaft twist is desired)} \tag{12.21a}$$

$$\text{or (degrees/mm of shaft length)} \leq 0.26 \times 10^{3} \tag{12.21b}$$

Units of shaft diameter and length alike.

Note: If $D > 7.5$ in. (190.5 mm) then Equation 12.20 governs.

US customary units (USCUs)

$$L_{\text{allowable}} \leq 80\,(D^2)^{1/3} \tag{12.22}$$

$$n \leq 0.67\, n_c \tag{12.23}$$

where

$$n_c = 4.732 \times 10^6 \frac{D}{L^2} \quad \text{(floating shaft)} \tag{12.24}$$

$$n_c = 1.55 \times 10^6 \frac{D^2}{L^2}\sqrt{\frac{L}{W}} \quad \text{(simply supported shaft with weight } W \text{ concentrated at centre)} \tag{12.25}$$

in which

D	shaft diameter (in.)
L	length (in.)
$L_{\text{allowable}}$	allowable length of shaft (in.)
n	shaft speed (rpm)
n_c	critical shaft speed (rpm)
W	weight (lb)

Metric units

$$L_{\text{allowable}} \leq 220\ (D^2)^{1/3} \tag{12.26}$$

$$n \leq 0.67\ n_c \tag{12.27}$$

where

$$n_c = 120 \times 10^6 \frac{D}{L^2} \quad \text{(floating shaft)} \tag{12.28}$$

$$n_c = 0.207 \times 10^6 \frac{D^2}{L^2}\sqrt{\frac{L}{m}} \quad \text{(simply supported shaft with weight concentrated at centre)} \tag{12.29}$$

in which:

$$m = \frac{W}{g} \tag{12.30}$$

and

D	shaft diameter (mm)
g	acceleration due to gravity (9.81 m/S^2)
L	length (mm)
$L_{\text{allowable}}$	allowable length of shaft (mm)
m	mass (kg)
n	shaft speed (rpm)
n_c	critical shaft speed (rpm)
W	weight (N)

The critical speed of a shaft, or any other rotating system for that matter, is that which corresponds to a resonance frequency. For shafts the resonance frequency is often taken as the natural frequency of lateral (beam) vibrations of the shaft (Ungar, 1964). Attention is called to the assumptions on which Equations 12.24 and 12.28 are based; geometry of shaft is perfect, elastic behaviour, self-aligning bearings spaced a distance L apart, infinitely stiff bearing supports and no axial torque in the shaft.

AASHTO (2007) Section 6.7.4.3 states that the maximum shaft speed shall not exceed 67% of the critical speed. This represents a factor of safety against resonance of only 1.5. Furthermore, it permits

L to be taken as the distance between flexible gear couplings. For the case of a string of alternate fixed and floating shafts that length may be considerably less than the distance between bearings. Industrial experience proves that taking L as the floating shaft length rather than the distance between bearings is unconservative.

Although neglected in the derivation of the common equation for critical speed of a uniform shaft, investigations have shown that the critical speed of such shafts is decreased as the applied torque is increased (Oberg *et al.*, 2012).

Also, in some cases, lateral (bending) deflection of the shaft becomes a consideration for alignment: for example, for shafts that support gears. For this case, the shafts not only have to meet the requirements for stress and fatigue design, they also are required to have limited deflection in order to maintain acceptable contact between the gearset teeth.

12.5. Design of plain and anti-friction bearings

Radial machinery bearings are required in the design of movable bridges to support shaft self-weight, and loads applied to the shafts, and allow them to rotate. Thrust machinery bearings are used to support dead load and applied loads in the axial direction and allow the components to rotate. There are two basic types of bearings typically used in movable bridge design: plain (journal or sleeve) and rolling element bearings (see Figure 12.9). Plain journal/sleeve or roller bearings can be either mounted in a separate housing or in a frame with other components.

The selection of bearing type needs to be done with care and understanding of loading, speed, lubrication and the environment to which the bearing will be subjected. Plain bronze bearings are good for high loads, impact loads, and harsher environments. However, their operating friction is higher than that for rolling element bearings. Plain bearings can be either radial or thrust bearings.

Rolling element bearings, also called roller bearings, or anti-friction bearings, have a low frictional resistance to rotation. Roller bearings can be spherical, cylindrical, tapered or ball bearings. These

Figure 12.9 Radial bearings for shafts

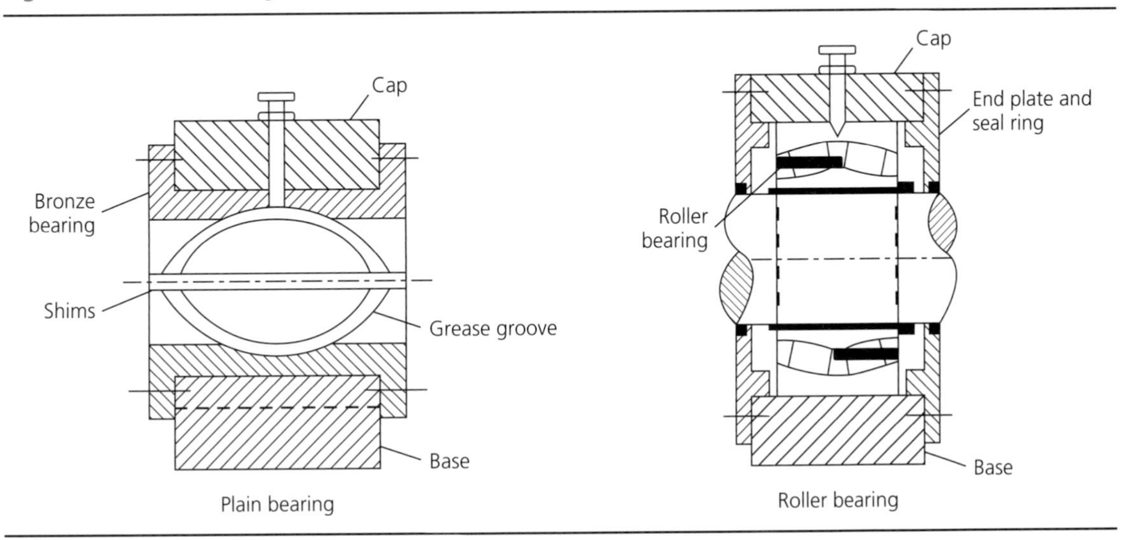

types of bearings typically have less resistance to impact loading than has a plain bearing, and they require that the internals of the bearing and lubricant stay in a clean condition. Roller bearings also can be either radial or thrust bearings.

Another important consideration in bearing selection is how the particular bearing type supports the loads. A plain journal bearing requires accurate alignment and installation to spread the load evenly across the bearing width. Of course, the deflection of the bearing support also influences the bearing stress distribution. A special case of a sleeve bearing is a spherical type which does compensate for some shaft misalignment. Manufacturing tolerances for this type of bearing for movable bridges and the greater friction compared to roller bearings, typically limits the use of such bearings to specific applications. Spherical, both plain and roller type radial bearings, rotate about a fixed point on the bearing to compensate for initial misalignment and shaft deflection as the shaft is loaded.

Examples of radial bearing loading for the plain journal and spherical types are shown in Figure 12.10.

When bearings are used to support dead and live loads of the structure they are typically sized based on the maximum load calculated for the structure. However, there are also additional specific requirements which the bearings must satisfy for design compliance which are given in AASHTO (2007) and AREMA (2014).

The bearings for operating machinery are generally sized based on a percentage of motor load. In many cases this is 150% of full load motor torque with the component weight added and is used as the service and fatigue load. Again specific project design specification or recommended practice must be reviewed and addressed in design for compliance.

A common design practice is to also consider a minimum percentage of primary bearing load in the non-primary direction. An example of this is (AASHTO, 2007; AREMA, 2014) for heavily loaded spherical radial bearings to be designed to include a minimum of 15% of radial load for axial load, or the actual axial load if greater. In some applications the bearing type does not have the ability to resist significant load in the non-primary direction. An example of this would be a cylindrical radial roller bearing application where there is a significant axial load. In such cases an additional means or a different bearing type would be needed to resist the axial load.

Figure 12.10 Examples of radial bearing loading

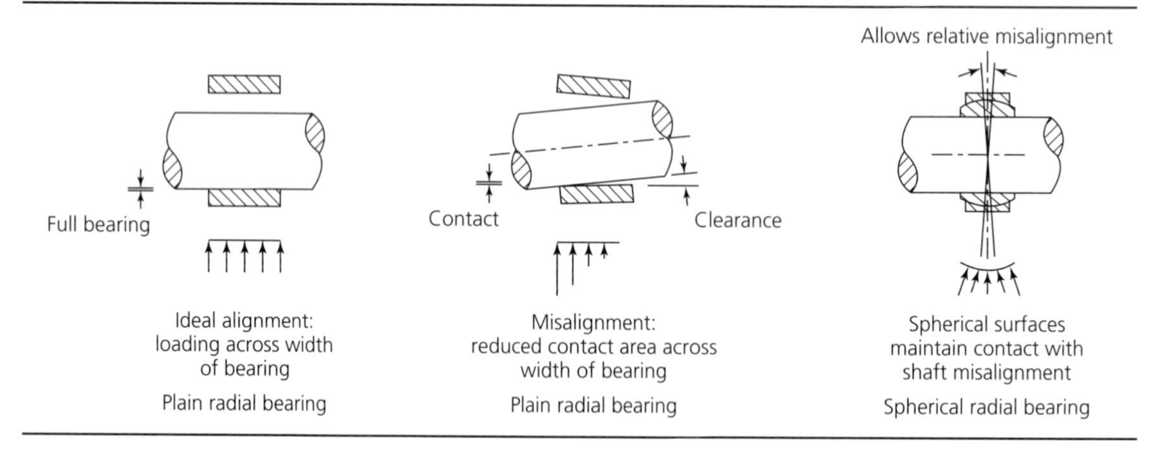

12.5.1 Plain bearings

Plain (or journal) radial bearing design requires determining the bearing pressure by dividing the radial load by the projected area of the bearing sleeve in contact with the shaft, minus the area for the grease or oil grooves. The allowable pressure is determined based on the materials used for the bearing, the shaft and the speed of the shaft journal at the sliding surface. It is also important that the bearing material be softer than the shaft journal. The difference in hardness helps avoid scoring or other shaft damage. Plain radial shaft bearings typically have a width parallel to the shaft equal to or greater than the journal diameter with the exception for special cases. Lubrication is usually by grease pumped through grease grooves. Typically, the grease grooves are a spiral figure 8 for smaller operating machinery bearings and straight (parallel to the shaft) on larger bearings that support the structure, such as trunnion bearings, where the shafts rotate slowly.

Plain radial sleeve bearings are typically detailed for a RC6 sliding fit between the journal and the sleeve. A common plain bearing finish would be 16 micro-inches (0.4 micro-metres) with a shaft journal finish of 8 micro-inches (0.2 micro-metres). The fits and finishes given are the same as those preferred for shaft journals and journal bushings listed in AASHTO (2007) and AREMA (2014). A commonly used design is a split plain-sleeve bearing. The major components of such a bearing are a base, cap, split bronze sleeve and fasteners. The bronze sleeves are typically designed with flanges to secure them in the base and cap, and they can also be used to resist thrust loads. In many cases the split plain sleeve bearings are detailed with shims at the split. The shims allow for future adjustment to compensate for wear of the sleeve. For large bearings, such as for trunnions, where the load always acts downwards into the base in many cases, the top half of the bearing assembly is just a copper alloy cap.

The basic bearing stress equation for plain radial bearing design is given below.

$$\sigma_{\text{brg}} = \frac{\text{Load}}{(dL - (\text{lubricant groove area}))} \tag{12.31}$$

in which

d shaft diameter
L bearing journal length
σ_{brg} bearing stress

The allowable bearing pressure for basic metallic bearing materials is given in AASHTO (2007) and AREMA (2014).

Plain thrust bearings, such as those for centre bearing swing bridges, are usually made of an upper bronze convex spherical disc and a lower concave hardened steel spherical disc with a slightly greater radius. Bearing pressure for a plain thrust bearing (full disc area, with no cut-out of an inner radius) is calculated by dividing the load by the projected area of the disc, minus the projected area for the lubricant grooves. The basic bearing stress equation for plain thrust (with a full disc area except for lubricant grooves) bearing design is given below.

$$\sigma_{\text{brg}} = \frac{\text{Load}}{(\pi r^2 - (\text{lubricant groove area}))} \tag{12.32}$$

Table 12.4 Suggested radii for pivot disc bearing surfaces

Diameter of disc in in.: mm	Lower steel disc concave surface radius in in.: mm	Upper bronze convex surface radius in in.: mm
12 (305)	30 (762)	29 (737)
18 (457)	48 (1,219)	46.75 (1187)
24 (610)	72 (1,829)	69.75 (1772)
30 (762)	96 (2,438)	93 (2362)
36 (914)	120 (3,048)	116.25 (2953)
48 (1219)	165 (4,191)	160.25 (4070)

Suggested radii in inches from Hovey (1926)

in which

r disc-projected bearing radius
σ_{brg} bearing stress

Some suggested spherical radii for common sizes of the upper and lower discs from Hovey (1926) are listed in Table 12.4.

Some centre bearing railway swing bridges erected in the late 19th and early 20th centuries in the USA have three element pivot bearings (centres). They comprise an upper and a lower concave bronze disc with a steel lenticular disc between them, with convex surfaces on the top and bottom. The motivation for this design was the idea that foundation settlement of the centre pier would have less effect on rotation of the span with the lenticular disc present in the centre bearing (Hovey, 1926, vol. 1).

There are also plain flat thrust bearings for shaft shoulders and plain flat bearings to support load from such components as lock bars and wedges. For shaft shoulder thrust bearings or flat load bearings the general bearing stress would be the load divided by the net area. The net area would be the contact area of the flat contact surfaces minus the area of the lubrication grooves.

Plain bearings, both radial and thrust, also need to be checked for heating and seizing due to rotational speed. Allowable pressure equations to avoid heating and seizing of a rolled or forged steel shaft on bronze are given in AASHTO (2007) and AREMA (2014).

12.5.2 Rolling element bearings

Roller bearings are available from various manufacturers as pre-engineered components. Roller bearing design typically follows the bearing manufacturer recommendations. A common practice is for operating machinery bearings to be sized for the service load and to have a minimum L_{10} life of 40 000 hours. In general terms, a L_{10}, 40 000 hours of life means that 90% of a population of the same size and type of bearing would last for 40 000 hours. Additionally, the overload case should be investigated. For larger bearings intended for supporting the structure, such as thrust bearings for swing bridges and radial bearings for bascule and vertical lift trunnions, the designer will also need to consult with the manufacturer for these special applications. Additional requirements are given in bridge design specifications and recommended practices such as those listed in AASHTO and AREMA.

The general bearing life equations given in Deutschman *et al.* (1975) for radial roller and ball bearing selection are listed below.

Radial roller bearing life

$$L_{10} = \frac{10^6}{60n}\left(\frac{C_{\text{basic rating}}}{P_{\text{equivalent}}}\right)^{10/3} \quad \text{(use greater value from equivalent load Equation 12.35 or 12.36)} \qquad (12.33)$$

Radial ball bearing life

$$L_{10} = \frac{10^6}{60n}\left(\frac{C_{\text{basic rating}}}{P_{\text{equivalent}}}\right)^{3} \quad \text{(use greater value from equivalent load Equation 12.35 or 12.36)} \qquad (12.34)$$

Some manufacturers may use an exponent higher than three for ball bearings.

$$P_{\text{equivalent}} = XVF_r + YF_a \qquad (12.35)$$

$$P_{\text{equivalent}} = VF_r \qquad (12.36)$$

in which

$C_{\text{basic rating}}$ bearing basic load rating
F_a applied axial load
F_r applied radial load
L_{10} bearing life in hours
n bearing rotational speed (RPM)
$P_{\text{equivalent}}$ equivalent load on bearing
V rotation factor
X radial load factor
Y thrust load factor

Note for V: According to Deutschman *et al.* 1975), V is taken as 1.0 for self-aligning bearing and inner ring rotation applications and 1.2 for non-self-aligning bearing outer ring rotation applications.

The equations for bearing life can be rewritten as an allowable load similar to the AASHTO formula by using a bearing life of 40 000 hours.

When selecting roller bearings, a designer should obtain data from the manufacturer for the V, X and Y factors and confirm specific selection requirements. These factors are used in determining equivalent loads to be used in the equations above.

12.5.3 Bearings in general

Typically, the radial bearing types listed above can be procured from various manufacturers as pre-engineered units that are complete assemblies, including housings. When selecting bearings as pre-engineered units it is important to verify the housing rating to confirm that it is adequate for the loads as well as the direction applied to the housing. This should also include checking that bolts of proper grade and strength are used. For roller bearings, housings can be provided with different seal options to keep the internal components clean and free of contaminates. The seals also function to keep the lubricant contained within the bearing housing. Careful seal selection is important and should be based on the application, the environment in which the bearings will be used and the lubricant. For large plain-sleeve bearings the size may be more limited for pre-engineered units than for roller bearings. A custom design is typically required for large plain-sleeve bearings. The bearing manufacturers commonly publish dimensions, loads, maximum speeds and alignment capacities. The literature available in most cases also includes installation, operation, maintenance and lubrication information.

12.6. Contact stresses and design of rollers

In the design of movable bridge machinery, rollers are used for different purposes. One use is for the rim bearings of swing bridges. Another is for large curved treads on flat tracks for rolling bascule bridges. The design of both of these types of components is dependent on contact stresses. There are various design formulae for calculating the allowable line load based on the yield strength of the material. The allowable load on a roller or curved surface depends on the diameter of the rolling element and the yield strength of the material.

For rolling lift bridges it is also noted that the thickness of the curved treads and tracks are an important consideration. If these elements are too thin, the material will yield and displace plastically under rolling contact stresses. The minimum thicknesses for these components are given in both the AASHTO specifications and AREMA recommended practices. The equations from AREMA (2014) for steel rollers on steel tracks (in USCAs) are listed below.

In USA units

for roller diameters up to 25 in.

$$\frac{\text{line bearing load (lb)}}{\text{linear in. of width}} \leq (400d)\,\frac{(F_y - 15\,000)}{20\,000} \tag{12.37}$$

for roller diameters of 25–125 in.

$$\frac{\text{line bearing load (lb)}}{\text{linear in. of width}} \leq (2000\sqrt{d})\,\frac{(F_y - 15\,000)}{20\,000} \tag{12.38}$$

in which

d diameter of roller (in.)
F_y minimum material yield strength (psi)

Note: For the static case the above allowables can be increased by 50%.

Metric units (converted from US unit equation):

for roller diameters up to 635 mm

$$\frac{\text{line bearing load } (N)}{\text{linear mm of width}} \leq (2.8d)\,\frac{(F_y - 105)}{140} \tag{12.39}$$

for roller diameter above 635 mm to 3175 mm

$$\frac{\text{line bearing load }(N)}{\text{linear mm of width}} \leq (70\sqrt{d})\ \frac{(F_y - 105)}{140} \tag{12.40}$$

in which

d diameter of roller (mm)
F_y minimum material yield strength (MPa)

Note: For the static case the above allowables can be increased by 50%.

The AREMA (2014) formula for steel rolling treads on flat steel tracks in USCUs is listed below.

USCUs

$$\frac{\text{line bearing load (lb)}}{\text{linear in. of width}} \leq (12\,000 + 80D)\ \frac{(F_y - 15\,000)}{20\,000} \tag{12.41}$$

where

$D > 120$ in.

in which

D diameter of curved surface (in.)
F_y minimum material yield strength (psi)

Note: For the static load case the above allowable can be increased by 50%.

Metric units (converted from US unit formula)

$$\frac{\text{line bearing load }(N)}{\text{linear mm of width}} \leq (2100 + 0.55D)\ \frac{(F_y - 105)}{140} \tag{12.42}$$

where

$D > 3048$ mm

in which

D diameter of curved surface (mm)
F_y minimum material yield strength (MPa)

Note: For the static case the above allowable can be increased by 50%.

Additional general information about contract stresses can be found in Young and Budynas (2002).

Figure 12.11 One example of wire rope construction

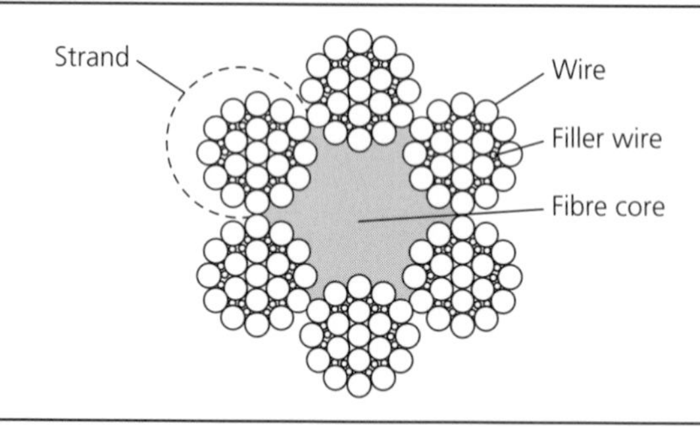

12.7. Selection of wire rope, drums and sheaves

12.7.1 Wire rope

Wire ropes are used in movable bridges to transmit force around an arc. A wire rope is a machine element made up of strands which are twisted around an inner core. The constructions of wire rope are described in Chapter 6. The strands are made of a multiplicity of wires helically twisted, usually about a wire core. The wires and strands move relative to one another as the ropes pass over an arc. Because of this movement, lubrication becomes an important factor in the life of the rope to reduce friction, wear and corrosion. Lubrication of wire ropes during manufacture and in operation is a major consideration. The wire rope cores for movable bridges are typically a fibre core (see Figure 12.11) or an independent wire rope. The rope shown in Figure 12.11 is known as 6 × 25 filler wire. The integer '6' refers to the number of strands, each having 19 main wires and 6 filler wires that occupy voids, as shown in Figure 6.9(a). The grade of wire steel is typically improved plow steel or extra improved plow steel. (AREMA, 2014) Section 6.6, 'Wire ropes and sockets', refers to ASTM Specification A1023 Standard specification for stranded carbon steel wire ropes for general purposes, which also gives data on wire rope construction.

The minimum wire rope strength (minimum breaking force, MBF) for a particular wire rope diameter and construction is given in ASTM A1023 Standard specification for stranded carbon steel wire ropes for general purposes. The size and selection of wire rope is based on safety factors given for the direct tension in the rope and for combined direct tension plus bending of the rope around a sheave or drum.

The AASHTO (2007) wire rope formulae for stress from direct loading and total stress are listed below.

$$\sigma_{\text{direct load}} = \frac{P}{A_{\text{effective}}} \quad \text{(based on AASHTO equation)} \tag{12.43}$$

$$\sigma_{\text{b}} = E_{\text{w}} \frac{d_{\text{w}}}{D_{\text{tread}}} \tag{12.44}$$

$$\sigma_{\text{t}} = \frac{P}{A_{\text{effective}}} + \sigma_{\text{b}} + \frac{P_{\text{o}}}{A_{\text{effective}}} \tag{12.45}$$

in which

$A_{\text{effective}}$ effective metal cross-sectional area of wire rope
D_{tread} tread diameter of sheave grooves

d_w	diameter of outer wire
E_w	tensile modulus of elasticity of steel wire
P	direct load on wire rope
P_o	operating load
σ_b	maximum bending stress in wire rope
$\sigma_{direct\ load}$	stress on wire rope due to direct load (based on AASHTO equation)
σ_t	total stress in wire rope

Note: See AASHTO (2007) for further information on loads and stresses. For a conservative approach, all operating loads acting on the wire rope would be included as part of the direct load, not separately.

The allowable load stress factors for wire rope given in AASHTO (2007) are listed in Table 12.5.

The main counterweight wire ropes are usually terminated at the ends with standard open spelter sockets with custom bored holes for pins or block sockets. Typically, block sockets for main counterweight ropes are custom designed. Operating ropes are also typically socketed with open spelter sockets on one end. The other end is usually terminated at the drum by a wedge assembly, block socket or clamps.

A common method for connecting main counterweight ropes includes an open spelter socket at both ends with a take-up for length adjustment on one end. One end is connected to a connection plate fixed to the counterweight and the other end to a take-up assembly on the lift span (see Figure 12.12). A take-up assembly is also sometimes used on one end of an operating rope for adjustment. With the take-ups, the length and tension can be adjusted for proper load distribution between multiple wire ropes.

Splay angles of ropes are the relative positions from the sheave or drum groove tangent points on one end to the position of the rope at the other end support or connection. Allowable splay angles in terms of ratio of length between points over relative displacement are listed in AASHTO (2007) and AREMA (2014).

12.7.2 Drums and sheaves

Drums and sheaves are used to support and direct the travel of wire ropes. They are made from steel weldments, forged steel or cast steel. The rim is grooved to support the ropes properly. Sheaves can range from ones that have one groove to ones with many grooves to support rope(s). The rim section is connected to the hub by spokes which are made up of webs, diaphragms and stiffeners leading to the hub. Hubs are either shrink-fitted to shafts or have internal bearings. A typical sheave rim section is shown Figure 12.13.

Table 12.5 Wire rope allowable load and stress factors (loads in rope not to exceed factor times MBF) divided by effective metallic cross-sectional area of rope for stress

	Direct load	Total stress
Counterweight rope	0.125	0.222
Operating ropes	0.167	0.300

Allowable wire rope factors from AASHTO (2007)

Figure 12.12 Typical wire rope take-up assembly

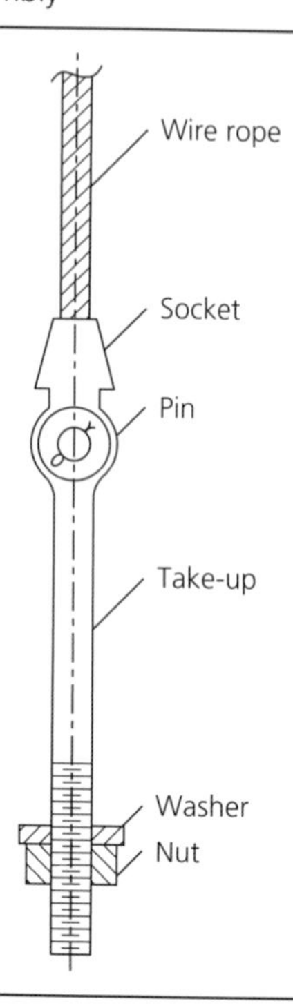

The minimum sheave or drum pitch diameter is typically specified as a function of rope diameter in a movable bridge design specification or recommended practice. As the pitch diameter is increased for a given rope diameter the resulting bending stress in the rope is reduced. See Chapter 6. The minimum ratios of pitch diameter to wire rope diameter required by AREMA (2014) are given in Table 12.6.

The rim sections of sheaves and drums are typically designed as beams loaded by the rope pressure. The rim should also be sized to limit deflection. Uneven deflection of the rim could cause some uneven loading in the ropes. The spokes are essentially columns that support the rim from the hub.

Conservative allowable working stress design may be used for initial sizing of the drums or sheaves. Analysing the rim as a shell may permit the designer to reduce the thickness of the rim. More accurate hand stress analyses or finite element analyses may be used to optimise designs. See AASHTO (2007) and AREMA (2014) for more detail on the design of sheaves and drums.

Large, and more frequently now, even smaller drums and sheaves are fabricated out of steel plate, rings, bars and tubes. A complete fabrication of such a component is commonly referred to as a weldment. More details and formulae used for the design of weldments such as drums and sheaves can be found in Blodgett (1963), AASHTO (2007) and AREMA (2014).

12.8. Linkage design

Linkages are used for many different machinery systems on movable bridges. In some cases they are used as part of the movable bridge support and counterweight systems. These types of bridges include heel trunnion bascules, Rall, Strauss and balance beam bascules. This section of the chapter will only

Figure 12.13 Counterweight sheave

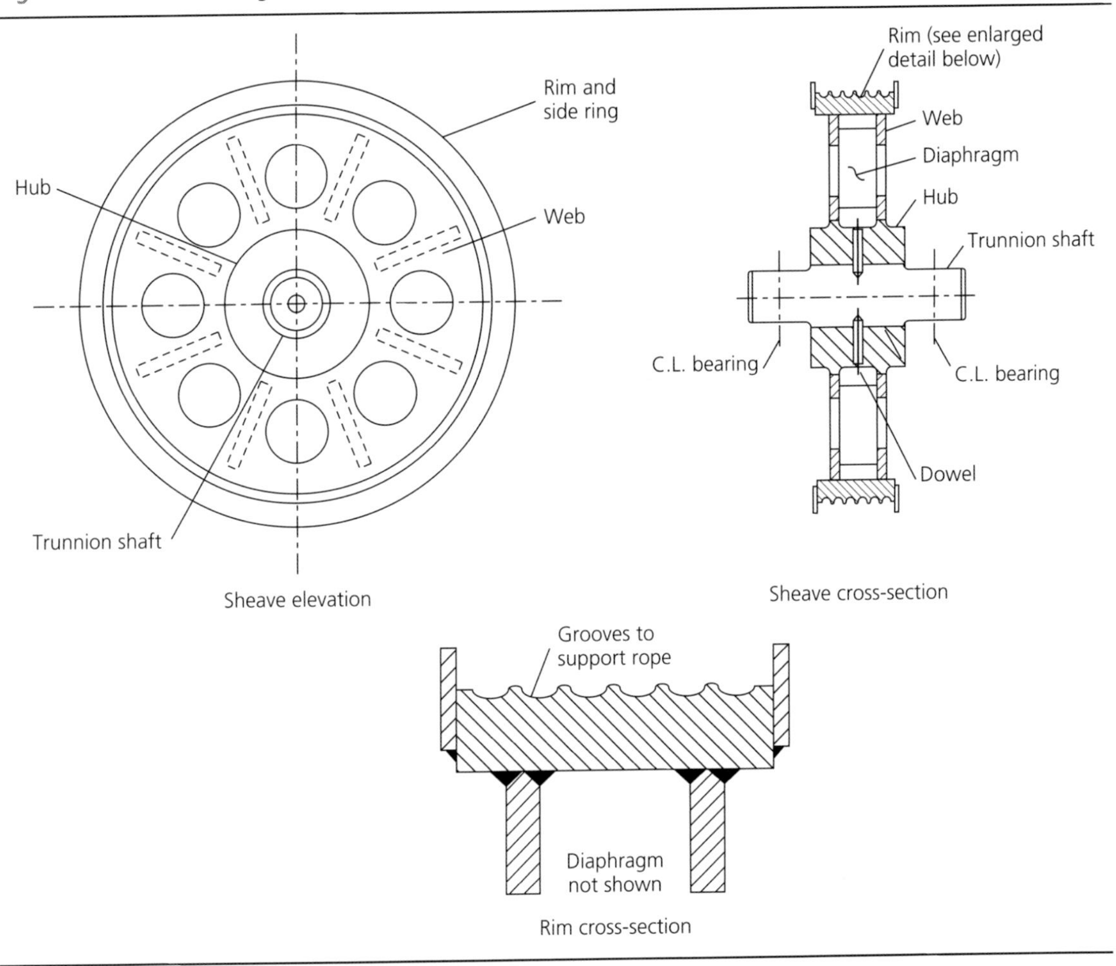

be focused on smaller linkage systems that are used frequently for the drives of lock bars, wedges and tail locks. This common linkage system is called the slider crank.

The basic components of a slider crank mechanism for a lock bar includes a sliding bar, connecting rod and crank. The connecting rod is pinned between the crank and the lock bar. The bar is supported in

Table 12.6 Ratio of sheave pitch diameter to wire rope diameter

	Preferred	Minimum
Counterweight wire rope sheave	80	72
Auxiliary counterweight wire rope sheave	–	60
Operating wire rope drum	48	45
Operating wire rope deflector sheave	48	45

Ratio of sheave pitch diameter to wire rope diameter from AREMA (2014)

guides and the crank shaft is supported by bearings. Both the bar guides and crank shaft bearings are located at fixed points relative to one another. As the crank rotates it causes the bar to translate in a linear motion by virtue of the connecting rod. The linkage for this case is the crank, connecting rod, bar and common base between the bar guides and the crank shaft bearings. Knowing the geometry of the components, the position and the driving force of the bar can be computed from linkage analysis based on the input torque. A general diagram for an in-line slider crank and example equations for position are shown in Figure 12.14. Typically, the length of the connecting rod is at least three times the crank radius.

A powerful tool for numerical analyses of linkages is the digital computer. Even with just a spreadsheet program, displacements, angles of linkages and forces can be computed for finite increments of lock bar motion. Linkage analysis of the slider crank and even of more complex cases is covered in detail in Mabie and Reinholtz (1987).

For a slider crank mechanism, the crank shaft needs to be designed for shear force, bearing area, bending and torsion. The connecting rod is subjected to a variable axial force (tension and compression) as the sliding bar is driven or retracted from the socket. It needs to be checked for buckling due to the compressive force in addition to the axial stresses. The linkage pins at the lock bar end are typically sized for

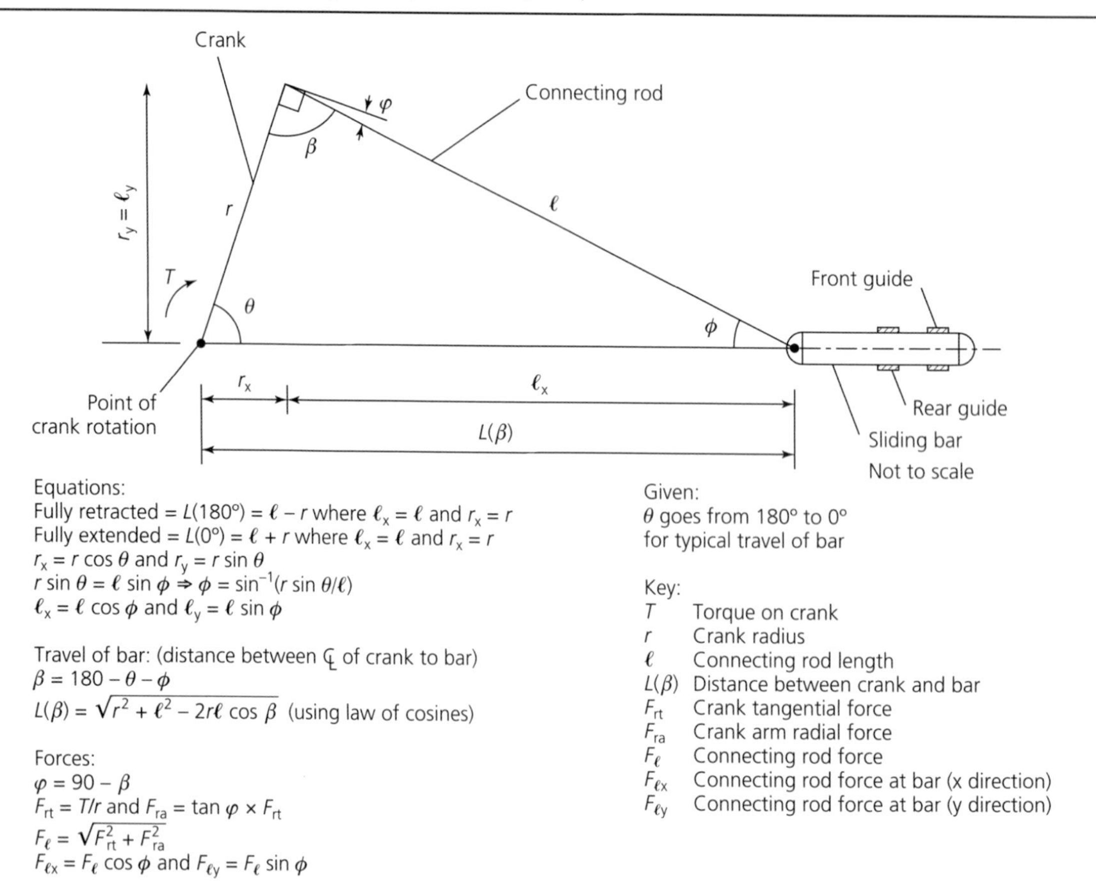

Figure 12.14 General slider crank geometry and example equations

shear forces, bearing stresses and bending. The crank, connecting rod, and pin are typically made from carbon or alloy steel. The crank is typically driven by a reducer and motor with brake.

12.9. Lock bar design

Lock bars are used in a few different forms for locking and aligning movable bridges. Some examples are

- double-leaf bascule – mid-span and tail locks
- single-leaf bascule – toe lock
- swing bridge centring device
- vertical lift bridge – span lock.

A diagram of a general lock bar arrangement which is driven by a crank and connecting rod is shown in Figure 12.15.

Lock bar assemblies comprise a bar sliding in two guides, a receiving socket and drive machinery. One of the more common and most stringent applications of a lock bar for movable bridges is for a double-leaf bascule. For this case the bar is subjected to load reversals as traffic passes over the mated leafs. The bar, guides and socket shoes are subject to high impact forces due to the moving live loads. For this reason this case is described below in detail. Note for other uses of lock bars on movable bridges such as a single-leaf bascule, swing bridge, and vertical lift bridge, the bars are only subjected to significant load when there is a severe misalignment of the spans or when they are used as a safety device for inadvertent powering of the span drive.

The span locks for double-leaf bascules serve to align the leaves and transfer vertical shear between them. The lock bar is usually modelled as a simply supported rectangular bar with one end cantilevered. Lock bars are typically either carbon or alloy steel forgings. On one leaf, the bar is supported by front and rear guides and on the other leaf, a receiving socket. The worst case loading is where one leaf is fully loaded with live load and the other is not. The fully loaded leaf transfers a shear load through the lock bar to the other leaf. Because the lock bar also aligns the spans, the deflection at the tips of the two leafs should be equal, or nearly so. This is the ideal, but actually the guide and socket shoes have clearances with respect to the bar and the bar bends, thus the toes of the mated leaves may be at slightly different levels. Considerations for lock bar design are

- lock bar material and design
- guide design

Figure 12.15 General lock bar assembly

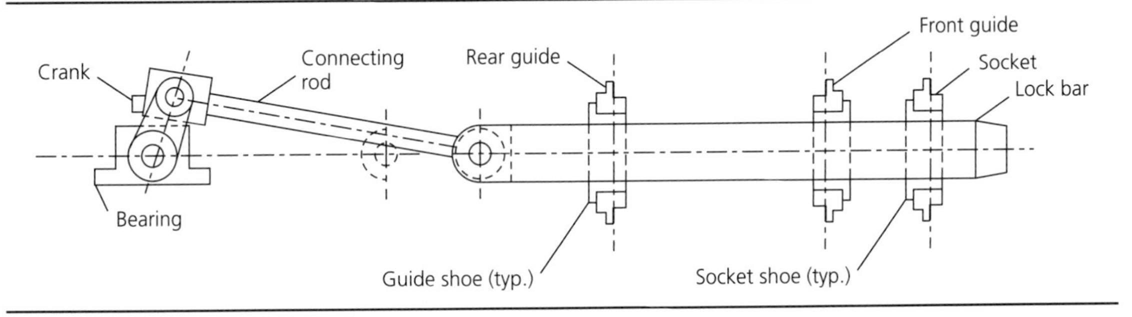

- socket design
- lock bar drive machinery
- lubrication.

Rectangular lock bar design is usually governed by bending stresses due to the load on the end of cantilevered beam. For the case of a double-leaf bascule, the bar is also typically analysed and sized for fatigue due to the many cycles of reversed loading from live load traffic. The guide and socket shoes are designed based on bearing area, or on allowable line contact if a cylindrical surface is used.

12.10. Wedge design

Wedges are usually used for live load supports at swing bridge rest and centre piers (of centre bearing swings). In some cases they are also used for the tail locks on bascule bridges and on hold-down systems for vertical lift bridges. The sloped surfaces of wedges create mechanical advantage to produce high separating forces when pairs are driven together.

In older and current swing bridge designs, the material used for the wedges, top guides/supports and bottom bases, is typically cast steel. In more recent designs, one or more sliding surfaces may be a bronze or nylon plate fastened to the steel wedge in order to reduce friction and promote smother sliding. A diagram of a general end wedge arrangement is shown in Figure 12.16.

The wedge design depends on the following

(*a*) forces created by the wedge as it is driven
(*b*) loads imparted on the wedges after it is driven and subjected to live load
(*c*) the surface area of the wedge and the speed of wedge driving and retracting.

An example free-body diagram of a wedge is shown in Figure 12.17.

The sliding area and vertical movement is a function of the wedge position and should be used in conjunction with the analysis of the driving mechanism to ensure that there is an appropriate driving force available at all positions of the wedge. Design should also include checking the pressures between the wedge, guide and base surfaces for both the dynamic and static load cases. Wedges are typically driven and retracted by machinery similar to that used for lock bars.

Figure 12.16 Wedge assembly

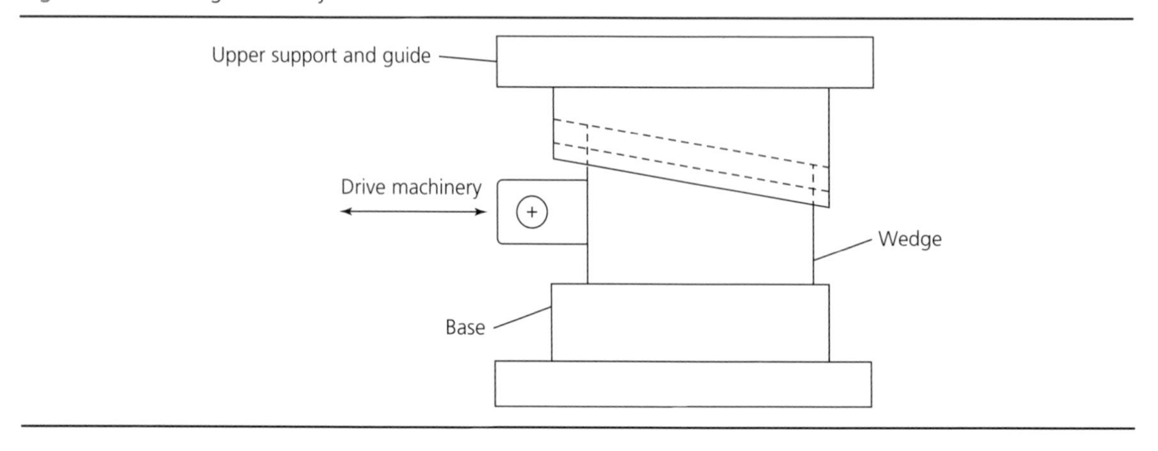

Figure 12.17 General wedge-free body diagram and example equations

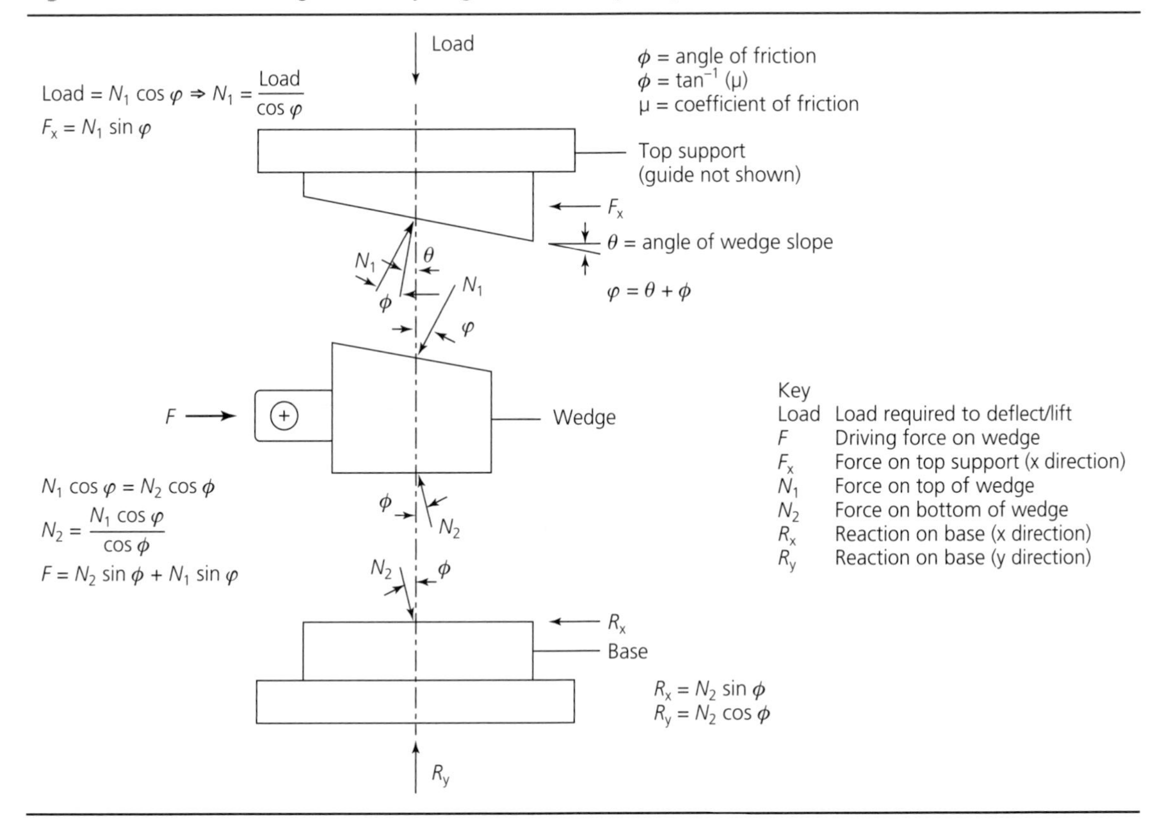

12.11. Selection and design of drive machinery for lock bars and wedges

The drive machinery arrangements for lock and wedges vary. Typical drive systems include a slider crank mechanical drive, rack and pinion, linear power screw actuators and hydraulic actuators. These systems provide power in a linear motion.

The slider crank mechanism was previously discussed in Section 12.8. A rack and pinion converts the rotating motion and torque of a pinion into a linear motion and force by engaging straight rack. Linear power screw actuators use a threaded power screw rod and matching nut to translate either the rotation of the nut and translating the rod, or rotating the rod and translating the nut into linear motion. Threaded power screw actuators are available as pre-engineered manufacturer units. The last system listed, hydraulic actuators, use hydraulic fluid under pressure to translate a rod and piston assembly inside a cylinder with pressure chambers on each side of the piston. This system is also available as pre-engineered manufactured units. Hydraulic machinery is also discussed in Chapter 14.

These systems have specific requirements and specific benefits for each movable bridge. An example, if the main drive system for a bridge is powered by hydraulics, using hydraulics for the sub-systems should be considered.

Another example would be possible considerations for span locks on double-leaf bascules where the machinery is located over water. For a hydraulic system, where a break in a pipe could create an

Figure 12.18 Coupling/shaft alignment

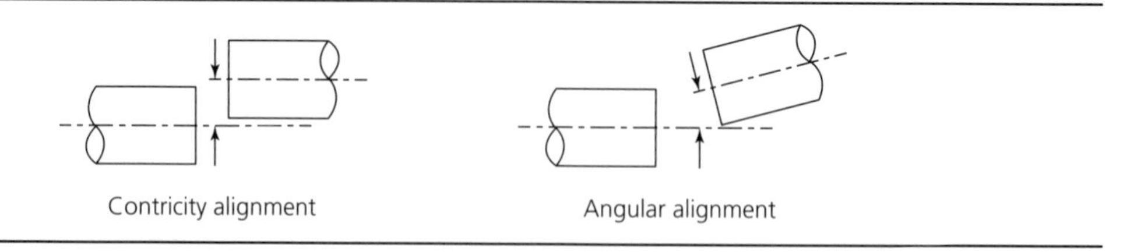

environmental spill, consideration is given to the type of fluid. An environmentally friendly hydraulic fluid could be used. Containment for possible leakage of fluid would also be another consideration.

12.12. Selection of shaft couplings

Machinery couplings connect two shafts and transmit torque during shaft rotations. Flexible couplings permit some initial shaft misalignment and misalignment due to deformation of the structure as it is moved and due to temperature change. The amount of permissible misalignment is dependent on the coupling type and design.

The main alignment requirements are the relative concentricity and angular orientation between the two shafts. Concentricity alignment is the relative positions of the centre of shafts relative to one another. Angular alignment is the relative positions of the shaft centreline axes relative to one another (see Figure 12.18). Some common types of couplings used in connecting movable bridges machinery are

- flex–flex gear coupling – generally used for connecting two fixed shafts (each shaft having at least two bearings)
- flex–rigid gear coupling – generally used for connecting floating shafts
- rigid–rigid coupling – used for connecting two shafts in special applications, often found on older bridges
- grid coupling – typically used as motor couplings (grid has greater resistance to shock loading from the motor than teeth of a gear coupling)
- jaw coupling – typically used to connect small motors to equipment or limit switches.

12.12.1 Gear couplings

A gear coupling comprises a hub with external coupling teeth that are mounted to the end of a shaft and an outer coupling sleeve which has internal gear teeth that mate with the external hub teeth. The typical coupling sleeve also has a flange that can either be bolted to another set of coupling hub and sleeve, making it a double-flex coupling or to a rigid hub which is just a hub with a bolting flange making it a single-flex coupling (see Figure 12.19).

12.12.2 Rigid couplings

Rigid couplings are typically just two rigid hubs bolted together at the centre between the two shafts. This type of coupling does not allow for any concentricity or angular alignment between the two shaft ends, either initially or while the shafts rotate.

12.12.3 Grid couplings

Grid couplings are comprised of a hub that is mounted on each of the two shafts ends that are to be coupled together. The hubs have axial slots cut in an outer periphery of the hub. A one- or

Figure 12.19 Coupling types – flexible gear type

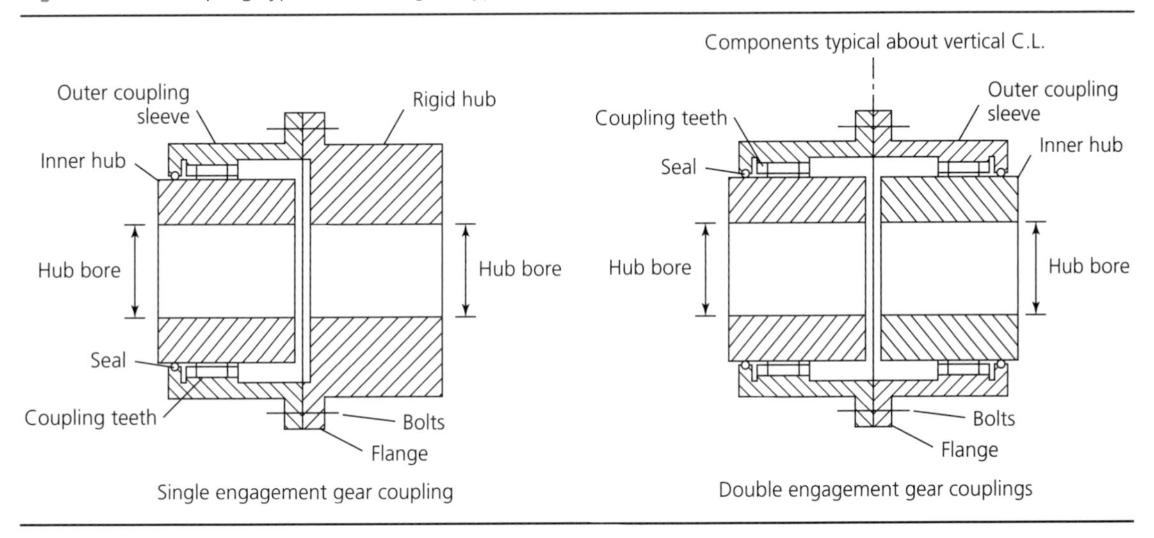

multiple-piece grid is inserted into the slots to transmit the torque and speed between the two shafts. The grid in the slots of the two hubs allows for misalignment between the two shafts. In some coupling designs the grid also is able to absorb some shock loading.

12.12.4 Jaw couplings

Jaw couplings are hubs that have protruding teeth which mesh with one another in the axial direction. The protruding teeth either mesh directly with one another, or otherwise with an insert placed between the coupling teeth. The use of bronze or non-metallic insert is typically used with modern couplings and improves misalignment and shock resistance between the two hubs.

12.12.5 Couplings in general

The coupling types listed above are available as pre-engineered manufactured units. The manufacturers typically publish dimensions, service factors, torque, speed, and alignment capacities in their literature. In most cases the literature also includes installation, operation, maintenance, and lubrication requirements. Coupling selection guidelines are given by coupling manufacturers. The basic minimum selection torque equations are listed below.

In USA units

$$\text{minimum selection torque (in} \times \text{lb)} = \frac{\text{service factor} \times \text{power (hp)} \times 63\,000}{\text{rpm}} \tag{12.46}$$

Metric units:

$$\text{minimum selection torque (mm} \times N) = \frac{\text{service factor} \times \text{power (kw)} \times 9\,550\,000}{\text{rpm}} \tag{12.47}$$

The power used in the above equations would need to be based on the design requirements of the machinery. The couplings for operating machinery are generally sized for a percentage of motor load. In many cases this is 150% full load motor torque. Coupling selection should also include investigation of the overload case. Specific project design specification or recommended practice must be reviewed and addressed along with confirmation of the manufacturer's selection requirements during the design of the machinery for compliance.

There are also special coupling designs that are also available as pre-manufactured units, which are too numerous to list. These types of couplings include low to zero backlash couplings used in limit switch applications.

12.13. Interference fits, key and mechanical friction locking devices

Interference fits are used to connect mechanical components by friction. This is done by making the inner component (a shaft) slightly bigger than the outer component (a hub). The two parts are assembled by a press fit or a shrink fit. Shrink fits are usually preferred so that mating surfaces do not gall during the part's assembly. Typically, an interference denoted as FN2 by ANSI is used. When the material of the two components is the same, the pressure between the two components can typically be calculated using the following equation (Shigley and Mischke, 2001). Note that the pressure also creates hoop stress in the hubs and compressive stress in the shaft, which should also be checked.

$$P_c = \frac{E\delta}{2R^3} \frac{(r_o^2 - R^2)(R^2 - r_i^2)}{(r_o^2 - r_i^2)} \tag{12.48}$$

in which

E modulus of elasticity
P_c shrink-fit pressure
R nominal outside radius of inner member
r_i inside radius of inner member
r_o outside radius of outer member
δ radial interference

The friction coefficient developed between the two parts depends on the types of material and the condition of the surfaces. Typically, for movable bridge machinery both the inner and outer members are steel.

The general equation for the torque capacity of a shrink fit is

$$\text{torque capacity} = 2\pi r^2 L P_c \mu \tag{12.49}$$

in which

L length of hub-bearing surface
P_c shrink-fit pressure
r radius of shaft at friction surface
μ friction coefficient

Keys are installed in almost all coupling, drums and gear connections. Even if an interference fit is used, a key is also typically installed in order to prevent slippage between the shaft and hub surfaces should

Figure 12.20 Shaft and hub key and keyways

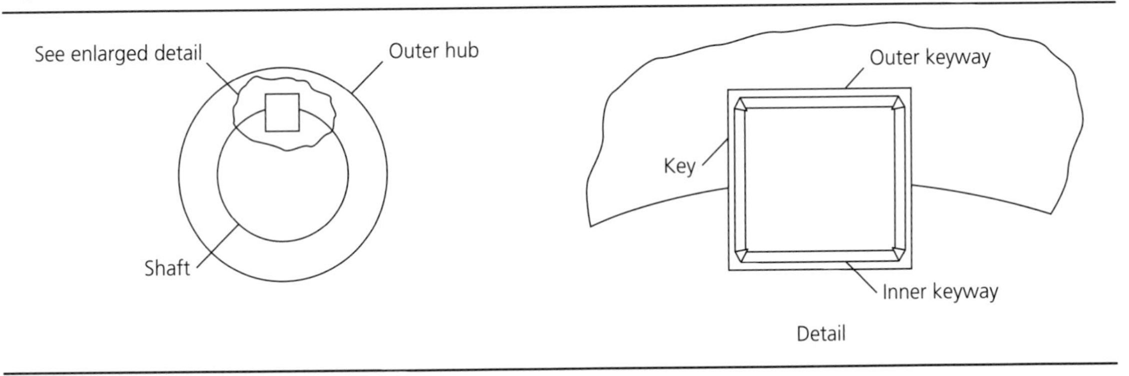

friction be inadequate for transmitting torque. The keys are set in keyways of both the inner and outer component. The key, therefore, is subject to shear and bearing stresses when transferring torque (see Figure 12.20). Keys are typically made of carbon or alloy steel. The corners of keyways are typically rounded as well as the ends in order to reduce stress concentration. The keys are also chamfered on the edges and rounded on the ends to fit the keyway.

When keys are used with shrink fits, they act as a secondary redundant means to transmit the torque. The general stress equations for shaft keys are given below.

$$\text{key shear stress} = \frac{\text{torque}}{r_s \times A_s} \tag{12.50}$$

in which

A_s key shear area
r_s radius at shear plane

$$\text{key bearing stress} = \frac{\text{torque}}{r_b \times A_b} \quad \text{(In some cases } r_s \text{ is used for } r_b\text{)} \tag{12.51}$$

in which

A_b keyway bearing area
r_b radius to centre of keyway bearing area

Note: Bearing stress needs to be checked for both the key/keyway inner and outer bearing surfaces (shaft key/keyway area and hub key/keyway area). Allowable key bearing and shear stresses can be found in AASHTO (2007) and AREMA (2014).

The principles of keys and keyways can also be applied to applications other than shafts and hubs. Such an application is for preventing sliding between two flat surfaces. In this case instead of resisting torque between a shaft and hub the key would be resisting a direct force between two flat surfaces over the key shear and bearing areas.

Mechanical shrink-fit assemblies are similar to interference fits in concept of torque capacity. The difference is that they use a mechanical means to generate the pressure. This is typically done by using opposing wedges drawn together by fasteners (see Chapter 9). The use of mechanical shrink-fit assemblies is typically limited to specific connections and conditions. The appropriate design specification should be consulted. This type of connection is not recommended where the assembly would be subjected to appreciable bending moments during rotation. In such cases slight movement of parts might loosen bolts and diminish the required radial pressure. Sizing and selection of mechanical shrink-fit assemblies is typically based on manufacturer's recommendations

12.14. Brake selection

Brakes serve an important safety function on movable bridges. They are used to stop and hold movable bridges during operation. In some modern machinery installations, the main drive electrical system accelerates and decelerates the drive machinery. However, brakes are still required to hold and also to stop the movable span in the event of power or control failure.

In the early years of movable bridges (1800s to early 1900s) brakes were usually custom designed for a specific application (see Hovey, 1926, for details of older brakes). Custom brakes have been phased out, for the most part, with the introduction of the pre-engineered units used today (see Chapter 15 for further details on modern brakes). This later type comprises a frame, linkages, springs and friction pads which press against either a drum or disk. They are usually spring set and are released by an actuator called a thrustor. In the cases of brakes on small motors, and sometimes for larger brakes for special applications, solenoids are used to release the brake. For auxiliary drive systems, a hydraulic system with foot pedal pump may be found at some installations.

Bridge design specifications generally specify the torque requirements and stopping time for brakes in the machinery system. The requirements are typically given for the motor brakes to slow down, stop and hold the bridge, in a specific time. Motor brakes are located close to the motor. Machinery brakes are typically installed to increase the holding capacity by an additional amount required by design specification or recommended practices. Machinery brakes are located as close to the final driven component as practical. This is typically on the input to a secondary gear reducer so as to minimise the brake size by taking advantage of the final ratio of gear reductions.

12.15. Bolts

Bolts are used for connecting machinery parts together and to their supports. Two common types of bolts are used on movable bridges. The first type is a high-strength finished body bolt or cap screw (actually, the body may not be mechanically machine finished if the shank dimensions meet an ANSI specification) and the second is the high-strength turned bolt. In current design, both types are typically made from the same or equivalent material (ASTM A449). See AASHTO (2007), AREMA (2014) and Bickford (1995) for additional information on bolts and design.

A high-strength finished body bolt or cap screw is a fastener that is manufactured to have a shank with a tight dimensional tolerance. Typically these types of bolts are used in tight tolerance holes so that the total diametral clearance is less than that of a structural connection. A high-strength finished body bolt in a machinery connection is typically designed as a connection similar to a friction-bolted structural connection, but with a tighter limitation on the distance that the connection can slip before the bolt shank goes into bearing. A bolt hole clearance not to exceed 0.010 in. (0.25 mm) is recommended by AREMA (AREMA, 2014). This type of fastener is typically torqued with a suitable nut to 70% of its minimum tensile strength.

Turned bolts are used where the amount of possible slip needs to be less than that of high-strength finished body bolted connection. A turned bolt is typically machined with a shank diameter 1/16 in. (1.6 mm) greater than the thread size. The shank diameter in most cases is held to within 0.000/ −0.001 in. (0.000–0.025 mm) or tighter diametral tolerance. The fit between the shank and hole is anywhere from a very slight clearance to a slight interference. The specific clearance used would be dependent on the connection and design requirements. AREMA (2014) specifies a LTI fit as per Ansi for this type of bolt. These types of bolts are typically used with double nuts or lock washers and tightened to produce a tensile bolt stress that is less than the yield strength.

12.15.1 Bolt preload

One important requirement for any bolt is the initial preload in the bolt at assembly (Bickford, 1995). Properly tensioned fasteners will limit the fatigue cycles a bolt experiences over its years in service. Establishing the preload is seldom a straightforward matter. For structural bolts, the turn of the nut or the calibrated torque wrench are typical methods used for tightening. The calibrated torque wrench method typically involves sampling fasteners of the same lot and storage condition and installing and tightening in a device that indicates tension in the bolt. An appropriate number of samples would be tightened to the required tension in order to establish the proper torque setting of the wrench. For threaded components a general equation for preload and tightening torque is sometimes used. The equation is dependent on fastener diameter, preload, friction, and the condition of the threads and bearings surfaces. The equation listed in Deutschman (1975) is as follows

$$\text{bolt tightening torque } (T) = c \times d \times F_i \tag{12.52}$$

in which

c bolt torque coefficient, also known as the bolt 'K factor'
d nominal bolt diameter
F_i bolt initial preload

The bolt K factor is dependent on thread geometry as well as friction coefficients. Suggested values from Deutschman (1975) for bolt torque coefficients are 0.20 for dry and 0.015 for lubricated conditions. Some anti-seize compound manufacturers list K values even lower for fastener applications. The K values (torque coefficients), including the ones listed here, should be considered approximate and should be checked and confirmed for the actual fasteners, conditions and lubricants.

12.16. Welding

Welding is commonly used to connect steel plates, forged pieces and structural shapes to make machinery components. In general terms, welds of either full or partial penetration, and fillet cross-section, are used to connect steel. The process fuses the two steel base metal pieces together with weld metal (welding rod) using one of a few different processes. For arc welding, a rod, also known as a consumable, has a current that passes through it, which generates the arc and heat necessary to fuse the base and rod metal together. Refer to Blodgett (1963), AASHTO (2007) and AREMA (2014) for further information on welding.

For full penetration welds the complete thickness of metal is fused. For partial penetration, like its name, only a partial thickness of the metal is fused together. Fillet welds are triangular welds at the inside corner of two plates. The short sides of the triangle fuse to the plates and loads are transmitted by shear through the minimum throat of the weld. Some common welding processes used for the fabrication of machinery components include

- manual shielded – arc welding
- automatic submerged – arc welding.

Standard weld procedures have been developed for common welds and special weld procedures may need to be developed for non-common welds. The procedures typically include specifying base metal thickness, type of welding rod, speed of and number of passes, preheat of material and amperage used to fuse the material.

The procedures, processes and requirements for welding for movable bridge components are typically specified in the bridge design and construction specifications or recommended practices. An example of this is the fabrication of sheaves for counterweight ropes to support the lift span and counterweight. AASHTO (2007) lists requirements for material, and other specifics. It also states that automatic submerged-arc welding should be used where practicable, and that full penetration welds be made from low-hydrogen electrodes. See Chapter 7 for deleterious effect of hydrogen on steels used in manufacturing machine parts.

During the welding process, the pieces should be held in rigid fixtures and a suitable weld procedure used to minimise distortion. Welded components for machinery are typically stress relieved after welding is complete and prior to machining. This is done to relieve residual stresses from the welding process and to prevent distortion of the weldment caused by relieving them when machined later (see Chapter 6). Machining processes are used either to finish surfaces used for mounting of the components, or as sliding and contact surfaces with other components. Sliding and contact surfaces include gear tooth surfaces, hubs and bores for shafts, and wire rope grooves. Sometimes the weld material is also required to be ground to a specific profile or to just be smooth. This is common at transition points and zones of high cycle stress.

12.17. General gearing

Open and enclosed gearing design is an important subject for movable bridge design. These components are most often used for mechanical systems of movable bridge drives.

Gear design is a broad topic and is covered in Chapter 13. This section is intended to review a simple design method used for high load, low speed, spur gearing where the gear tooth is analysed as a cantilevered beam. For a simple analysis the Lewis equation (Hovey, 1926; Spott *et al.*, 1985; Shigley and Mischke, 2001) is sometimes used with a velocity factor. A variation of the Lewis equation for spur gear design is given in the AREMA recommended practice. These equations are also listed in Chapter 13. They are used for preliminary sizing and, in some cases, for final design, as with AREMA recommended practices. However, in any case the wear and fatigue should be considered in the final design.

12.18. Machinery inspection and maintenance

Inspection plans and programme are important to maintain a movable bridge in good working order for the full design life. The intent of these plans and programmes are to identify small problems before they become big problems.

Inspection categories include maintenance, routine, biennial and special. The inspection tasks include checking for proper lubrication, paint systems, fasteners, alignments, wear, bearing clearances, loose or broken components, brake function, any overstress condition or excessive vibration in components. Inspections are also used to monitor conditions previously noted during inspection or maintenance.

They typically include observation of the components at rest as well as in motion from a safe area. Specific requirements and frequencies for inspections are developed for a movable bridge and also conform to a required specification. An example is the inspection of movable highway bridges in the USA which follow the applicable portions of AASHTO (1998). Another reference that includes additional information on inspection specific to movable bridges is Parsons Brinckerhoff (1993).

Proper maintenance is also required to maintain a movable bridge in good working order for its full design life. Maintenance programmes not only address regular tasks that need to be maintained, such as proper lubrication and adjustment of components; they can also help identify minor problem areas before they progress into major ones. Maintenance should include

- checking, and changing when required, and maintaining proper lubrication
- checking and repairing the paint system for protection from corrosion
- checking, maintaining, and adjusting brakes
- checking for loose fasteners
- performing tests (including imbalance measurements) to identify potential problems
- checking for loose or broken components
- monitoring any adverse conditions previously noted during inspection or maintenance.

12.19. Other machinery design considerations

Neither AASHTO (2007) nor AREMA (2014) deal explicitly with the topics of single-mode failure analysis and machinery vibration (except for shafting). Yet these two topics are important considerations in machinery design.

Both the AASHTO design specifications and the AREMA recommended practice do not specifically discuss single-mode machinery failure analysis. It is implied that if the design specification requirements are satisfied, there will not be any failure causing an uncontrolled motion. But machinery components do fail. The operating machinery design should be such that the failure of one component should not cause the bridge to undergo uncontrolled motion of the movable span. The bridge machinery should be analysed for single-point failure (Sööt, 1990).

A movable bridge as a whole is a machine and unless there has been successful experience with a design of similar type and scale, it may be advisable to consider the dynamics of the structural and machinery system. Adverse vibration effects have occurred at resonance between structural, electrical and mechanical components. It may be desirable to analyse the whole design from a vibration viewpoint with regard to equipment damage or malfunction, structural fatigue failure, noise, or human discomfort. The most detrimental dynamic effects occur at resonance, between machinery units or between a machinery unit and the structure. Resonance occurs when the natural vibrating frequency of a component is close to the frequency of the driving force. For situations where resonance occurs, or is likely to occur, between the machinery and the structure, such as at a machinery support it is preferable to stiffen the structure. Adding stiffness with little added mass results in shifting of resonance to higher frequencies. In general, the higher the natural frequency of the component or system compared to the frequency of the driving force, the better.

If resonances are found to occur within the range of the excitation frequencies, and redesign is not feasible, it may be necessary to introduce damping into the system. Methods to reduce vibration at resonance are discussed in (Ungar, 1964).

12.20. Summary

Movable bridge machinery as well as structural designs are seldom exactly the same from bridge to bridge. However, valuable experience and knowledge can be gained from studying the designs of existing movable bridges. Past experience is also very important in the development of new designs and concepts.

Each movable bridge design project should start with establishing the governing design criteria, specification, recommended design practice, or owner requirements. Note that specifications and recommended practices are revised and updated over time. They include more details and requirements than discussed in this chapter. Therefore, the applicable specifications and recommended practices should be reviewed completely and frequently, so that all requirements will be satisfied. They become the basis of the design and should be checked throughout the design process for compliance.

Design is an iterative process. From the start, a design is formulated and then is optimised in order to develop a final design that satisfies all of the established objectives. Experience and skill of the engineer is needed to determine when design meets all of the requirements and is ready for bid or fabrication and construction.

REFERENCES

AASHTO (American Association of State Highway and Transportation Officials) (1988) *Standard Specifications for Movable Highway Bridges*. AASHTO, Washington, DC, USA.

AASHTO (1998) *Movable Bridge Inspection, Evaluation and Maintenance Manual*. AASHTO, Washington, DC, USA.

AASHTO (2007 with interim revisions) *Standard specifications for movable highway bridges*, 2nd edn. AASHTO, Washington, DC, USA.

AREMA (American Railway Engineering and Maintenance-of-Way Association) (2014) Movable bridges – steel structures. Chapter 15 in *AREMA Manual for Railway Engineering*. AREMA, Lanham, MD, USA.

Bannantine J, Comer JJ and Handrock JL (1990) *Fundamentals of Metal Fatigue Analysis*. Prentice Hall, Upper Saddle River, NJ, USA.

Bickford JH (1995) *An Introduction to the Design and Behavior of Bolted Joints*, 3rd edn revised. CRC Press, Boca Raton, FL, USA, http://dx.doi.org/10.1115/1.3269076.

Blodgett OW (1963) *Design of Weldments*. James F Lincoln Arc Welding Foundation, Cleveland, OH, USA.

Childs PRN (2014) *Mechanical Design Engineering Handbook*. Butterworth-Heinemann, Oxford, UK.

Deutschman AD, Michels WJ and Wilson CE (1975) *Machine Design: Theory and Practice*. Macmillan, New York, NY, USA.

Hool GA and Kinne WS (1943) *Movable and Long-Span Bridges*, 2nd edn. McGraw-Hill, New York, NY, USA.

Hovey OE (1926) *Movable Bridges*, in two vols. Wiley, New York, NY, USA.

Koglin TL (2003) *Movable Bridge Engineering*. Wiley, Hoboken, NJ, USA.

Mabie HH and Reinholtz CF (1987) *Mechanisms and Dynamics of Machinery*, 4th edn. Wiley, New York, NY, USA.

Oberg E, Jones FD, Horton HL, and Ryfel HH (2012) *Machinery's Handbook*, 29th edn. Industrial Press, New York, NY, USA.

Parke GAR and Hewson N (2008) *ICE Manual of Bridge Engineering*, 2nd edn, Thomas Telford, London, http://dx.doi.org/10.1680/mobe.34525.

Parsons Brinckerhoff (1993) *Bridge Inspection and Rehabilitation – A Practical Guide*. Wiley, New York, NY, USA.
Pilkey WD (1997) *Peterson's Stress Concentration Factors*, 2nd edn. Wiley, New York, NY, USA, http://dx.doi.org/10.1002/9780470172674.
Rankine WJM (1877) *Machinery and Millwork*, 3rd edn. revised by E.F. Bamber. Charles Griffin and Company, London.
Shigley JE and Mischke CR (2001) *Mechanical Engineering Design*, 6th edn. McGraw-Hill, New York, NY, USA.
Sööt O (1990) The need for single failure proof design for movable structures. *Proceedings of the 3rd Biennial Symposium of Heavy Movable Structures*. St Petersburg, Florida, 12–15 Nov.
Spotts MF (1985) *Design of Machine Elements*, 5th edn. Prentice Hall, Englewood Cliffs, NJ, USA.
Spotts MF, Shoup TE and Hornberger LE (2004) *Design of Machine Elements*, 8th edn. Pearson Prentice Hall, Upper Saddle River, NJ, USA.
Ugural AC (2004) *Mechanical Design: An Integrated Approach*. McGraw-Hill, New York, NY, USA.
Ungar EE (1964) Mechanical vibrations. In *Mechanical Design and Systems Handbook* (Rothbart HA). McGraw-Hill, New York, NY, USA.
Waddell JAL (1916) *Bridge Engineering*. Wiley, New York, NY, USA.
Young WC and Budynas RG (2002) *Roark's Formulas for Stress and Strain*, 7th edn. McGraw-Hill, New York, NY, USA.

Movable Bridge Design
ISBN 978-0-7277-5804-0

http://dx.doi.org/10.1680/mbd.58040.297

Chapter 13
Gearing and speed reducer design

Robert L Cragg and Robert J Tosolt

13.0. Introduction

Movable bridge operating systems have historically used mechanical power transmission components to couple the prime mover to the movable span and generate movement. From the outset, the power train of choice utilised gearing drives in the USA.

Gears are shaft-mounted machine elements taking the form of toothed (or cogged) wheels that operate in pairs (i.e. sets) to transmit motion and power between shafts by successively engaging teeth on the mating wheels. The evolution of gears was related by Dudley (1969) (see Figure 13.1 and Chapter 9). The point of engagement of the mating wheels is denoted the gearmesh. For each gearset, the driving gear is typically denoted as the pinion and is on the motor side of the gearmesh; the driven gear is denoted as the gear and is on the load side of the gearmesh. The gearing ratio is the ratio of gear teeth to pinion teeth and impacts the system torque and speed in an inverse proportion. With each subsequent gear reduction in a gear train, the system torque increases and the speed decreases by the gear ratio. Careful selection of the total gearing ratio in the design process is required to achieve the necessary torque to overcome design loads while producing a rate of bridge movement that is acceptable.

Gearing drives on movable bridges utilise both open and enclosed gearing. Shaft-mounted gears that are supported either on individual pedestals or in common frames, and that are exposed to the

Figure 13.1 Basic gear nomenclature

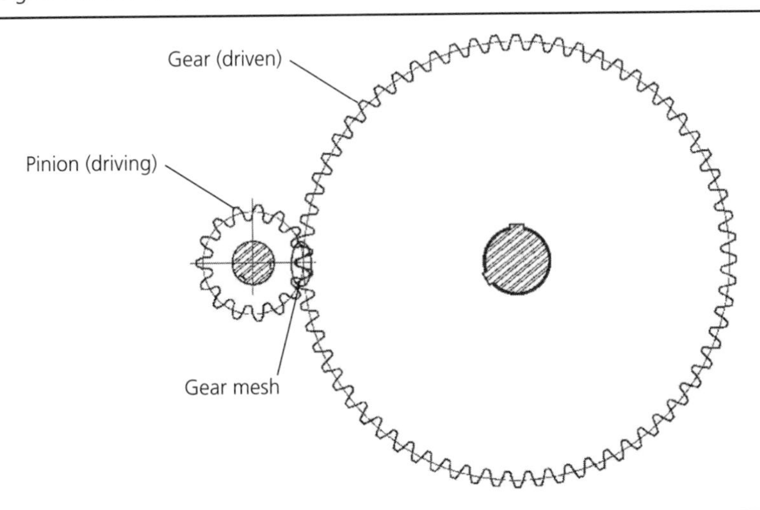

Figure 13.2 Open gear span drive. (Courtesy of Stafford Bandlow Engineering)

environment as depicted in Figure 13.2, are considered open gearing. Shaft mounted gears that are supported in sealed housing are considered enclosed gearing, speed reducers or, in general, gearboxes (see Figure 13.3). With advancing technology, open gearing has been replaced with enclosed speed reducers where practicable. However, the final connection of the drive train to the bridge structure has remained, usually because of size, an open gearset comprising a large drive pinion mating with a large rack or circular gear. In addition, where limited rehabilitation is performed on existing bridges, sometimes the practicable alternative is to utilise open gearing throughout. Therefore, any general discussion of movable bridge design is incomplete without a discussion of gearing.

Figure 13.3 Parallel shaft speed reducer. (Courtesy of Stafford Bandlow Engineering)

13.1. Symbols and abbreviations

Add gear tooth addendum (in.)
B/L gearset backlash (in.)
C gearset centre distance (in.)
C_f surface condition factor for pitting resistance
C_H hardness ratio factor
C_p elastic coefficient for steel
d pinion pitch diameter (in.)
D gear pitch diameter (in.)
Ded gear tooth dedendum (in.)
Eff power transmission efficiency
f, F tooth net face width (in.)
FLT full load torque
I geometry factor for pitting resistance
J geometry factor for bending strength
K_B rim thickness factor
K_f stress correction factor
K_m load distribution factor
K_{my} load distribution factor for overload condition
K_o overload factor
K_R reliability factor
K_S tooth size factor
K_T temperature factor
K_v dynamic factor
K_y yield strength factor
m_G gear ratio
n number of teeth
N_G number of teeth in the gear
PA pressure angle (degree)
P_c circular pitch (in.)
P_{nd} normal diametral pitch (in.)
p circular pitch (in.)
p_d pitch diameter
S_{ac} allowable contract stress (psi)
S_{at} allowable bending stress (psi)
S_{ay} allowable yield stress (psi)
S_F bending safety factor
S_H pitting safety factor
s permissible unit stress (psi)
T full design torque load (in.–lb)
T_i input torque (in.–lb)
T_o output torque (in.–lb)
V velocity of pitch circle (ft/min)
W allowable tooth load (lb)
WD tooth whole depth (in.)
W_{max} maximum factored resistance (lb)
W_t transmitted tangential load (lb)
W_{tac} factored surface durability resistance (lb)

W_{tat} factored flexural resistance (lb)
y tooth form factor
Y_N stress cycle factor for bending strength
Z_N stress cycle factor for pitting resistance

13.2. Gear design

Selection of gear size, gear type and gear material is predicated on ensuring that gear stresses remain within allowable levels for the imposed design load. In the USA, movable bridge machinery design is presently governed by AREMA (American Railway Engineering and Maintenance-of-Way Association, 2014) and AASHTO (American Association of State Highway and Transportation Officials, 2007). Both require allowable working stress design for machinery even though AASHTO (2007) is entitled a load and resistance factor design (LRFD) specification. In fact, AASHTO states in the commentary provision C1.3.1 on page 1.5 that

> The design of bridge machinery in the United States is based on allowable working stress design, therefore, this Specification follows the accepted industry design practice in this regard. As of this writing (2006), reliability-based design at the strength limit state is not possible given the dearth of necessary data.

The following section describes the types of gears commonly used on movable bridges and the fundamental calculations that have historically been utilised to size the gearing.

13.2.1 Open gearing fundamentals

Gearsets are used to transmit force and motion between paired shafts. All gearsets have two members – a driving component and a driven component. The driving component, called the pinion, usually has fewer teeth than the driven component, known as the gear. The ratio of the gear teeth, or gear ratio (m_G), is the number of teeth in the gear (N_G) divided by the number of teeth in the pinion (N_P): this ratio is inversely proportional to the gearset speed, but directly proportional to the increase in torque that the gearset transmits. Thus a 48-tooth gear meshing with a 16-tooth pinion ($m_G = N_G/N_P = 3$) will operate at a third of the speed but will transmit three times the torque of the pinion. Because power losses occur in all gear sets, an efficiency factor (Eff) should be included when calculating torque. The AASHTO LRFD design specification (AASHTO, 2007) paragraph 5.8.4.1 recommends an Eff design value of 0.96 for each open gear set. Therefore, for a known input torque (T_i), the output torque may be calculated as $T_o = T_i \times m_G \times \text{Eff}$. Two types of gears that are usually used for open (exposed) gear sets on movable bridges are spur and straight bevel gears.

13.2.1.1 Spur gear fundamentals

Spur gears are cylindrical in form with teeth on the outer circumference that are straight and parallel to the shaft axis and transmit power and speed between parallel shafts. Mating gears can have differing sizes. However, the gear teeth must share similar geometry to ensure that they will mesh properly. Tooth standards were established to ensure interchangeability of gears of all tooth numbers that are cut to the same pitch and pressure angle. Each gear must have an integer number of teeth and the gearset must be set at the proper centre distance (C). Figure 13.4 illustrates the following features and nomenclature.

- Pitch, which defines the proportions of the teeth, may be stated as circular (P_c) or normal diametral (P_{nd}). Circular pitch is the distance from a point on one tooth to the corresponding point on the next tooth, measured along the pitch circle. Diametral pitch designates the ratio of

Figure 13.4 Spur gearset nomenclature

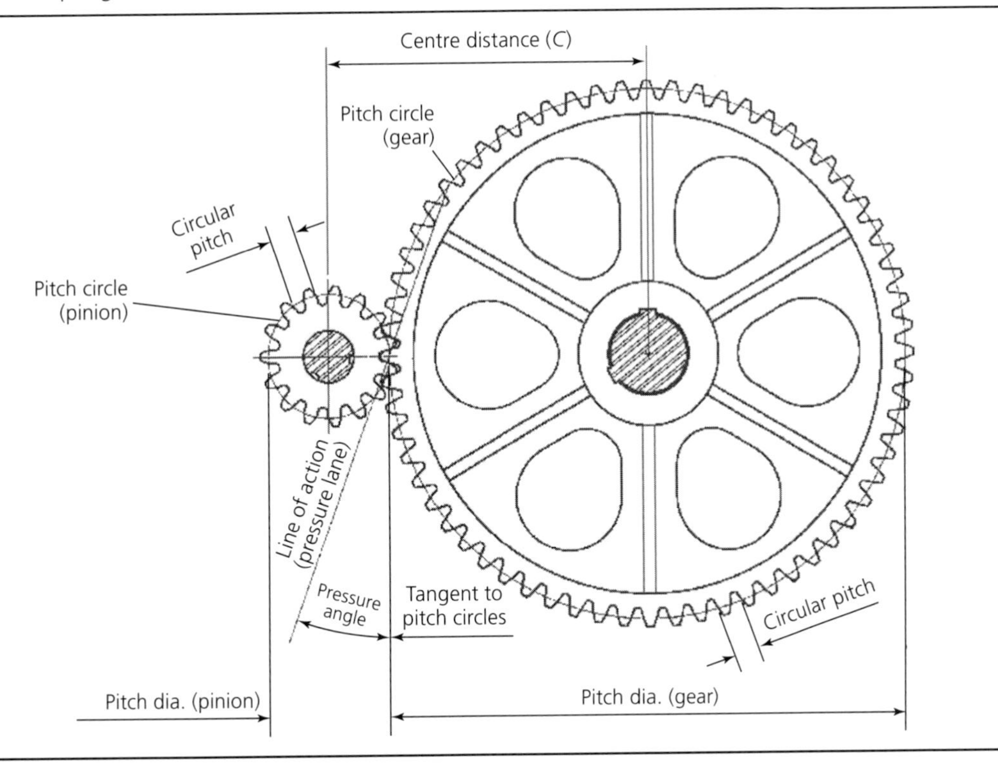

the number of teeth in the gear to its pitch diameter – that is, the number of teeth for each inch of pitch diameter; pitch diameter = number of teeth/diametral pitch. Accordingly, the pitch diameter of a $P_{nd} = 4$ pinion with16 teeth is $d = N_P/P_{nd} = 4$ in.. The relationship between circular and normal diametral pitch is expressed as $P_{nd} = \pi/P_c$.

- Pressure angle (PA) is defined as the angle at a pitch point between the line perpendicular to the involute tooth surface and the line tangent to the pitch circle.
- Centre distance (C) is the sum of the pitch diameters of the pinion (d) and gear (D) divided by 2.
- The pitch circle is a circle having a circumference of π times the pitch diameter. (Every open gear used in movable bridges should have its pitch circle machined into both ends of the teeth during manufacture.)

The objective of gear teeth is to transmit load and motion between adjoining parallel shafts, and to do so in a uniform manner. When the gear tooth profiles transmit uniform angular velocity from one shaft to another, the tooth profiles are said to be conjugate. The preferred solution for obtaining conjugate action, as demonstrated through its widespread acceptance in industry, is through use of an involute profile. The involute concept predates Rankine (1876); it was probably due to Leonard Euler because a thorough mathematical treatment of the involute was presented in a handbook on statics (Eytelwein, 1808, vol. 3, p. 50) along with other transcendental curves studied by Euler. An involute is the curve traced by a point on a taut cord unwinding from a cylinder, which, applied to gearing, is the base circle of the gear; the involutes on the opposite sides of each tooth are inclined toward each other (Buckingham, 1988; Hindhede *et al.*, 1983; Shigley, 2001; Spotts *et al.*, 2004; Childs, 2014). The

Figure 13.5 Spur gear tooth nomenclature

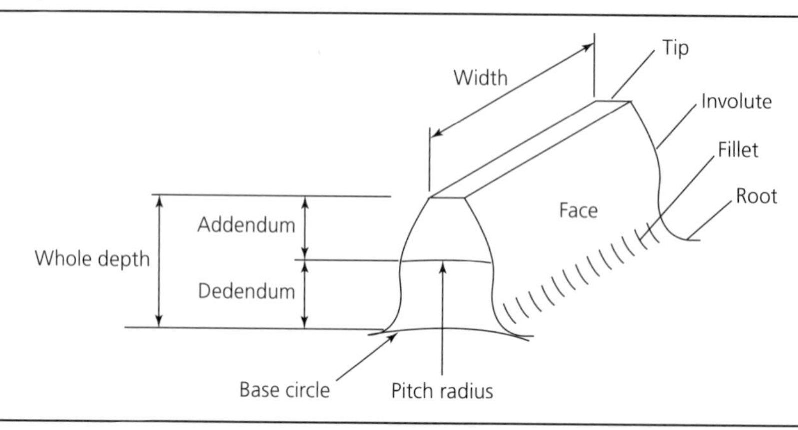

actual tooth proportions are governed by the tooth system selected. The involute profile and other characteristics that are used to define spur gear tooth systems are illustrated in Figure 13.5 and are defined as follows:

- pitch radius = pitch diameter/2
- addendum (Add), the radial distance from the pitch circle to the outside diameter, $1/P_{nd}$
- whole depth (WD), the radial distance from the root circle to the outside diameter, $2.25/P_{nd}$
- dedendum (Ded), the radial distance from the pitch circle to the root diameter, WD minus Add
- tooth face, the active tooth surface in contact during operation
- face width (F), the width of the active face
- root, the surface between the dedenda of adjacent teeth
- root circle, the circle which describes the root
- base circle, the circle from which the involute profile is derived
- fillet, the smooth curved surface between the root and tooth face
- axis, the central line about which the gear rotates.

The terminology described above complies with ANSI (American National Standards Institute)/ AGMA (American Gear Manufacturers' Association) 1012-G05, which provisions are those utilised by AASHTO. The relationships specified by other organisations' standards for movable bridges (AREMA, 2014 for instance) may vary slightly from those given, but the basic definitions and considerations are the same.

Teeth may be full depth or stub forms, which terms refer to modifications to the tooth proportions including the addendum, dedendum, working depth and tooth thickness. Stub teeth are shorter than full depth ones and are somewhat stronger at their root; full-depth teeth operate more smoothly. Early bridges used stub teeth gears, but today the established standard of almost universal use is the ANSI 20° standard spur gear form, which has proved reliable through many years of service. Where loading demands exceed the full-depth gear tooth ratings, modified addendum gearing can be employed to improve the strength of the most heavily loaded gearsets. Early designs frequently used 'as-cast' gears, particularly for the coarse pitch racks and pinions, without further finishing or machining the teeth. Today, teeth on all gears are accurately machined to close tolerances to enable smooth, conjugate tooth meshing during operation. In many cases the pinion is integral with the shaft, having its teeth cut into the shaft, and the gears are separate pieces, mounted on their supporting shafts with an interference or keyed fit. Surface finish of the tooth faces is also very important to assist in providing long reliable service. When gearsets are properly mounted to the correct centre distance, the axes of the pinion and gear will be parallel and the tooth faces of the teeth must also be parallel so that full face contact exists as the gears rotate, and the load is evenly distributed across the faces of the driving and driven gear. Correct setting is achieved when the pitch circles of the pinion and gear are tangent at both ends of the teeth; and a line drawn through the centres of rotation of the pinion and gear intersects the pitch lines at their point of tangency. To provide a slight clearance for a lubrication film to separate the tooth faces, and avoid metal-to-metal contact during operation, the teeth of one or both gears in the set are intentionally cut a little narrow; the resulting assembled clearance is known as backlash (B/L). Precisely defined, it is the amount by which the sum of the circular tooth thicknesses of two gears in mesh is less than the P_c. Practically, it may be considered as the amount by which a space between two teeth is greater than the thickness of one tooth. There are no standards on the amount of backlash to be provided in any gearset; but AGMA gives suggested amounts, based on ($P_{nd)}$ and (C) in its publication ANSI/AGMA 2002-B08, Appendix A (ANSI/AGMA, 2014).

For more in-depth treatment of gearing basics and tooth modification, the reader is referred to (Dudley (1984) and Shigley *et al.* (2004).

13.2.1.2 Rack fundamentals

A rack is a spur gear section having an infinite pitch radius – accordingly the section has a straight pitch line and is flat, as illustrated in Figure 13.6. The tooth forms, dimensions and descriptions are identical to those associated with spur gears. Specifically, according to gearing terminology, a rack is flat, but the colloquial use of the term in movable bridges also includes large-diameter gear segments, as used in trunnion bascules and large-diameter full ring gears, such as those on vertical lift counterweight sheaves and swing span pivot piers. Briefly, the final drive gear on a movable bridge may be called a 'rack'. Of course, because racks are the final connection between the moving span and pier, they are subjected to the greatest forces and should have the sturdiest teeth.

Scherzer rolling lift bascule and Strauss heel trunnion bascule bridges normally have flat racks, which result in a linear motion between the rack and pinion, so applied as to cause the movable structure to open and close. One such rack for a Strauss heel trunnion bridge is depicted in Figure 13.7. Full ring gear racks used on tower drive vertical lift bridge counterweight sheaves are mounted directly on the end of the sheave and swing span racks are usually mounted on the top of the pivot pier. Span drive vertical lift bridges have the racks mounted on the ends of the operating rope drums. Racks on simple trunnion bascules are usually mounted on the bascule girders and may have either external teeth on the outside diameter or internal teeth on the inside diameter. Figure 13.8 shows a simple trunnion bascule girder having a rack with external teeth, in a partially open position.

Figure 13.6 Straight rack

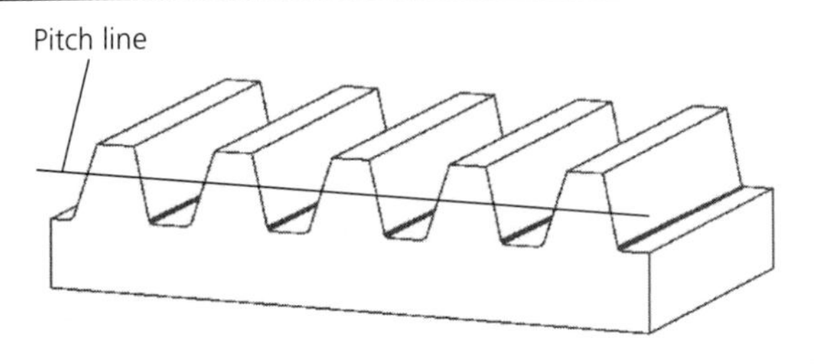

Figure 13.7 View along length of straight rack. (Courtesy of Stafford Bandlow Engineering)

13.2.1.3 Bevel gear fundamentals

Straight bevel gears are used to transmit motion and power between shafts that are at an angle to one another. Basically, a bevel gear is a frustum of a cone having tapered teeth cut into the conical surfaces. A set includes a driving and a driven component, and may be used to reduce the shaft speed or, at times,

Figure 13.8 Curved rack. (Courtesy of Stafford Bandlow Engineering)

Figure 13.9 Bevel gearset nomenclature

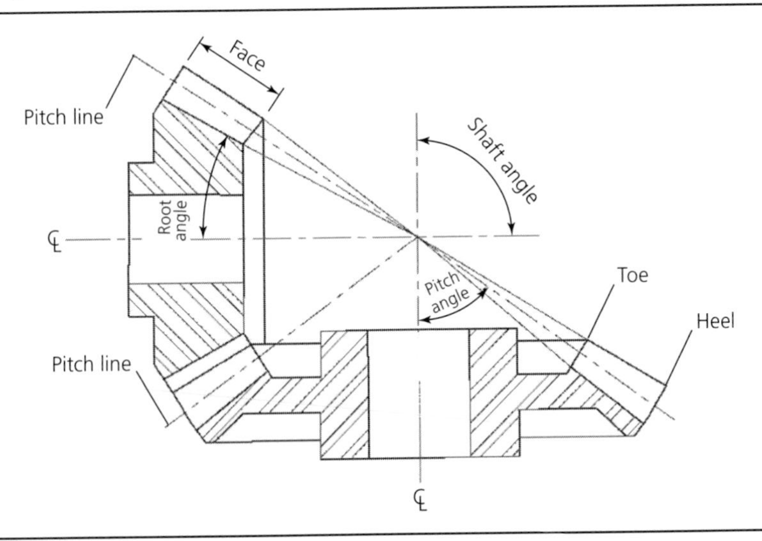

merely change the direction and maintain the same speed (1 : 1 ratio). The terms used to describe spur gears apply to bevel gear teeth: the pitch diameter is at the outside, the large end of the teeth, and the specified size dimensions are given for the large end of the tooth. Additional terms describing bevel gears are shown in Figure 13.9.

Virtually all bevel gear sets on movable bridges are mounted on shafts at 90° to one another. A typical example is the drive on a swing span, where the output shafts of the primary drive are positioned horizontally and the low speed shafts are vertical as depicted in Figure 13.10.

Figure 13.10 Bevel gearset at output of primary reduction on swing span. (Courtesy of Stafford Bandlow Engineering)

Figure 13.11 Open differential gearset, manufactured c. 1925. The bevel pinions are located between the chocks, on either side of the assembly, which resist the separating force and maintain the position of the bevel gears. (Courtesy of Stafford Bandlow Engineering)

13.2.1.4 Gear differential fundamentals

Load sharing between the final drive pinions is now considered desirable for most bascule and two pinion drive swing bridges (see also Chapter 9). A special adaption of parallel shaft gears and bevel gears, called a differential, is the most satisfactory way of achieving load sharing in mechanical systems. Figure 13.11 shows an open differential gearset used to equalise loading to the drive pinions on a Strauss heel trunnion bascule bridge.

The typical differential consists of a ring gear having no fewer than two, typically three, and sometimes more, equally spaced bevel pinions mounted within the web of the gear. These pinions engage bevel gears mounted on the two transverse, output shafts, which drive the rack pinions, either directly or through additional reduction. Torque is transmitted from the ring gear, through the bevel pinions to the bevel gears. Because the output shafts can rotate independently, any tendency of one drive pinion to require increased torque demand will slow that shaft and proportionally increase the speed of the other shaft, thus quickly resolving the situation and equalising the torque to each output shaft.

13.2.1.5 Open gearset design and rating

Gearsets of movable bridge span drives are individually designed to sustain the stresses at 150% of rated full motor torque (FLT) for their life expectancy. Although there is no accepted standard for life expectancy, 10 000 operating hours is a frequently used value. Of course, the owner and designer should agree on a life, considering the anticipated frequency of operation. The calculation method followed to determine the rating, or suitability, of a gearset varies, depending on the standards designated in the contract documents and special provisions prepared by the owner.

13.2.1.6 Spur gear design

Although the matter of spur gear tooth form received much attention in the 19th-century literature it appears that the matter of gear tooth thickness did not. In Rankine (1876) the formula cited for required cast-iron tooth thickness was that attributed to Thomas Tredgold (1788–1829). Tredgold's derivation assumed that the whole contact force between a pair of teeth due to torque was transmitted between mating teeth at a corner of the tooth. A conservative approach, but reasonable in situations of severe shaft misalignment, especially for cast-iron gears. Tredgold did not account for variations of tooth form or the velocity of contact between the gears (pitch line velocity).

Many alternative formulae were proposed in America for gear tooth thickness after 1875 but it was not until Wilfred Lewis read his paper on the topic of gear strength to the Engineer's Club of Philadelphia on 15 October 1892 that a rational method of spur gear tooth design was proposed (Lewis, 1893). His method is the basis of present gear design procedures.

In contrast to Tredgold's corner contact assumption, Lewis concluded that the gear teeth may be considered as cantilevers tangentially loaded near the ends for their full width. He considered the radial component of the contact force negligible. In his paper, he treated four tooth forms, including a 20° involute. The Lewis formula is

$$W = spFy \tag{13.1}$$

in which

F = face width of gear (in.)
p = circular pitch (in.)
s = allowable stress (psi), a function of the material and pitch line velocity (V)
W = transmitted tooth force (lb)
y = tooth form factor

Form factors (y) are listed in his Table 1 and allowable stresses for cast iron and steel in his Table 2. Some Lewis form factors may be found in Spotts (1985) and his table of allowable stresses as a function of V in Buckingham (1988).

The allowable tooth loads for spur gears in AREMA (2014) are based on a modified Lewis formula as shown in Article 6.5.19 as follows.

For spur gears and bevel gears with full-depth involute teeth with a 20° pressure angle:

$$W = \text{psf}\left(0.154 - \frac{0.912}{n}\right)\frac{600}{600 + V} \tag{13.2}$$

in which

n = number of teeth in the gear
V = velocity of pitch circle (fpm)

For stub involute teeth with a 20° pressure angle

$$W = \text{psf}\left(0.178 - \frac{1.033}{n}\right)\frac{600}{600 + V} \tag{13.3}$$

The torques corresponding to Equation 13.2 and Equation 13.3 are

$$T = W\cos 20^\circ\left(\frac{d}{2}\right) \tag{13.4}$$

in which

$d =$ pitch diameter (in.)

AREMA has many requirements for gearing which are stated in Articles 6.5.18 and 6.5.19.

The AASHTO design procedure for gears considers both modes of spur gear tooth failure: by flexure and by contact stress at the tooth surface. For flexural analysis, AASHTO specifies an empirical equation that is based on the Lewis formula. Whereas Lewis used two modifying factors in developing his relationship, one based on tooth geometry and the other on pitch line velocity, AASHTO provides for many more, at least 10, to account for variations in manufacturing and operating conditions. Most of these factors are listed in Section 13.1 and are presented in AASHTO (2007) and in diagrams and tables by Shigley *et al.*, 2004). AASHTO requires that the factored flexural resistances, W_{tat}, of the spur gear be computed from:

$$W_{\text{tat}} = \frac{FJS_{\text{at}}Y_{\text{N}}}{P_{\text{d}}K_{\text{o}}K_{\text{v}}K_{\text{s}}K_{\text{m}}K_{\text{B}}S_{\text{F}}K_{\text{T}}K_{\text{R}}} \tag{13.5}$$

The second primary mode of gear tooth failure, pitting, caused by contact stress exceeding the surface endurance strength, is also treated by AASHTO using an empirical equation based on the Lewis formula. Recall that the Lewis formula considered bending only, but a similar approach was adopted by AGMA to develop a relationship for the resistance of a gear to pitting failure. Again many factors are involved in order to account for variations in manufacturing, operation, and environment. They are listed in Section 13.1 and AASHTO (2007) and Shigley *et al.* (2004). AASHTO requires that the factored durability resistance, W_{tac}, of the spur gear be computed from:

$$W_{\text{tac}} = \frac{FdI}{K_{\text{o}}K_{\text{v}}K_{\text{s}}K_{\text{m}}C_{\text{f}}}\cdot\left(\frac{S_{\text{ac}}Z_{\text{N}}C_{\text{H}}}{C_{\text{p}}S_{\text{H}}K_{\text{T}}K_{\text{R}}}\right)^2 \tag{13.6}$$

As per AASHTO a spur gear must satisfy

$$W_{\text{tat}} \geq \frac{2T}{d} \tag{13.7}$$

and

$$W_{\text{tac}} \geq \frac{2T}{d} \tag{13.8}$$

AASHTO also requires that the teeth be checked for possible yielding of the gear material at occasional overloads. The maximum factored resistance (W_{max}) must be equal to or greater than the maximum tangential force (W_{t}) that could occur during an overload condition, where

$$W_{\text{max}} = \frac{K_{\text{y}}FK_{\text{f}}JS_{\text{ay}}}{P_{\text{nd}}K_{\text{my}}} \tag{13.9}$$

13.2.1.7 Bevel gear design

Bevel gears are not rated by AASHTO. It is suggested the designer follows the procedures given in ANSI/AGMA 2003-B97 (2003). AREMA does provide a method for rating bevel gears using the same formulae as for spur gear teeth.

13.2.2 Enclosed gearing

Where multiple gearsets are configured together they are referred to as a gear train. Speed reducers refer to custom or commercially manufactured gear trains mounted in enclosed, oil-tight, sealed housings. Specifically, the term speed reducer refers to an enclosed reducer whose purpose is to provide speed reduction, and a corresponding increase in torque, from the input to the output shaft. It should be noted that enclosed gearing units which provide an increase in speed from input to output are properly denoted as speed increasers. However, due to the fact that the primary function of enclosed reducers in span drives is to provide speed reduction in the movable bridge industry, the colloquial use of the term speed reducer serves as a catch-all which encompasses all enclosed gearing. Speed reducers provide many advantages to the user over open gearing since

- generally, they are more compact and smaller, with attendant weight savings
- they provide a safer environment for workers because the gears are enclosed
- wear and corrosion are reduced as the components are protected inside a housing
- they usually ensure continuous lubrication during movable span operation
- they offer flexibility of shaft positions and mounting arrangements
- all bearings may be rigidly mounted in a substantial housing
- shaft misalignment is virtually eliminated as the bearing bores may be accurately located
- the power transmission system operates more efficiently.

13.2.2.1 Speed reducer types

Rarely, if ever, will two bridges have identical power, speed, torque and space requirements; accordingly, speed reducers for span drives are usually designed and manufactured for specific projects: they are seldom off-the-shelf products. While there are many types of speed reducers the ones regularly installed on movable bridges are: parallel shaft, right angle, planetary and worm units.

Parallel shaft reducers are the most frequently used reducer type, and may be found in the main drive machinery, auxiliary drive machinery, span lock machinery and wedge drive systems. As the name implies, all shafts are parallel.

Right-angle reducers are utilised where the output shaft is oriented at a right angle to the input shaft, whether due to space constraints or practicability. Right-angle units may find usage in the same systems as the parallel shaft reducers.

Worm reducers are a type of right-angle unit that may be used for end lift and wedge drive systems. However, AASHTO prohibits the use of worm reducers for main drive machinery and AREMA indicates a preference against their use as well. If a worm reducer is utilised for a main drive system, it must be verified that it is not of a self-locking design, in order to prevent catastrophic failure under back-driving loads.

Planetary reducers are called for in designs dictating concentricity of the input and output shafts or where a very large reduction ratio is required in a small space. An elementary discussion of planetary reducers appears in Chapter 9 and in Lynwander (1983).

Figure 13.12 Parallel shaft reducer in shop with cover removed. (Courtesy of Stafford Bandlow Engineering)

13.2.2.2 Speed reducer gearing

Spur gears are infrequently used in movable bridge speed reducers; both AASHTO and AREMA require helical or herringbone gears. Helical gears have tooth forms similar to spur gears, but the teeth are helically cut at an angle to the axis. The load capacity of helical gears, for a given face width, is greater than that of a spur gear, and the helical overlap of the teeth contributes to smoother operation. Helical teeth introduce an axial thrust load into the shaft during operation, which must be considered in addition to the radial load when selecting the shaft support bearings and designing their housing. This axial load can be almost eliminated by mounting two helical gears, their teeth sloping in opposite directions, exactly side by side, so the net axial load is zero. Such an arrangement is often called a double helical, or herringbone gear. A special type, a continuous tooth, double helical gear, cut into a single cylinder is the Sykes Herringbone shown in Figures 13.12 and 13.13.

Spiral bevel gears bear the same relationship to straight bevel gears that helical gears have to spur gears; therefore, where a right angle unit is necessary in a reducer, spiral bevels are used. The reducers may be a single, spiral bevel reduction, or a multiple reduction with the spiral bevel set used in concert with parallel shaft reductions as shown in Figure 13.14 (see also Figure 9.17).

Similar to open gearing, where loading sharing between the output shafts is required, differentials gearsets can be incorporated into parallel shaft and right-angle reducers. Figure 13.15 depicts a differential

Figure 13.13 Helical and herringbone gearsets in enclosed speed reducer. (Courtesy of Stafford Bandlow Engineering)

Figure 13.14 Spiral bevel gearset at input for right-angle enclosed speed reducer. (Courtesy of Stafford Bandlow Engineering)

Figure 13.15 Differential gearset within output gearset of enclosed speed reducer. One of three differential bevel pinions is partially visible at top left of gearset. (Courtesy of Stafford Bandlow Engineering)

gearset mounted on the output gear of a parallel shaft reducer. One differential bevel gear is in the foreground of the photo, and one of the differential pinions, which is internally mounted in the output gear, is partially visible at the top left of the gear.

The ratio for a single reduction helical gearset is limited to about 10 : 1; the ratio in a single reduction worm reducer can be many times as great. In some centre and end wedge applications where space is at a premium, worm reducers are used because they can provide a very high reduction ratio within a single stage unit.

13.2.2.3 Speed-reducer material considerations

Gears for movable bridge reducers are manufactured from steels, ranging from plain carbon steel to high alloy steel (see Chapter 6). Material properties are an important consideration in the selection and design of gearing. Commercially manufactured speed reducers utilise gearing that has been heat treated to a much greater hardness than the open gearing typically found on movable bridges. Heat treating serves two purposes: to improve machinability, and to develop the necessary hardness, strength and wear resistance for the intended use Rakhit (2000). Heat treatment may take the form of through hardening or case hardening. Through hardening produces a substantially uniform increase in hardness throughout the whole gear tooth. Case hardening, whether by carburising, nitriding or induction heating, produces an increase in hardness of the outer layer of metal at the tooth surface but does not produce a corresponding increase in the hardness or strength of the core material. Case-hardened gears are more ‘power-dense’ than through-hardened gears and are the preferred option of many reducer manufacturers as they provide greater output for a given size unit. However, if the ‘case’ layer on the gear teeth is inappropriately selected and breaks down, there will be a rapid acceleration in deterioration due to the inferior properties of the core material. This concern is not a problem with through-hardened gears, which have uniform properties throughout. For this reason, some owners, and the AREMA standard in particular, have mandated that through-hardened gears be used in span drive speed reducers. This issue needs to be evaluated by the designer and owner for every project to ensure that the benefits and limitations of each heat treatment are understood prior to a decision being made as to which option to pursue.

13.2.2.4 Speed reducer design

Both AASHTO and AREMA require that speed reducers be designed in accordance with the current ANSI/AGMA 6013.A06 (ANSI/AGMA, 2011b), including certain additional provisions those organisations (AASHTO and AREMA) believe necessary. The ANSI/AGMA document treats the complete reducer – not only the gears, but also the bearings, shafts, keys, threaded fasteners and housings.

AASHTO specifies that reducers be designed to sustain the stresses resulting from 150% of rated full load motor torque, with an AGMA service factor of 1 without exceeding 75% of the yield strength of any component.

AREMA requires that reducers be designed to sustain the stresses resulting from a momentary 300% overload without any component reaching its yield strength; and, that the gears be rated using an AGMA service factor of 1 for surface durability and 1.5 for bending strength, based on full load motor torque. Similarly to open gears, those in speed reducers are evaluated for surface durability and bending strength. While the basic relationships remain the same, many of the variable modifying factors (I, J, K, Q, etc.) differ due to the improved quality, accuracy, lubrication, and reliability of a shop-assembled speed reducer.

Rolling element bearings are required and selected to provide for a 40 000 hour, L-10, minimum life expectancy. The method of calculation used is that designated by the bearing manufacturer. Other components, not specifically addressed by AASHTO or AREMA, are designed in accordance with ANSI/AGMA (2001).

While the overall size of the reducer casing, or housing, is dependent on the number of reductions and sizes of gears necessary to deliver the design power, the requirements for configuration and type of mounting are quite flexible. For instance, in a parallel shaft unit, the plane of the input and output shaft may be horizontal, vertical or at some angle in between. Also, the reducer may be foot- or side-mounted to the pier or structure, and even straddle-mounted between two structural members, such as stringers on a rolling lift bridge. Nowadays, the housing is usually a steel weldment, designed to support the loads imposed by the gears and to restrict the vertical and horizontal deflections from adversely affecting the alignment of the gear teeth. Likewise, the foot mounting bolts and the threaded fasteners securing the cap and base are selected to ensure rigidity during operation. Lubricant leakage around shafts protruding through the reducer housing is contained by the use of lip-type seals. While dual lip seals have served that purpose well, some designs call for double seals, with the lips opposed and a grease chamber in between the slightly separated seals. End covers are used to seal the bores in the housing at all non-extension shaft locations. Removable covers are located on the housing top and or the sides to permit visual inspection of the gear teeth without having to remove the housing top. Lifting lugs should be located at balanced positions on the housing to enable ease of handling during shipment and installation.

The gears and bearings are usually oil lubricated. Generally, a splash system is used. A splash system is one in which the static oil level in the base, known as the sump, is above the outside diameter of one or more gears, at a height to assure oil is reaching every bearing. Then, as the gears rotate, oil is picked up by the submerged teeth and slung throughout the housing, thereby lubricating all the components. However, some applications require a circulating pump to ensure adequate lubrication of all components. This is an important consideration for reducers that are mounted on bascule leaves and rotate with the leaf, and for reducers where the input shaft is at a higher elevation than the oil level. In either case, a sight gauge is located on the outside of the housing base for the purpose of monitoring the oil level. The correct static oil level is marked and a drain is positioned in the base at the lowest point to facilitate draining before adding new lubricant. The type and grade of oil used should be as specified by the reducer manufacturer. The oil should be drained, and the interior flushed and filled with fresh oil annually.

Overall, the torque transmission efficiency of speed reducers, including losses in the gear meshes, bearings, seals and lubricating system, is significantly higher than that of open gearing. A rough comparison for a triple reduction installation would be 94% for an enclosed drive and 87% for open gearing, both with shafts supported by rolling element bearings. The efficiencies would be lower with sleeve bearings.

13.3. Inspection fundamentals

It is common to observe wear or distress on a gearset that has been in service and to have no reference point against which to evaluate the rate of deterioration, which is of paramount importance in evaluating the remaining service life. Periodic inspection of gearing should be conducted to monitor the integrity of the gears as well as to chart the progression of deterioration.

13.3.1 Inspection of open gearing

Ideally, the as-fabricated dimensions of the gear teeth will be documented, and serve as a baseline against which to evaluate future wear. This information should be mandated for all new installations.

Figure 13.16 Chordal tooth thickness measurement using gear tooth caliper. (Courtesy of Stafford Bandlow Engineering)

Unfortunately, it is often unavailable for existing installations. Where original information is unavailable, gear tooth wear can be compared against historical inspection information, where an active inspection programme is in place. Gear inspection should encompass the following areas.

13.3.1.1 Gear tooth measurements

Gear tooth thickness can be evaluated either by a chordal thickness measurement using a special gear vernier caliper as shown in Figure 13.16, or by a span thickness measurement using a standard vernier caliper as depicted in Figure 13.17. It is preferable to take span thickness measurements wherever

Figure 13.17 Span tooth thickness measurement over three pinion teeth. (Courtesy of Stafford Bandlow Engineering)

Figure 13.18 Backlash measurement with thickness gauge. This measurement identified inadequate backlash due to a new pinion being manufactured with the incorrect tooth form to mate with an existing gear. (Courtesy of Stafford Bandlow Engineering)

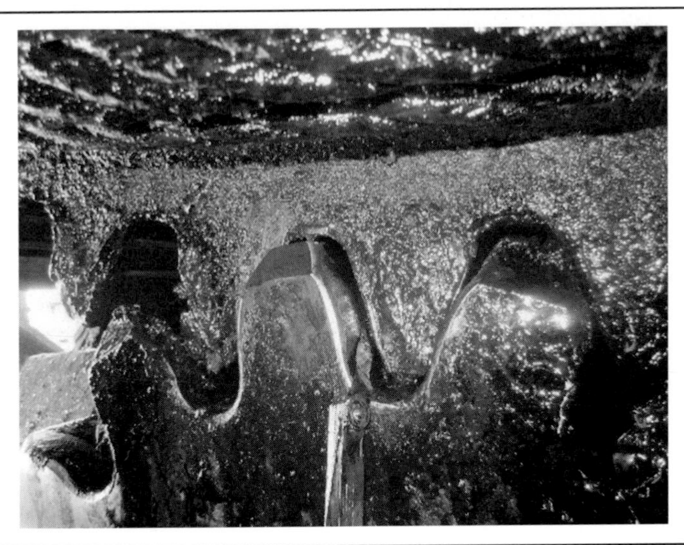

practical as these measurements are not affected by wear at the tips of the teeth which can result in fins that directly affect the chordal measurements.

13.3.1.2 Backlash and tip clearance

Backlash is the amount by which the width of a tooth space exceeds the thickness of the engaging tooth on the pitch circle. Tip clearance is the radial distance between the tip of one tooth and the root of the mating tooth. Backlash and tip clearance are necessary for the gear teeth to properly engage, but they may also be used as a secondary indicator of tooth wear. For a given tip clearance, an increase in backlash is a direct indicator of tooth wear. However, backlash and tip clearance are interrelated in that if the gearset is not installed at the proper centre distance, the tip clearance and backlash will either increase or decrease correspondingly. Therefore, it is important to measure both backlash and tip clearance to properly assess the gearset installation. Figure 13.18 shows a feeler gauge being used to measure backlash.

13.3.1.3 Documentation

All measurements should be documented for historical reasons. It is also preferable to punch mark or otherwise record which teeth have been measured so that the same teeth can be monitored from cycle to cycle of inspections as common references. It should also be noted that the most heavily loaded teeth are typically located in the accelerating zone for span opening (raising) and span closing (lowering). Deciding which tooth to measure may sometimes be influenced by which teeth can be readily accessed. However, regardless of which teeth are initially, or historically, chosen for measurement, all teeth should be visually examined periodically to ensure that there is not a tooth (or teeth) which exhibits substantially greater wear than the location chosen for measurement. If substantial wear is noted at a different location than was originally selected, the scope of work should be expanded to include these teeth for measurement.

Figure 13.19 Open gear tooth wear. The gear teeth exhibit heavy plastic flow resulting in a pronounced ridge at pitch line, large fins at the sides of the teeth and general breakdown of the contact surface. (Courtesy of Stafford Bandlow Engineering)

13.3.1.4 Visual inspection

A proper inspection of gearing should assess the physical condition of the teeth to indicate the type and location of wear, condition of lubricant and apparent misalignment. Open gearing has historically been fabricated from materials softer than that for enclosed gearing. As a result, the teeth have a tendency to wear and deform when subjected to high loads as opposed to fracture. Therefore, wear of open gearing occurs primarily in the form of plastic flow resulting in fins at the tips and edges of the teeth with indentations and ridges at the tooth pitch lines as evidenced in Figure 13.19. This type of wear can be quantified through measurement to provide an assessment of decreasing tooth strength with increasing wear so that an appropriate recommendation can be made for gearing replacement. Open gears, due to their exposed mounting, are also subject to abrasive wear from lubricant contamination. Severe contamination of lubricant is unfortunately a common occurrence, particularly due to blasting media spread during painting projects, and can severely foreshorten gearing life, as depicted in Figure 13.20. Active monitoring of the lubricant condition as part of routine inspections can identify if or when the lubricant is contaminated and can facilitate recommendations to replace the lubricant to prevent subsequent abrasive wear. Open gearing can also be subject to pitting and scoring given the right combination of high loading, poor lubrication, and poor alignment. Where pitting and scoring is first noted, an evaluation of the gear tooth loading and lubricant type may be warranted.

As part of the visual inspection, the gear teeth should be evaluated for adequate lubrication. Lubrication should be sufficient to maintain a film of lubricant on the tooth surface in the contact areas. A sure sign of inadequate lubrication is where spotty surface corrosion is evident on the tooth surface in the contact areas. Evaluation of the lubricant contact pattern on the teeth can also provide additional information such as if the gear teeth are aligned satisfactorily (i.e. full face contact? cross-bearing? end loading?) as well as if the bridge is appropriately balanced (i.e. do gear teeth only exhibit contact in one direction of rotation? Is wear significantly heavier in one direction of rotation?).

Figure 13.20 Abrasive damage to open teeth. The gear teeth exhibit severe wear following contamination of lubricant with sandblasting media. Note the pronounced steps from the unworn ends of the teeth to the worn contact area. (Courtesy of Stafford Bandlow Engineering)

Another element of visual inspection is to observe the integrity of the gear body itself (i.e. hub, spokes, rim) as well as the mounting of the gear on its shaft. Areas that have been historically problematic and warrant particular attention include crack initiation in the roots of teeth and at keyways, crack initiation through the rims of gears, and loss of fit at gear hub and shaft resulting in relative movement at this interface. Figure 13.21 illustrates extensive cracking at the root of a curved rack tooth. Issues that

Figure 13.21 Cracks in gear. Cracks have developed and extend across the full width of the root of the rack teeth. (Courtesy of Stafford Bandlow Engineering)

Figure 13.22 Speed reducer. Access to gearing is through inspection port following removal of cover. (Courtesy of Stafford Bandlow Engineering)

have proven especially problematic for straight and curved racks include accumulation of debris in and around mounting bolts contributing to severe deterioration and loss of integrity and, for rolling lift bridges, bottoming conditions developing between the main pinions and racks (when racks are located below the pinions) due to wear at the tread and track on which the bascule span rolls.

13.3.2 Enclosed reducer inspection

Gearing design for enclosed reducers differs from open gearing, and the inspection methods differ as well. Whereas there is typically unobstructed access to open gearing, access to the gearing in an enclosed reducer is typically limited to one or two inspection covers or ports which are provided in gear boxes as depicted in Figure 13.22. In addition, the gearing in enclosed reducers may be partially or fully submerged in the oil bath. This restricted access precludes taking wear measurements on the gearing inside enclosed reducers as part of a normal inspection programme. Because the material selected for enclosed reducer gearing is much harder than that selected for open gearing, the gearing typically does not wear as much as it does for open gearing, which mitigates the inability to take wear measurements. A thorough visual inspection of the enclosed gearing usually provides an adequate assessment of gear tooth integrity whether due to polishing from acceptable contact or damage from destructive wear. Tooth degradation in hardened gearing typically takes the form of pitting, scoring, frosting or ridging (see Figures 13.23 and 13.24). The source of the damage could be attributable to several factors: therefore the presence of any of these wear modes should be documented and monitored by an ongoing inspection programme. Some pitting is acceptable at initial installation and is regarded as corrective pitting. However, continuation of pitting throughout the service life will degrade the tooth surface and result in loss of tooth profile and ultimately may lead to tooth breakage. Pitting, scoring and scuffing are all conditions which could be indicative of excessive loading, inadequate lubrication and inadequate tooth hardness, and if left uncorrected may also lead to tooth breakage. As previously stated, it is uncommon to see appreciable wear on hardened gear teeth. When wear is noted, it should be investigated as it is likely to be indicative of a problem, whether due to overloading, inadequate tooth hardness, or misalignment as shown in Figure 13.25.

Figure 13.23 Speed reducer tooth wear. Teeth exhibit moderate to heavy pitting damage. (Courtesy of Stafford Bandlow Engineering)

Figure 13.24 Speed reducer teeth wear due to inadequate lubrication. Teeth exhibit moderate localised scoring. Damage resulted from misdirection of forced lubrication stream, resulting in lubricant starvation during operation. (Courtesy of Stafford Bandlow Engineering)

Figure 13.25 Speed reducer tooth wear due to heavy plastic flow at flank of teeth. Inadequate housing stiffness allowed deflection under load resulting in extreme end loading of teeth. (Courtesy of Stafford Bandlow Engineering)

To guard against lubrication issues, it is prudent to maintain an oil sample analysis programme throughout the life of a reducer. Historic oil sample test results may aid in identifying if the oil is inadequate for the application or if there has been a sudden increase in the amount of wear particles in the oil at any given time, which could be indicative of a problem requiring further investigation.

While some older reducers utilised a mix of sleeve bearings for the output shafts and roller bearings for all other shafts, modern enclosed reducers typically are equipped only with roller bearings. Where sleeve bearings are used, inspection techniques are the same as for standard sleeve bearings. Where roller bearings are used, the bearings are generally inaccessible for inspection. Basic external checks of the reducer can be performed during span operation to determine if there is an apparent problem with a bearing, such as abnormal noise, excessive heat, or vibration. If a problem is suspected with an internal roller bearing, specialised vibration testing can be performed to isolate the roller frequencies and identify if, in fact, a problem exists. Many reducer bearings are equipped with grease lubricated seals. Routine inspection should verify that the seals are adequately lubricated.

Some owners have chosen to incorporate partial disassembly of enclosed gear units in their preventative maintenance programme. This in-depth inspection task is typically performed only after the initial warranty period has expired and is typically predicated on interrogating a deficiency previously identified through the routine maintenance programme or to inspect gearing that is otherwise inaccessible, either due to location or due to submersion. While disassembly can provide valuation information regarding reducer condition, reducer disassembly should not be attempted without skilled labour support. In lieu of disassembly, boroscope inspection can supplement normal visual inspection of otherwise inaccessible locations.

13.3.3 Lubrication and maintenance

Lubrication of gear teeth is necessary for several reasons: to minimise metal-to-metal contact of the tooth faces as they go through the mesh, to maintain a low coefficient of friction during operation, and to serve as a rust and corrosion inhibitor. High-grade grease is the lubricant normally used for open gearing. The very fact that open gears are exposed, suggesting that rain, dirt, roadway debris and other foreign materials can easily contaminate the grease and promote corrosion and wear, dictates the importance of regularly checking the condition of the lubricant and cleaning old, dirty grease from the gears and applying fresh, clean lubricant. Safety measures require that protective shields be installed over open gear sets; but, even with partial enclosures the lubricants do become contaminated and should be monitored.

Before installation the gears should be coated with an anti-corrosive compound to protect against rust build-up. Once installed the teeth need to be lubricated with the grease specified by the manufacturer. It is frequently applied using a grease gun or spatula to insert grease in between the teeth and then rotating the set to distribute the lubricant across the tooth faces and roots. Proper lubrication also aids in visually checking the alignment and contact pattern of the rack and pinion teeth during operation. It is recommended that the lubricant condition be checked at least semi-monthly for condition and adequacy, and cleaned and re-lubricated as required.

13.4. Gear design considerations

The design standards for movable bridges specify that machinery should be simple, and of substantial construction. The configuration and arrangement of the components should permit easy erection, adjustment, inspection, lubrication, cleaning, painting and replacement of worn or defective parts. Contemporary design has shifted to a clear preference for enclosed reducers in lieu of open gearing wherever possible. Typically, it is not practical to eliminate the final drive open gearset which couples the machinery to the structure due to the large final ratio usually needed at this gearset. However, it should be noted that some designs have, in fact, pushed this limit by coupling the drive reducers directly to the main trunnions (rotational axis of bridge).

Enclosed reducer usage satisfies many of the design objectives, including simplifying alignment and handling, decreasing maintenance, ensuring an adequate lubrication supply through a continuous oil bath and improving safety by enclosing the rotating machinery. However, the reducer arrangements which provide the above benefits typically also limit access for thorough inspection and increase the complexity of replacing defective or worn parts. While enclosed reducers are touted for their improved reliability and as long-term closed box solutions, practical experience has demonstrated that reducers are not immune to failures, whether due to excessive tooth damage or breakage, seized bearings, failed bearing seals, excessive leakage or due to an improperly functioning oil circulation system. Where internal reducer problems are identified, the corrective action typically requires that the reducers be removed from the structure and shipped to a shop for rehabilitation. Therefore, access to reducers for future maintenance and removal should be considered in the design process.

Open gearing in contemporary design is typically limited to the final drive gearset, for example, where the drive pinion mates with the rack that is attached to the structure. This final gearset typically provides a substantial ratio which would otherwise result in a substantial increase in the size of the enclosed reducers. From a historic perspective where open gear trains were used as the primary power-train, there are many instances where the gear trains have provided satisfactory service over a long life (i.e. over 50 years). However, there are also widespread problems with open gearing being subjected to extensive damage over a much shorter service life. Contributing factors to a poor service life include overloading, fabrication from inadequate materials, poor initial alignment, inadequate lubrication and contamination.

While it is likely that overloading of reducers will contribute to accelerated damage as at the open gearing, the manufacturing techniques and controls which produce high-quality gearing for the enclosed reducers can also be applied to open gearing to ensure the quality of the gearing. Unlike the reducers, the open gearing will require frequent maintenance to ensure adequate lubrication during service and to monitor the lubricant to ensure that it is not compromised by contaminants.

If open gearsets are to be considered as part of a design, gearing alignment needs to be considered. Experience has demonstrated that relying on field alignment of multiple individual machinery bases is an onerous task and decreases the likelihood of obtaining satisfactory alignment. Wherever possible, mating gears should be mounted on a rigid common frame of substantial design so that they can be shop installed and aligned to ensure satisfactory field installation.

Several bridge machinery 'innovations' were based specifically on this principle. Leonard Hopkins patented a particular type of common frame-mounted machinery assembly for bascule bridges intended to replace conventional floor mounted assembly which utilised multiple discrete components. The Hopkins design saw widespread usage as an alternative to conventional design for a many years. However, over time several flaws in the Hopkins design became evident which highlight the need to ensure that the machinery support has adequate stiffness to maintain the position of the gears under operating loads.

Regardless of whether or not an enclosed reducer or gear frame is selected, shop assembly, inspection and testing is warranted to ensure that the assembly meets the intent of the design prior to it being

Figure 13.26 Reducer shop load test. Two drive assemblies consisting of small primary parallel shaft reducers driving large secondary right-angle reducers are coupled back-to-back for load testing. (Courtesy of Stafford Bandlow Engineering)

shipped to the field. Contemporary practice dictates that enclosed reducers be subjected to shop tests which encompass a leak test, a no-load spin test at rated speed, and a load test at varying percentages of rated power (see Figure 13.26). The objective of the load tests is to ensure that the reducers are manufactured and assembled to produce acceptable alignment and can withstand the design loads without damage. The actual load test requirements vary with differing owners. These must be worked out between the owner and the designer with the understanding that the manufacturers have advised that, in order to pass the load test, increasingly stringent test requirements require the reducers to be designed for a higher effective service factor than would otherwise be necessary – with the result that the test increases the cost of the reducers. However, the load tests have identified issues such as inadequate contact and resultant gear tooth damage that would otherwise have gone undetected until after the reducers were placed into service (see Figure 13.27). In selecting the load test requirements, the owner and designer must find an acceptable balance between cost versus assurance. Shop testing of open gear frames is typically limited to a no-load spin test of the machinery. However, given the results of the enclosed reducer testing, combined with historical issues with open gearing alignment, consideration should be given to performing load testing of open gear frames wherever possible.

Figure 13.27 Reducer shop load test results. Dark areas on teeth are remains of blueing, indicating lack of contact between mating teeth. Blueing contact check reveals less than 50% tooth contact under load, which is inadequate. (Courtesy of Stafford Bandlow Engineering)

13.5. Future considerations

Movable bridges range in age from recent construction back to the late 1800s, and many of the older bridges remain in service with no plans for immediate replacement. Therefore, the problem the designer often faces is upkeep and rehabilitation of existing machinery rather than the design of new machinery. For many structures, the existing machinery has been designed integral with, or closely tied to, the structure. Therefore, when planning the rehabilitation of machinery on an existing structure, the designer must determine what is the best machinery design for the structure as a whole, and not just for the machinery *per se*. The designer must ask:

- Will the reducer fit in the available space?
- Are the available support beams and structure adequate to support the load of the reducer?
- Will new support beams fit in the available space?
- Will the reducer and/or new support beams fit in the available clearance envelope throughout the range of span operation?
- Will the reducer and support beam arrangement adversely alter the loading of the structure?
- Is there access to maintain the machinery and has it been improved by the rehabilitation?
- Will the rehabilitation improve the reliability and upkeep of the system?
- Is the design cost effective for the owner?

With careful consideration of the above factors, the designer can design a gear drive that is durable, that is compatible with the structure and that provides for reliable long-term service.

REFERENCES

AASHTO (American Association of State Highway and Transportation Officials) (2007) *LRFD Movable Highway Bridge Design Specifications*, 2nd edn. AASHTO, Washington, DC, USA.

ANSI (American National Standards Institute)/AGMA (American Gear Manufacturers Association) (2001) 2001-D97: Design and selection of components for enclosed gear drives. American National Standards Institute, Washington, DC, USA.
ANSI/AGMA (2003) 2003-B97: Rating the pitting resistance and bending strength of generated straight bevel, zero bevel, and spiral bevel teeth. ANSI, Washington, DC, USA.
ANSI/AGMA (2010) 2001-D04: Fundamental rating factors and calculation methods for involute spur and helical gear teeth. ANSI, Washington, DC, USA.
ANSI/AGMA (2011a) 1012-G05: Gear nomenclature, definition of terms with symbols, USA.
ANSI/AGMA (2011b) 6013.A06: Standard for industrial enclosed gear drive. ANSI, Washington, DC, USA.
ANSI/AGMA (2014) 2002-B88: Tooth thickness specification and measurement (Appendix A).
ANSI/AGMA (2013) Standard 2013.A06: Industrial gear drives, ANSI, Washington, DC, USA.
AREMA (American Railway Engineering and Maintenance-of-Way Association) (2014) Movable bridges – steel structures. Ch. 15 in *AREMA Manual for Railway Engineering*, AREMA, Lanham, MD, USA.
Buckingham E (1988) *Analytical Mechanics of Gears*. Dover, Mineola, NY, USA.
Childs PRN (2014) *Mechanical Design Engineering Handbook*. Butterworth-Heinemann, Oxford, UK.
Dudley DW (1969) *The Evolution of Gear Art*. American Gear Manufacturers Association, Washington, DC, USA.
Dudley DW (1984) *Handbook of Practical Gear Design*. McGraw-Hill, New York, NY, USA.
Eytelwein JA (1808) *Handbuch der Statik fester Körper* in three vols (in German). Realschulbuchhandlung, Berlin, Germany.
Hindhede W, Zimmerman JR, Hopkins RB, Erisman RI, Hull WC and Lang JD (1983) *Machine Design Fundamentals: A Practical Approach*. Wiley, New York, NY, USA.
Lewis W (1893) Investigation of the strength of gear teeth. *Proceedings of the Engineer's Club of Philadelphia*, vol. X, 19–23.
Lynwander P (1983) *Gear Drive Systems: Design and Application*. Marcel Dekker, New York, NY, USA.
Rakhit AK (2000) *Heat Treatment of Gears: A Practical Guide for Engineers*. ASM International, Materials Park, OH, USA.
Rankine WJM (1876) *A Manual of Machinery and Millwork*, 3rd edn, revised by EF Bamber. Charles Griffen, London.
Shigley JE (2001) *Mechanical Engineering Design*, 6th edn. McGraw-Hill, New York, NY, USA.
Shigley JE, Mischke CR and Budynas RG (2004) *Mechanical Engineering Design*, 7th edn. McGraw-Hill, New York, NY, USA.
Spotts MF (1985) *Design of Machine Elements*, 6th edn. Prentice Hall, Englewood Cliffs, NJ, USA.
Spotts MF, Shoup TE and Hornberger LE (2004) *Design of Machine Elements*, 8th edn. Pearson Education, Upper Saddle River, NJ, USA.
Steward Machine Company (2008) *Earle Speed Reducers*. Steward Machine Company, Birmingham, AL, USA.

Movable Bridge Design
ISBN 978-0-7277-5804-0

http://dx.doi.org/10.1680/mbd.58040.325

Chapter 14
Hydraulic span drive systems

James M Phillips III

14.0. Introduction

Hydraulic systems are a viable alternative to electro-mechanical drives for movable bridges. They are specified for a large percentage of European bridges. In the USA they have only been installed on a small percentage of movables. When properly designed, constructed and maintained, hydraulic systems offer the following characteristics that make them well suited for bridge drive applications

- load sharing: inherent load sharing between actuators without the use of differential gear assemblies
- constructability: relatively large tolerances on installation and alignment compared with equivalent mechanical drives
- flexibility: numerous options are available for mounting power units and actuators providing flexibility to simplify the design of the movable span structure and supporting piers; power units can be located remote from actuators to reduce environmental exposure
- elimination of brakes: with cylinder drives, maintenance-prone brakes are eliminated
- cushioning: built-in cushioning is available with hydraulic cylinders, providing additional safety without use of buffer cylinders
- compact power: hydraulic systems offer a high power to weight (or size) ratio; they are well suited to bridge applications where space is limited.

This chapter discusses the application of industrial hydraulics (sometimes called fluid power) in movable bridge design. A detailed discussion of hydraulic system design, the internal workings of hydraulic components and design of hydraulic power units are beyond the scope of this text. Instead, the focus herein is on the selection of hydraulic equipment for movable bridge drive systems with emphasis on those aspects which are unique to movable bridges. Specific equipment and systems that have been proven to perform successfully in movable bridge applications are also stressed. Design of hydraulic systems for bridge auxiliary equipment, such as span locks or wedges, is not addressed in this chapter as it is more closely related to general industrial hydraulic system design.

There are several standards that govern design of movable bridge hydraulic systems, depending upon the location of the bridge. The intent of this chapter is to provide guidance that may be used when following any specific design standard. However, for the purpose of consistency and clarity, the terminology and design examples presented herein correlate with those of AASHTO (American Association of State Highway and Transportation Officials, 2007), the AASHTO load resistance factor design (LRFD) movable highway bridge design specifications. The units of measurement are the conventional US units (USCUs), used here for the reasons explained in Chapter 5.

This chapter starts by defining the terms and symbols used throughout the chapter. This is followed by a review of hydraulic system fundamentals with emphasis on the typical components and systems utilised on movable bridges. The most common types of bridge drive systems are presented, including comparison of the advantages and disadvantages of each to specific applications. At the end of the chapter a design example is presented for a hydraulic cylinder-operated bascule bridge.

14.0.1 Definitions

The following symbols and definitions are used throughout this chapter.

- Holding load – the loading against which the bridge machinery must be capable of holding the movable span stationary. It includes unbalanced dead loads and wind loads; also referred to as the holding torque when referenced to a rotating element.
- Maximum constant velocity load – maximum loading against which the bridge drive system must be capable of operating at full speed; includes dynamic friction, unbalanced dead loads and wind loads. Also referred to as the maximum constant velocity torque when referenced to a rotating element.
- Maximum starting load – maximum loading against which the bridge drive system must be capable of starting; includes static friction, unbalanced dead loads, wind loads and ice loads. Inertia may be included at the owner's discretion. Also referred to as the maximum starting torque when referenced to a rotating element.
- Maximum working pressure – the maximum hydraulic pressure in a circuit or branch of a circuit as limited by a non-adjustable relief valve.
- Normal working pressure – the maximum hydraulic pressure in a circuit or branch of a circuit as determined by the setting of an adjustable relief valve.
- Pilot operation – a fluid path (i.e. ports in a device and/or fluid conductors) connecting one section of a hydraulic circuit or device to another section of the circuit or device. Pilot lines may also link one chamber of a component to another chamber within that component or an adjacent coupled component. A pilot line provides fluid connectivity so that the pressure in one section of the circuit can be used to control or assist in operation of another device. The controlled device may be coupled to the source of the pilot pressure, contained within a common manifold, or piped to a remote device. The pressure applied by way of a pilot line is referred to as *pilot pressure*. A valve is said to be *pilot operated* when pilot pressure is used to control operation of the valve. Pilot operation also allows large valves to be operated by relatively small solenoids. Where pilot pressure controlled by a solenoid valve is used to shift a valve, the valve is referred to as pilot operated or *pilot assisted.*

14.0.2 Symbols

A_b hydraulic cylinder bore area (in.^{2})
A_r hydraulic cylinder rod end area (or rod end area) (in.^{2})
C_d orifice coefficient (unitless)
d displacement of a pump or motor (in.^{3}/rev.)
D_b diameter of cylinder bore (in.)
D_r diameter of cylinder rod (in.)
D_o diameter of an orifice (in.)
E modulus of elasticity, psi (taken as 29×10^6 for steel)
F force: lb
F_b design buckling load for a hydraulic cylinder (lb)
F_E Euler bucking load (lb)
FS factor of safety (unitless)
K effective length factor for buckling (unitless)

n rotational speed (rpm)
η_m efficiency, mechanical (unitless)
η_o efficiency, overall (unitless)
η_v efficiency, volumetric (unitless)
P pressure (psi)
P_b pressure acting on the blind end of a cylinder piston (psi)
P_r pressure acting on the rod end of a cylinder piston (psi)
ΔP differential pressure; pressure drop across a component; branch of a circuit (psi)
P_{hp} power (horsepower)
Q flow, gpm (gallons per minute)
S specific gravity of hydraulic fluid
T torque (in.–lb)
v cylinder rod velocity (in./sec)
ω angular velocity (rad/sec)

14.1. Hydraulic system fundamentals

Hydraulic systems (aka fluid power systems) utilise flow of pressurised fluid to perform work. By controlling flow and pressure these systems are capable of controlling the speed and motion of machines of all types and sizes and are particularly well suited for heavy movable structures such as navigation lock, gates, dam gates and movable bridges. Many different types of hydraulic components are available and nearly limitless arrangements can be composed, giving hydraulic systems great flexibility in application. Hydraulic systems have the capability to provide the motive power and control necessary to operate movable bridges of many different types. They are also well suited for operation of movable bridge auxiliary devices such as span locks and swing bridge centre and end lifts.

A rudimentary hydraulic system, as presented in the hydraulic schematic in Figure 14.1, consists of a pump (item 1.0), prime mover (item 2.0) (electric motor shown), reservoir (item 3.0), hydraulic fluid,

Figure 14.1 Rudimentary hydraulic schematic

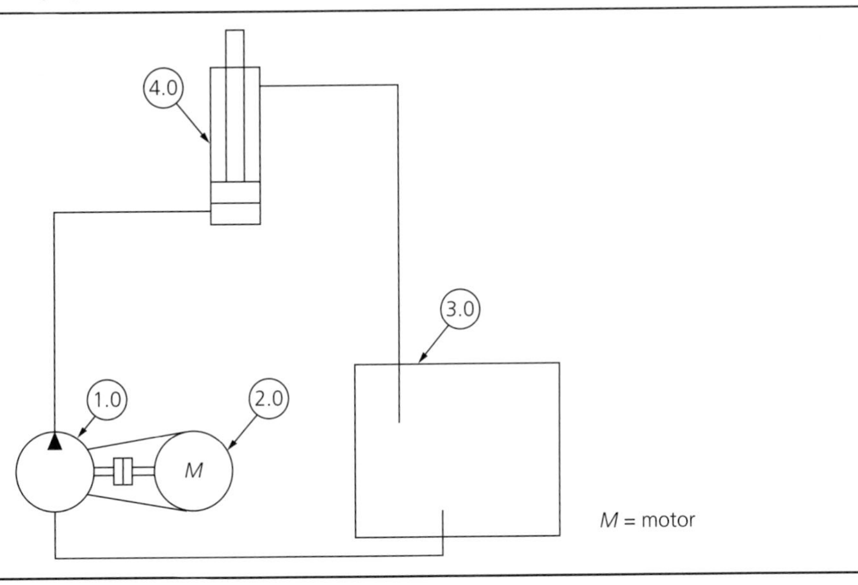

piping and actuator (item 4.0) (hydraulic cylinder shown). The prime mover could be a combustion engine, but is most often a standard AC induction motor. The actuator could be a hydraulic motor or hydraulic cylinder as is shown. For the purpose of this discussion, the use of mineral oil, the most common type of fluid in bridge applications, is assumed.

A hydraulic pump is a rotary device that converts power from the prime mover into fluid power. It draws in hydraulic fluid through an inlet port and forces it out through an outlet port. If the flow from the outlet port is restricted, the exiting fluid is pressurised and thus is available to perform work. The primary characteristics that define a pump are the type, displacement, pressure rating and speed rating. Pump displacement is the volume of fluid processed through the pump with each rotation of the input shaft.

Hydraulic motors are devices that produce rotary action, when provided a supply of pressurised hydraulic fluid. Hydraulic motors convert fluid power to rotary power, measured as torque and angular velocity (e.g. in.–lbs and rpm). The primary characteristics that define a hydraulic motor are its type, displacement, pressure rating and speed rating. Motor displacement is the volume of fluid passing through the motor with each rotation of the output shaft.

Hydraulic cylinders are devices that produce linear action when provided a supply of pressurised fluid. Cylinders convert fluid power into linear power, measured by force and velocity (e.g. lb and in./sec). Hydraulic cylinders are characterised primarily by their type, mounting, bore diameter, rod diameter, stroke and pressure rating.

The fundamental working process of a hydraulic bridge drive can be summarised as follows

(*a*) prime mover turns the pump
(*b*) pump draws fluid from the reservoir, where it is not pressurised, and pushes it into the circuit
(*c*) pressurised fluid flows through piping to the actuator
(*d*) pressurised fluid acts upon the actuator creating force (cylinder) or torque (motor)
(*e*) the resulting force or torque acts upon the movable span or auxiliary device to perform the desired action
(*f*) fluid exiting the actuator returns to the reservoir.

This simplified circuit and idealised operation is presented as a starting point for discussing hydraulic system design. As will become apparent in further discussion, this circuit must be significantly modified to become a suitable hydraulic drive capable of controlling the motion of a movable bridge or movable bridge auxiliary device.

In the design of many hydraulic systems, including those used to operate movable bridge auxiliary devices, such as bascule bridge span locks or swing bridge end lifts, the dynamic nature of the load can be ignored. The mass and inertia of these devices is small compared to the applied loads due to friction or weight. Furthermore, there are generally no load conditions, or only limited load conditions, which tend to overhaul the load. In cases where no significant overhauling loads are present, it is still necessary to provide means to control motion, but dynamic effects can generally be neglected. Figure 14.2 presents a simple but viable circuit for operating an auxiliary movable bridge device such as a pair of wedges. In addition to the basic motor, pump and cylinder of Figure 14.1, this circuit features a relief valve (item 2.1) to limit the maximum pressure in the system, a directional control valve (item 2.2) to direct fluid to either end of the cylinder so that the cylinder can drive and pull the wedge,

Figure 14.2 Hydraulic cylinder drive schematic

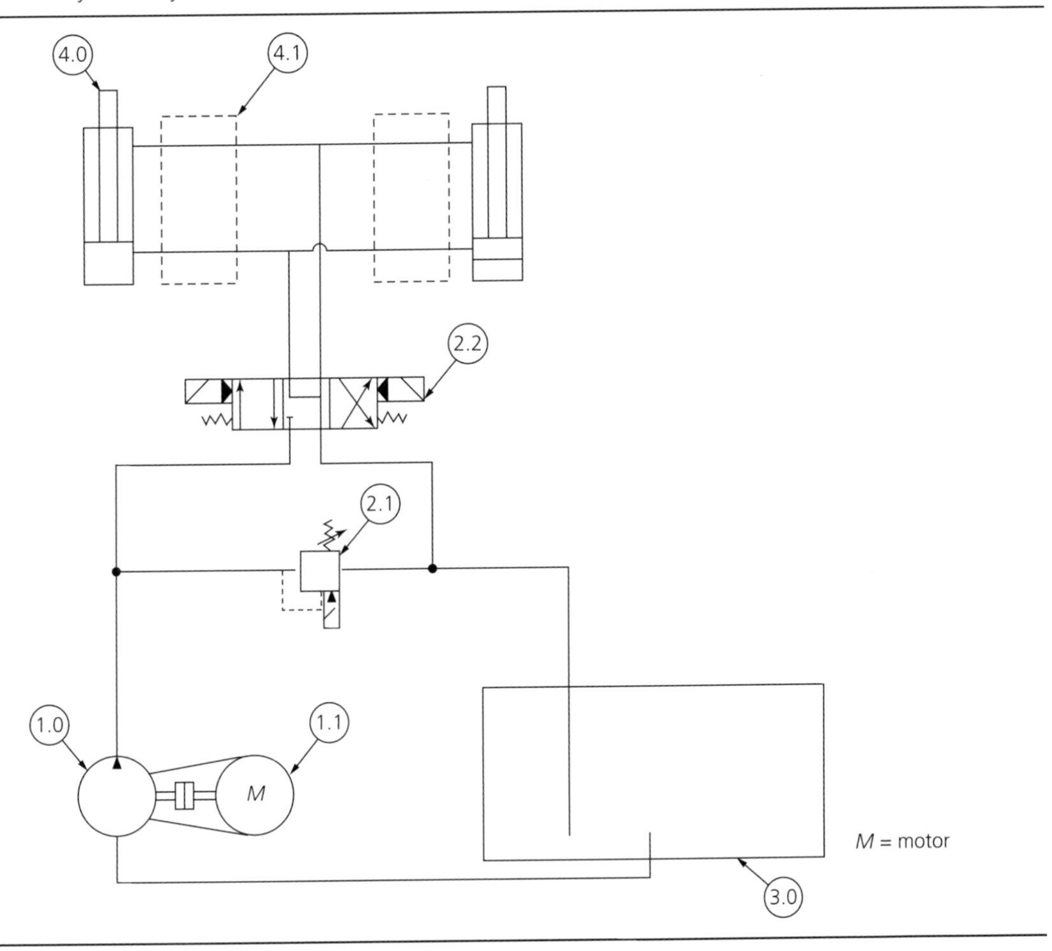

and a cylinder manifold (item 4.1) that contains check valves to hold the cylinder in position when not being actuated.

For span drives, dynamic effects, including inertia, wind, and varying span unbalance, are a significant component of the design loading and must be given primary consideration in selecting actuators and designing a hydraulic system. Similarly, with the exception of unbalanced bridges, such as uncounter-weighted vertical lift bridges for which the load is always acting in one direction, overhauling loads are dynamic in nature and must be adequately damped to avoid system oscillations. For these reasons, the basic hydraulic schematic as shown in Figure 14.2 is not valid for use as a bridge drive. The modifications required to develop a viable bridge drive are discussed later in this chapter.

Before delving into the details of movable bridge hydraulic systems, review of some fundamental principles is in order, including presentation of several basic formulae. The principles at work in a hydraulic system are described in detail by Frankenfield (1984), Stewart and Philbin (1984) and Rabie (2009). In the context of movable bridge system design, the following parameters and formulae are of primary interest:

- Hydraulic fluid: for bridge drive hydraulic systems the fluid is typically petroleum-based mineral oil or synthetic oil formulated with anti-wear additives and corrosion inhibitors for heavy industrial use. Mineral oil has proven to be the most reliable hydraulic fluid for industrial and bridge applications and provides long lasting performance even in harsh conditions. In some cases, vegetable oils, such as rapeseed oil, are used when an environmentally friendly (biodegradable) fluid is necessary because of the potential of a leak to contaminate an environmentally sensitive area. These fluids can perform well, but generally require careful monitoring and have a shorter service life than conventional mineral oil. Unless there is a specific need for an environmentally friendly fluid, mineral oil is recommended for bridge applications.
- Fluid conditioning: hydraulic fluid is the lifeblood of any hydraulic system. The condition of the hydraulic fluid in a circuit is critical to the system's performance and long-term serviceability. The term 'fluid conditioning' is used in this text to describe the devices or subsystems incorporated into a hydraulic system to maintain the hydraulic fluid, including filters, heat exchangers and heaters. These devices are frequently included in the main hydraulic system to perform a specific function, such as filters for filtration. For bridge drives that feature a large reservoir, as is often the case for cylinder drives, it is recommended to package heaters and heat exchangers into an independent fluid conditioning circuit. Typical fluid-conditioning circuits feature a small gear pump, pressure relief valve, heater, heat exchanger and additional filters to supplement those in the main circuit. Conditioning circuits are 'off-line'; in other words, they do not rely on operation of the drive system pumps to function. Off-line configuring provides the advantage of operation independent of the hydraulic drive system that operates the movable span. This enables filtration and either heating or cooling to be performed prior to or between bridge operation cycles, not just during bridge operation.
- Hydraulic power: the power transmitted at any point in a hydraulic system is a function of flow and pressure. The basic equation is as follows

$$P_{\mathrm{hp}} = \frac{Q \cdot P}{1714} \tag{14.1}$$

in which

P_{hp} = power (horsepower)
Q = flow in gallons per minute (gpm)
P = pressure (psi)
1714 = unit conversion factor

- Motor torque: torque, T (in.–lb), produced by a fluid power system utilising a hydraulic motor is the product of the motor displacement, d (in.^{3}/rev), and the pressure drop across the motor, ΔP (psi).

$$T = \frac{\Delta P \cdot d}{2\pi} \tag{14.2}$$

- Motor power: the power, in horsepower, produced by a hydraulic motor is the product of torque, T (in.–lb), and speed, n (rpm):

$$P_{\mathrm{hp}} = \frac{T \cdot n}{63\,025} \tag{14.3}$$

- Pump flow: the flow, in gpm, Q produced by a pump is the product of displacement, d (in.^{3}/rev), speed, n (rpm), and volumetric efficiency, η_v:

$$Q = \frac{n \cdot d}{231} \cdot \eta_v \tag{14.4}$$

- Motor flow: the flow, Q in gpm, required to operate a motor at a given speed is the product of displacement, d (in.^{3}/rev), and speed, n (rpm), divided by the volumetric efficiency, η_v:

$$Q = \frac{n \cdot d}{231 \cdot \eta_v} \tag{14.5}$$

- Cylinder force: the force produced by a cylinder, F (lb), is the difference between the product of the pressure acting on the blind side of the piston, P_b (psi) and the cylinder bore area, A_b (in^2), and the product of the pressure on the rod side of the piston, P_r (psi) and rod annulus area, A_r (in.^{2}).

$$F = P_b \cdot A_b - P_r \cdot A_r \tag{14.6}$$

- Cylinder torque: when a cylinder is used to rotate a body, such as a bascule or swing span, the torque about the centre of rotation of that body is the product of the force in the cylinder and the effective moment arm (the distance along a line perpendicular to the line of action of the cylinder force and passing through the centre of rotation.) It is important to note that for many configurations, the effective moment arm changes as the movable span rotates.

- Cylinder power: the power produced by a hydraulic cylinder is the product of the force in the hydraulic cylinder, F (lb), and speed of the cylinder rod, v (in./sec).

$$P_{hp} = \frac{F \cdot v}{6600} \tag{14.7}$$

- Hydraulic system efficiency: as with all power transmission systems, hydraulic systems are subject to inefficiencies. In movable bridge hydraulic systems inefficiency can be significant and must be accounted for in design. Hydraulic system inefficiencies are generally categorised as volumetric or mechanical. Volumetric inefficiencies result from internal leakage within components, in particular hydraulic pumps and motors. This leakage cannot be eliminated as it is necessary to lubricate components and cool them by removing heated fluid from within the component. Volumetric losses must be accounted for in sizing pumps and motors. Mechanical inefficiencies result from friction in the internal working parts of the components and from friction in the fluid as it is pushed through components and fluid conductors. Mechanical inefficiencies produce heat that is lost to the environment, thereby reducing the work that can be performed. Mechanical losses are typically accounted for in hydraulic systems by calculating the pressure loss that results as fluid is forced through any component or fluid conductor. The following is the basic equation for the pressure loss that results from forcing fluid through a fixed orifice:

$$\Delta P = S \cdot \left[\frac{Q}{29.81 \cdot C_d \cdot D_o^2}\right]^2 \tag{14.8}$$

In this equation ΔP is the pressure drop in psi, S is the specific gravity of the fluid, C_d is the orifice coefficient, D_o is the diameter of the orifice in inches, and Q is the flow in gpm. Examining this equation reveals an important relationship – the pressure drop increases as the square of the flow. This is also generally true of the pressure drop through other elements of the system, including fluid conductors and most valves.

In design, pressure losses in conductors, including pipe, tube, hose and fittings are typically derived from published tables keyed to flow rate. For valves, the pressure drop information must be obtained from the manufacturer. Published pressure loss data are based on specific tests and must be adjusted to account for the properties (notably viscosity) of the hydraulic fluid specified for use.

An important consequence of the relationship between pressure losses and flow is that hydraulic system efficiency can generally be improved by reducing flow rates and increasing pressure to maintain power transmission. In other words, systems that are configured to operate at higher pressures are more efficient than those that operate at higher flow rates.

14.2. Hydraulic system components

Hydraulic systems for movable bridges are composed of a combination of standard manufactured products, such as pumps and valves, customised products, such as hydraulic cylinders, and custom fabrications, such as field piping, and reservoirs. The following section introduces some of the key components used in movable bridge hydraulic systems.

14.2.1 Hydraulic pumps

Hydraulic pumps are the workhorse of a hydraulic system. Two primary pump types are used for movable bridges: axial piston pumps for bridge drives and gear pumps for auxiliary drives. Axial piston pumps are also used for auxiliary drives where specific performance criteria are not achievable using a gear pump.

Axial piston pumps are available in a number of configurations and with a large variety of control options and features. Some of the key features and options relevant to movable bridge drives are discussed herein. As with all manufactured hydraulic components, it is important to fully understand the specific performance characteristics of a given pump product. Products that appear to be equivalent often differ in ways that can adversely affect bridge performance.

Axial piston pumps offer high efficiency and durability. In its basic form an axial piston pump is a fixed displacement pump capable of producing flows of over 200 gpm and pressures above 5000 psi, depending upon make and model. Fixed displacement axial piston pumps are commonly used in open-loop circuits in which proportional valves are used to provide control of fluid flow. Fixed displacement pumps can also be used in conjunction with a variable speed motor to provide flow control.

Variable displacement axial piston pumps are available in bent axis or swashplate designs. Figure 14.3 shows the internal configuration of an axial piston pump of the swashplate design. Key elements of the pump, as shown in the figure, include the drive shaft (item 1), swashplate (item 2), piston (item 3), stroke piston (item 4), opposing piston (item 5), suction side (item 6), high-pressure side (item 7), and control valve (item 8). Variable displacement pumps can be used to regulate flow so that actuator speed can be controlled without the use of a proportional valve or variable speed motor. This allows flow control of a pump coupled with a standard constant speed motor, such as an AC squirrel cage induction motor. The displacement of an axial piston pump is varied by changing the length of stroke of the pistons. Increasing the displacement is referred to as 'stroking' the pump. Conversely, reducing the displacement is called 'de-stroking' the pump. In bent axis pumps the angle of the pistons to the

Figure 14.3 Axial piston pump: swashplate design adapted from Bosch-Rexroth (2013)

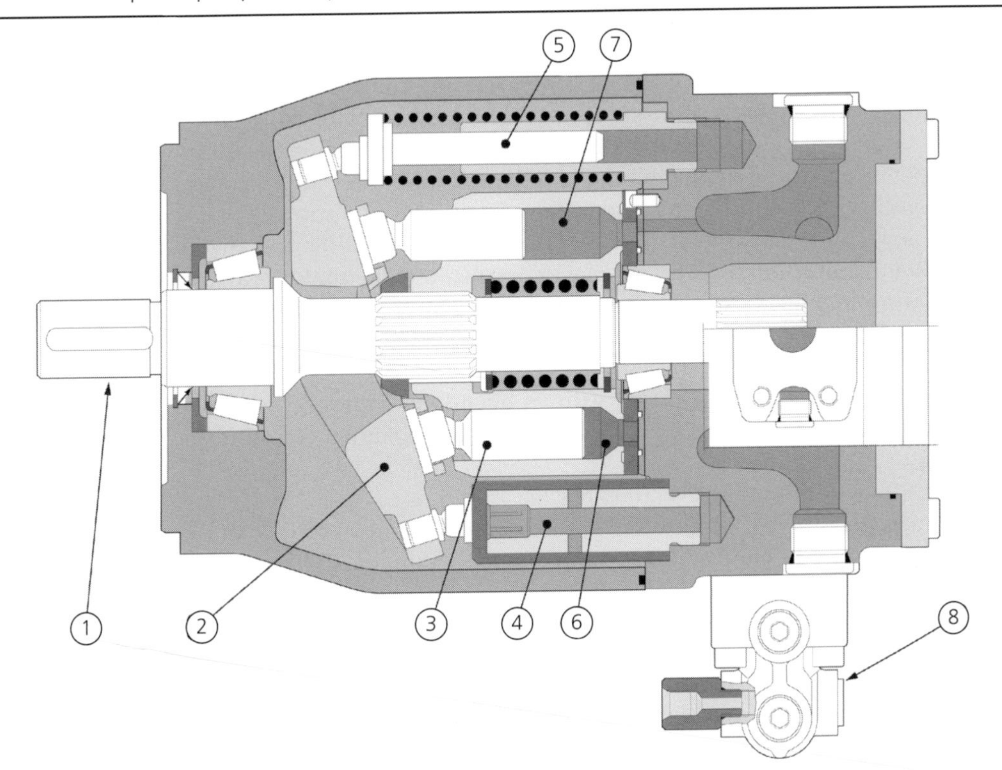

pump axis is varied. In swashplate pumps the angle of the swashplate to the pump axis is varied. Common means of displacement control are by way of a servo motor or hydraulic positioning piston. Servo motor control provides the most accurate control. However, the accuracy of servo motor control is generally only required for precision industrial hydraulic applications and is neither required nor recommended for most bridge applications. The following are some of the potential benefits of variable displacement pumps.

- Maximum flow stop: an adjustment that can be used to limit the maximum pump displacement of a variable displacement pump and thereby limit the maximum flow. This is helpful when the pump size selected is capable of producing more flow than is required, as is often the case because pumps are only available in a limited number of sizes and it is unlikely that the required flow exactly matches the full displacement of the pump. The maximum flow stop is also a means of preventing actuator speed from exceeding a design maximum.
- Pressure compensation: this controls the maximum pressure produced by the pump. It is arranged such that the pump stroke is automatically reduced to zero if the pump outlet pressure reaches a set limit, which is adjustable at the pump. Pressure compensation is an overriding control that governs over other pump displacement controls. In bridge drives, pressure compensation of the pump can serve several purposes. Most importantly, it can be set to protect the pump and hydraulic circuit from excessive pressure should the relief valve not function properly. It can also be used to protect the electric motor from current overdraw. This is accomplished by setting the compensating pressure such that at full pump displacement (or the

maximum-flow stop displacement) the power drawn by the pump does not exceed the motor rating.

- Horsepower limiting: variable displacement axial piston pumps with hydraulic positioning pistons are also available with constant power limiting control options that can be of benefit to certain bridge drive systems. Constant power limiting, also referred to as horsepower limiting, is a control option that automatically adjusts the pump displacement based on pressure, such that the power produced by the pump (i.e. the product of flow and pressure) remains equal to a pre-set constant. When designing bridge drives, this control option can be used to reduce the required power of the hydraulic drive by taking advantage of the fact that at operating loads greater than the maximum constant velocity torque, a reduction in bridge speed is acceptable. The following simplified example demonstrates the difference in a system with and without horsepower limiting.

  - Example: Trunnion bascule bridge with an open-loop hydraulic cylinder drive
  - Maximum pressure in blind end of cylinder under maximum constant velocity loading = 600 psi
  - Average flow to raise bridge = 100 gpm
  - Maximum pressure in blind end of cylinder at maximum starting loading = 1300 psi
  - Average pressure in blind end of cylinder under maximum starting torque load conditions (over full range of opening angle), except using dynamic frictional resistance = 1000 psi
  - Pressure drop in system raising the bridge at full speed = 400 psi
  - Pressure drop in system raising the bridge at half speed = 320 psi.

Without horsepower limiting and assuming 85% system efficiency, the required horsepower of the system is computed as follows:

$$\text{hp} = \frac{Q \cdot P}{1714 \cdot \eta_o} = \frac{100 \cdot (1300 + 400)}{1714 \cdot 0.85} = 117 \text{ hp} \tag{14.9}$$

In comparison, with horsepower limiting, the required horsepower can be calculated using the maximum constant velocity load pressure rather than the maximum starting load pressure.

$$\text{hp} = \frac{100 \cdot (600 + 400)}{1714 \cdot 0.85} = 69 \text{ hp} \tag{14.10}$$

Although the power available to move the bascule leaf is less, the pressure available to start the span has not changed. Furthermore, wind loading, as applied as per AASHTO (2007), is a minimum with the leaf in the closed position and increases as the leaf is raised. Ice loading exhibits an inverse effect and is a maximum with the leaf closed. The net of these varying loads is that the average load is less than the maximum load. Therefore, to confirm that operation is still acceptable with this reduced power, the average flow produced at 69 hp over the full operating cycle is calculated using the average pressure resulting from the loading associated with maximum starting load, except that dynamic frictional resistance is used rather than static frictional resistance.

$$Q_{\text{average}} = \frac{\text{hp} \cdot 1714 \cdot \eta_o}{P} = \frac{69 \cdot 1714 \cdot 0.85}{(1000 + 400)} = 72 \text{ gpm} \tag{14.11}$$

In the above calculation the full 400 psi pressure loss was used conservatively as although the system were running at 100 gpm. In fact, as the pump de-strokes to maintain constant power, the losses in the system will decrease because the flow will be less. This is evident by examining the calculated pressure loss in the system at half flow. With regard to bridge speed, at an average flow of 72 gpm the bridge will operate slower than full design speed. If the loading remains at the maximum starting load, the speed will average 72% of full design speed. The actual speed of the leaf will vary as a function of the load, with the pump stroke continually adjusting to produce constant power. When the loads are lower than the maximum, the bridge speed will increase until the point is reached where the loads are less than or equal to the maximum constant velocity load, at which point the bridge will run at full speed, limited by the maximum stroke setting of the pump. Therefore, as long as it is acceptable for the bridge to operate slower under conditions that exceed the maximum constant velocity load conditions (i.e. dynamic friction, 2.5 psf wind load and unbalance), a horsepower limiting system can be utilised and the power provided can be significantly reduced as seen above (i.e. 69 hp compared to 117 hp).

Caution must be employed regarding horsepower limiting systems for bridge drives. The horsepower limiter setting should be selected such that under normal operation (pressures up to those associated with the maximum constant velocity load) the pump does not de-stroke as a result of the power limiting settings. Pump de-stroking during normal operation can produce undesirable dynamic oscillations. This can be prevented by confirming that the pump/motor is capable of producing full speed flow at the pressure associated with the maximum constant velocity load.

14.2.2 Hydraulic cylinders

Hydraulic cylinders used on bridges are typically either single-acting, or most commonly, double-acting. Single-acting cylinders are constructed such that fluid pressure only acts upon one side of the piston. Single-acting cylinders are generally only used where the load is always applied in one direction, such as lifting an unbalanced bridge. Double-acting cylinders provide fluid pressure to both sides of the piston, allowing the cylinder force to act in either direction of the stroke. Because double-acting cylinders can apply or resist forces in both directions, they are the most commonly applied hydraulic cylinder in movable bridge drives and movable bridge auxiliary equipment. Each of these cylinder types is available in a number of mounting configurations.

The key parts of a cylinder are the body or tube (1), head (2), rod end port (3), cap (4), blind end port (5), rod (6), piston (7) and associated bushings and seals as shown in Figure 14.4. The tube is capped at one end, commonly referred to as the blind end, and has a head at the other end, commonly referred to as the rod end, that is bored to allow the rod to pass through it. In tie-rod type cylinders the cap and head are connected by high-strength rods that hold them securely to the tube. In mill-type (welded

Figure 14.4 Mill-type hydraulic cylinder

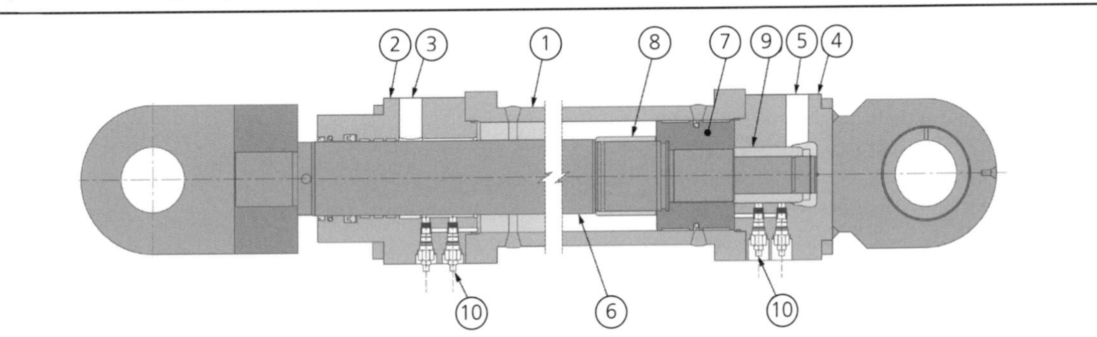

construction) cylinders, both ends of the tube are welded to a flange that is bolted to the head or cap. Inside the tube, the piston is connected to the rod. The piston has bearing rings and seals that fit tightly in the bore of the tube so that pressurised fluid on one side of the piston will not leak to the other side of the piston. Similarly, the head contains a bearing and series of seals to prevent leakage around the rod. The area between the outside diameter of the rod and the inside diameter of the tube is referred to as the rod annulus or rod-end area. The area on the blind end of the piston is referred to as the blind-end area. Other elements of custom cylinders shown in Figure 14.4 and discussed later in this chapter are the rod-end cushion bushing (8), blind-end cushion bushing (9) and cushioning control valves (10).

14.2.3 Hydraulic motors

Hydraulic motors for bridge drives are classified as either high speed or low-speed–high-torque (LSHT). High speed hydraulic motors operate in the same speed range as electric motors used on bridges and can be one-for-one replacements for electric motors. They are commonly employed to operate up to 1800 rpm and are configured with a mechanical drive train just as electric motors are. For bridge drive systems it is recommended to use high-speed motors of the fixed displacement axial piston type. Physically, fixed displacement axial piston motors are similar to hydraulic pumps. Like their pump counterparts they are available in bent axis or swashplate design.

LSHT motors produce higher torque output than is possible with a high-speed motor, but have lower operating speed limits. Although LSHT motors are available that operate at 400 rpm and above, in most bridge drive applications they are operated below 160 rpm and are commonly operated below 40 rpm. LSHT motors are of the radial piston design and derive their torque capability from having large pistons oriented radially around the drive shaft. In designing with LSHT motors it is important to configure the drive machinery such that the motor operates in a speed range for which it is capable of smooth, consistent rotation. Many radial piston motors, particularly those with a limited number of pistons (many models are available with only five pistons) will exhibit cogging at slow speeds. Even motors with a dozen or more pistons may exhibit cogging or require higher viscosity oil to operate at speeds below 3 rpm. Figure 14.5 depicts a radial piston hydraulic motor.

14.2.4 Directional control valves

Directional control valves are used to control the direction of flow in the hydraulic circuit. These valves allow the use of a unidirectional pump to operate bridges and auxiliary bridge equipment bi-directionally. That is, the direction of fluid flow through the actuator can be reversed at the valve.

Proportional directional control valves control flow direction and rate. Figure 14.6 shows the internal workings of a typical pilot operated proportional directional control valve. The valve body (1) contains ports (3) and houses a main spool (2) that is notched. Pilot pressure is applied to either end of the spool by way of the solenoid (5) operated spool valve (4). Shifting the main spool opens or blocks fluid paths between main ports (P, A, B and T). The spool notches provide variable openings that meter the flow rate in proportion to the spool movement.

14.2.5 Pressure relief valves

Pressure relief valves are used to control the pressure within a section of the hydraulic system. All movable bridge hydraulic systems must have relief valves to prevent pressures from exceeding those for which the system is designed. Pressure relief valves respond to pressure within the circuit, dumping fluid back to the reservoir or low-pressure side of the circuit, as required to prevent pressure from exceeding a pre-set value. Relief valves may be fixed, adjustable or variable. Fixed relief valves are factory pre-set to a maximum pressure and are best specified as tamper resistant to prevent field adjustment. Adjustable

Figure 14.5 Hydraulic radial piston motor. (Reproduced from Birnstiel, 2008, courtesy of Bosch-Rexroth)

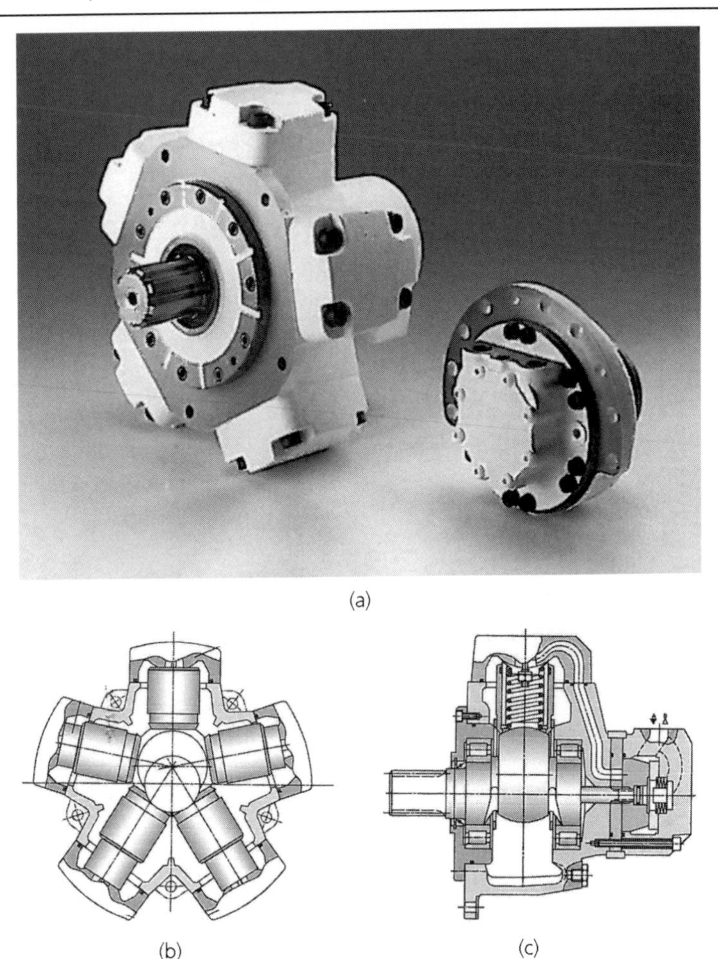

(a)

(b)

(c)

relief valves are manually adjusted to various pressure settings. Variable relief valves are equipped to be dynamically adjusted to change the pressure-relief setting while the system is in use. All relief valves should be specified with maximum pressure protection at a limit equal to or below that for which the circuit is designed. Pressure-relief valves open gradually to deter sudden changes in pressure. Most valves have a cracking pressure, the pressure at which the valve begins to open, that is 100–200 psi below the relief valve setting. Therefore, relief valves must be set above the desired working pressure if full system flow is to be maintained at the desired working pressure. Otherwise, some of the fluid flow will bypass the circuit through the relief valve, thereby reducing the flow to the actuator. In some cases this may not be a concern. For example, under maximum starting load conditions it is generally acceptable for the bridge speed to be reduced. However, during normal operation it is not advisable for system pressure to exceed the relief valve cracking pressure as this could create oscillations in the motion of the movable span.

The pressure available at the pump outlet is typically limited by a main system pressure relief valve. This valve should be factory set to limit maximum system pressure. If an adjustable or variable pressure

Figure 14.6 Proportional directional valve with pilot valve mounted on top. For guidance on reading hydraulic schematics and component diagrams, see Frankenfield (1984). (Adapted from Bosch-Rexroth, 2013, reproduced by kind permission)

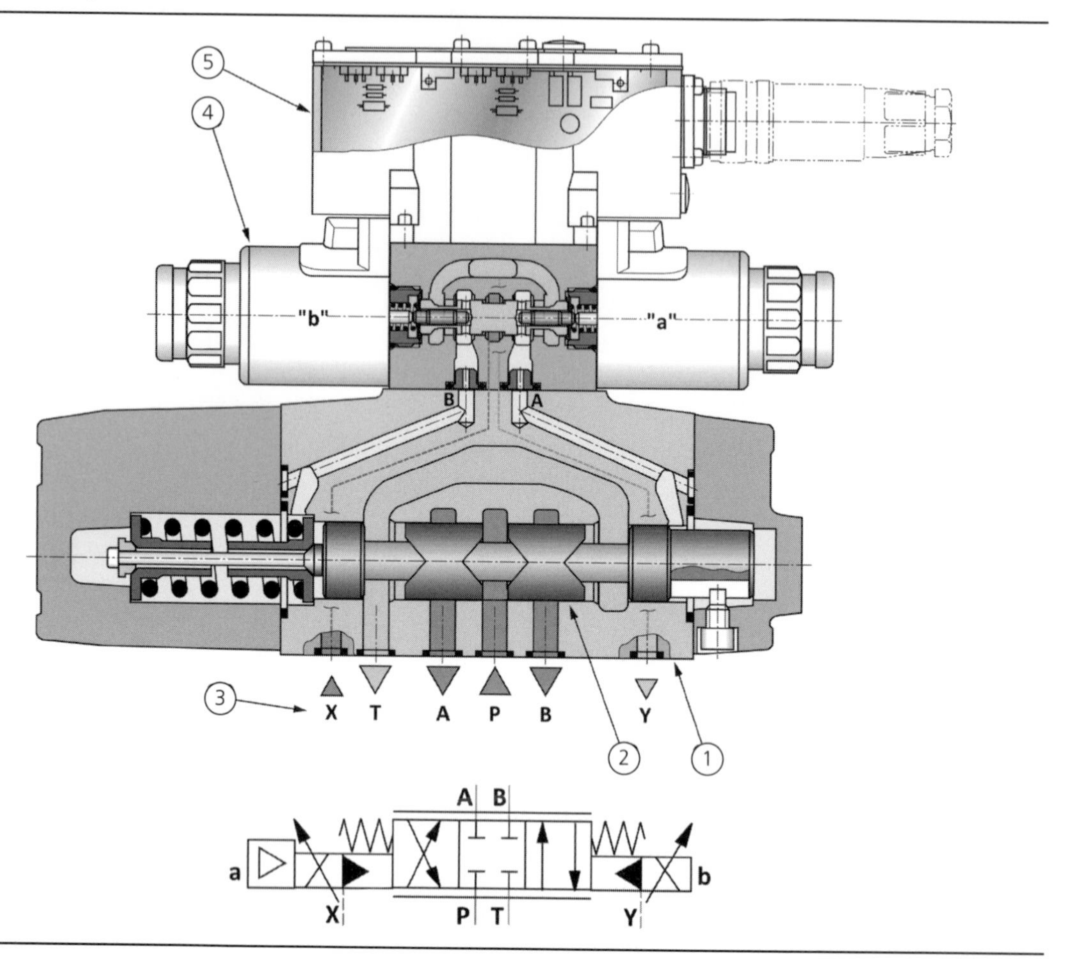

relief valve is used, a second fixed relief valve may be desirable or even necessary. Otherwise, the system must be designed for the maximum setting of the adjustable or variable relief valve.

14.2.6 Control of hydraulic pumps, valves and other components

Hydraulic components are available in a variety control options, including manual, hydraulic, servo, electronic and combinations thereof. Using a combination of these means can result in a bridge hydraulic system that can be operated from a remote or local control panel as well as manually in the event of control system failure.

For most modern movable bridge drive applications, pumps and valves used to control the motion of a bridge are controlled electronically by use of a control processor (such as an amplifier card) provided by the component manufacturer or by a remote motion controller (such as a programmable logic controller). This allows a hydraulic drive to be interconnected with, or made integral with, the bridge control system. Each control function can be executed from simple contact closures, such as those originating from a position limit switch, or variable commands such as those issued by a programmable logic controller or dedicated motion controller.

Directional control valves and relief valves are typically discrete in their function. That is, they are either shifted or not. For these valves electronic control is also discrete. However, proportional directional control or relief valves feature variable control of the valve. By controlling the amount by which the spool of a proportional direction valve shifts, the effective size of the orifice created within the valve can be adjusted and flow through the valve regulated with sufficient precision.

Where pump control is used to vary fluid flow, pump stroke can be controlled either by use of a manufacturer-supplied control processor or by a remote motion controller. Pump control processors accept discrete or variable (i.e. analogue) inputs and provide a command signal to a solenoid that controls the pump stroke cylinder. The pump control functions vary with manufacturer, make and model, but generally provide motion control functions such as stroke ramping for acceleration and deceleration and pre-set stroke levels for constant flow levels associated with full speed or seating speed. By matching these adjustable settings to a series of discrete inputs the bridge acceleration ramp, full speed, deceleration ramp, seating speed and stop ramps can be established. Pump stroke control can also incorporate feedback to allow for use of proportional–integral–derivative (PID) control.

14.2.7 Check valves

Check valves control the direction of flow in a circuit. A standard check valve allows fluid to flow in one direction only, blocking flow in the reverse direction. Pilot-operated check valves (PO checks) limit flow to one direction unless the valve is piloted open to allow flow in the reverse direction. Check valves are available with or without springs that seat the valve. Spring-set check valves require a specific pressure to unseat the valve and allow flow.

14.2.8 Counterbalance valves

Counterbalance valves are arguably the most important valve in achieving uniform, stable span operation of open-loop movable bridge circuits. Located downstream of the load, this valve provides resistance to actuator motion for both dynamic braking and dynamic stability. Counterbalance valves modulate to regulate flow in response to pilot pressure. Counterbalance valves are manufactured in a variety of configurations. Some counterbalance valves also respond to load pressure. Configured properly in the circuit, the counterbalance valve restricts flow in inverse proportion to the load pressure. Counterbalance valves are configured ‘cross-piloted’ so that pilot pressure is sourced from the opposite side of the actuator. In this arrangement, load pressure on the pump side of a cylinder pilots the counterbalance valve open. Therefore, when the actuator is working against a load, the counterbalance valve is piloted open to minimise its restriction to flow. This is desirable as it reduces the pressure drop through the valve while the actuator is performing work.

In the event of an overhauling load, a load acting to move the actuator away from the pump faster than desired, load pressure on the pump side of the actuator will drop. This will reduce the pilot pressure acting on the counterbalance valve. The counterbalance valve will respond by modulating (shifting towards the closed position of the valve) to restrict flow through the valve. This restriction of flow will create back pressure to effectively apply the brakes and counter the overhauling load.

Counterbalance valve response to load variation is dynamic. Under normal operating loads and conditions, the valve should modulate smoothly between full open and partially closed. The valve should never close completely during operation, nor should it be used to hold the load in a fixed position (see PO check valves). Counterbalance valves must not be oversized in an attempt to reduce the pressure drop and increase the efficiency of the system. Even when operating against maximum constant velocity load, the counterbalance valve must provide some resistance to actuator motion to prevent oscillation of the movable span during actuation. Recommended practice is to size the counterbalance valve so that when piloted full open, at full flow, it provides a pressure drop at least equal to the pressure drop in the circuit between the flow control device (pump or proportional valve for example) and the actuator.

A properly selected counterbalance valve will be reactive enough to prevent overhauling loads from increasing span speed beyond the desired speed and yet provide sufficient damping to prevent span oscillation under normal operating conditions. In general, counterbalance valves with a pilot ratio of 5 : 1 or less are preferred for bridge applications. Similar to setting the gain too high with a PID controller, use of counterbalance valves with high pilot ratios can result in unstable operation as the valve continually overcompensates rather than seeking a stable position.

14.2.9 Accumulators

Accumulators are used to store pressurised fluid so that it can be released to perform work without running a pump or otherwise applying external power to the system. Accumulators commonly used in bridge drives are of the bladder or the piston type. These accumulators are pre-charged with nitrogen to establish the minimum pressure at which the accumulator begins to fill with fluid. Frankenfield (1984) and Rabie (2009) both provide a description of accumulators and formulae for design.

14.2.10 Needle valves

Needle valves are used to close off, or partially close off, a passage to fluid flow. They can be used to isolate a section of a circuit, or when partially closed off, as a means of restricting flow.

14.2.11 Filters

Filters are a critical component of hydraulic systems. They function to remove contaminants from the hydraulic fluid that could otherwise cause increased wear or even damage to the hydraulic components. Filters are generally categorised by their location in the system as either a pressure filter or a return filter. Pressure filters are located near the outlet of the pump to remove contaminants that may be generated by pump wear before they are spread into the circuit. Return filters are located on the low-pressure side of the circuit where fluid is returning to the reservoir. Return filters are positioned to remove contaminants that may have entered the circuit before they are returned to the reservoir and eventually sucked into the pump.

Recommended practice for bridge drive systems is to include both pressure and return filters. The extra filtration is appropriate because of the fluid cleanliness needs of the components used in these systems, the cost of replacement components, and the disruption to operation that can result from system failure. The ISO (International Organisation for Standardisation) standard ISO 4406 (1999) establishes fluid cleanliness level definitions for hydraulic systems. Manufacturers publish the required minimum fluid cleanliness levels required for proper performance of their components. The fluid in a hydraulic system should be kept at or below (i.e. cleaner than) the ISO code level for the component in the circuit with the most stringent requirement. A minimum ISO solid contaminant code level of 17/14 is recommended for hydraulic bridge drives.

14.3. Heating and cooling of hydraulic systems

Maintaining the temperature of hydraulic fluid in a system is critical to serviceability and performance. If fluid temperatures are too low, fluid viscosity will increase and system performance and efficiency will suffer. Therefore, in cold regions, a means of heating the hydraulic fluid must be provided. If fluid temperatures are too high, fluid viscosity will decrease, leakage will increase and the ability of the fluid to lubricate the components diminishes. At extremely high temperatures, fluid and seals can suffer permanent damage. Therefore, hydraulic systems must be designed so that excess heat can be efficiently removed from the system, either by passive means such as thermal conductance (radiant heat released through the walls of the reservoir) or by active means such as use of a heat exchanger. Heat exchange in hydraulic bridge drives is similar to that of other industrial hydraulic systems. Refer to Frankenfield (1984) for guidance in designing heat exchangers and heaters.

14.4. Reserve power factor

Design of hydraulic systems for bridge drives involves selection of components that must function in unison for the system to work as intended. Many components have adjustable settings that serve to provide flexibility and allow field adjustment of hydraulic systems to optimise performance. The designer establishes settings for these adjustable components and bases calculations for system efficiency and power on those settings. Inherent in the flexibility of hydraulic systems is the risk that in adjusting system performance, actual performance characteristics may deviate from the design assumptions. Additionally, designs are often implemented under construction contract provisions that allow substitutions for specified design components. In most cases, field adjustments of components and substitution of alternate manufacturer's make and model for specified components can be accommodated without sacrificing the design intent or performance. The frequent exception to this generalisation is that most substitutions or adjustments tend to result in an increase in pressure drops in the system. This is particularly true of adjustments made to dampen a system to eliminate oscillations. In almost all cases this involves increasing the pressure drop in the system.

To accommodate future system adjustment, component substitution and even future component replacement, an appropriate 'reserve power factor' should be applied to sizing the prime mover. The amount of reserve power necessary varies with the specifics of design, contracting and maintenance. AASHTO (2007) specifies a reserve power factor of 20% for open-loop hydraulic systems. This value is appropriate for typical highway bridge projects in the USA, for which the construction is awarded to the low bidder. It also is appropriate for open-loop systems which utilise counterbalance valves for control of overhauling loads and to provide dynamic stability. Reserve factors of less than 20% are appropriate for closed-loop systems and/or systems for which the designer has control over the selection of specific component make and model. A minimum reserve power factor of 5% is recommended in all cases.

14.5. Hydraulic movable bridge drive types

Hydraulic circuits are classified as either 'open-loop' or 'closed-loop'. Open-loop systems draw fluid from a reservoir, impart energy to the fluid, and conduct pressurised fluid to an actuator where work is performed. Once the fluid has performed its work, it is returned to the reservoir where it is 'open' to atmospheric pressure. This fluid mixes with other fluid still in the reservoir and remains there until drawn into a pump to start the cycle again. In open-loop systems the direction of fluid flow is typically controlled by use of directional valves. Open-loop systems are characterised by large reservoirs. Conversely, in a closed-loop system, the fluid is contained within the circuit. The pump draws fluid from one side of the circuit and forces it into the other side. In closed-loop systems the direction of fluid flow is controlled by reversing the direction of fluid flow through the pump. When run continuously, the same fluid is circulated over and over around the circuit. Closed-loop systems only require a small reservoir needed to store and discharge fluid that is drawn from, or leaks from, the closed circuit. Closed-loop systems are commonly referred to as hydrostatic drives.

The two types of hydraulic actuators used for movable bridge drives are cylinders and motors. Each of these can be coupled with either an open-loop or a closed-loop circuit, although closed-loop cylinder drives are rare. In addition, control of the drive can be accomplished in a number of ways, thus creating a plethora of possible system configurations. Historically, the following are the most common and successful types of bridge drives

- open-loop hydraulic cylinder drives
- open-loop hydraulic motor drives
- closed-loop hydraulic motor drives.

14.5.1 Hydraulic cylinder-operated bridges

Hydraulic cylinders are commonly used to operate bascule and swing bridges. They are also occasionally used to operate vertical lift bridges that require limited lift. Cylinders can be positioned to actuate motion in a variety of ways and are adaptable for use on most types of bridges.

Trunnion bascule bridges are often configured with hydraulic cylinders positioned such that the cylinders extend to push the leaf open and retract to pull the leaf closed, as shown in Figure 14.7. Although leaf motion is angular, cylinder motion is linear. As a result, constant fluid flow and cylinder velocity results in non-uniform angular velocity of the leaf. Similarly, for a given torsional load on the leaf, acting about the trunnion's axis, the cylinder force and resulting pressure vary depending on leaf position.

In the arrangement shown in Figure 14.7, the effective moment arm of the cylinder force acting to rotate the leaf is at a minimum, with the leaf in the full closed and full open positions and at a maximum with the leaf approximately halfway open. For a nominally balanced bascule leaf this is generally the most effective configuration for push-to-open cylinder arrangements because the variation between cylinder velocity and leaf velocity is minimised. Similarly, cylinder pressure for maximum constant velocity loading remains nearly constant.

Trunnion bascule bridges have also been constructed with hydraulic cylinders configured in push–pull tandem pairs such that one cylinder pushes the leaf open, while the other simultaneously pulls the leaf open. An example of a bridge with push–pull tandem cylinders is shown in Figure 14.8.

Rolling-lift bascule bridges have been configured with the cylinders arranged horizontally such that they pull the bridge open. In most cases the cylinder rod end is attached to a pintle at the centre of roll and the blind end of the cylinder is anchored to the back wall of the bascule pier. Other rolling-lift bridges have been configured with the cylinders mounted to the bascule leaf below the track level and oriented to the push toward the channel. This is counterintuitive, but creates rotation to open the span because the line of force is below the pivot point of the span on the track.

Both rolling-lift and trunnion bascule bridges have also been configured with hydraulic cylinders located such that they pull down on the counterweight to open the bridge.

Swing bridges are often configured with a pair of cylinders that work in tandem, either as a push–pull pair or with both cylinders pushing together and pulling together. For swing bridges the cylinders are typically oriented horizontally and attached to either side of the pivot.

Some of the above arrangements have produced less than desirable results. Typical problems include positioning the cylinders such that they have excessively long strokes, which results in systems in which

Figure 14.7 Bascule bridge with hydraulic cylinder drive

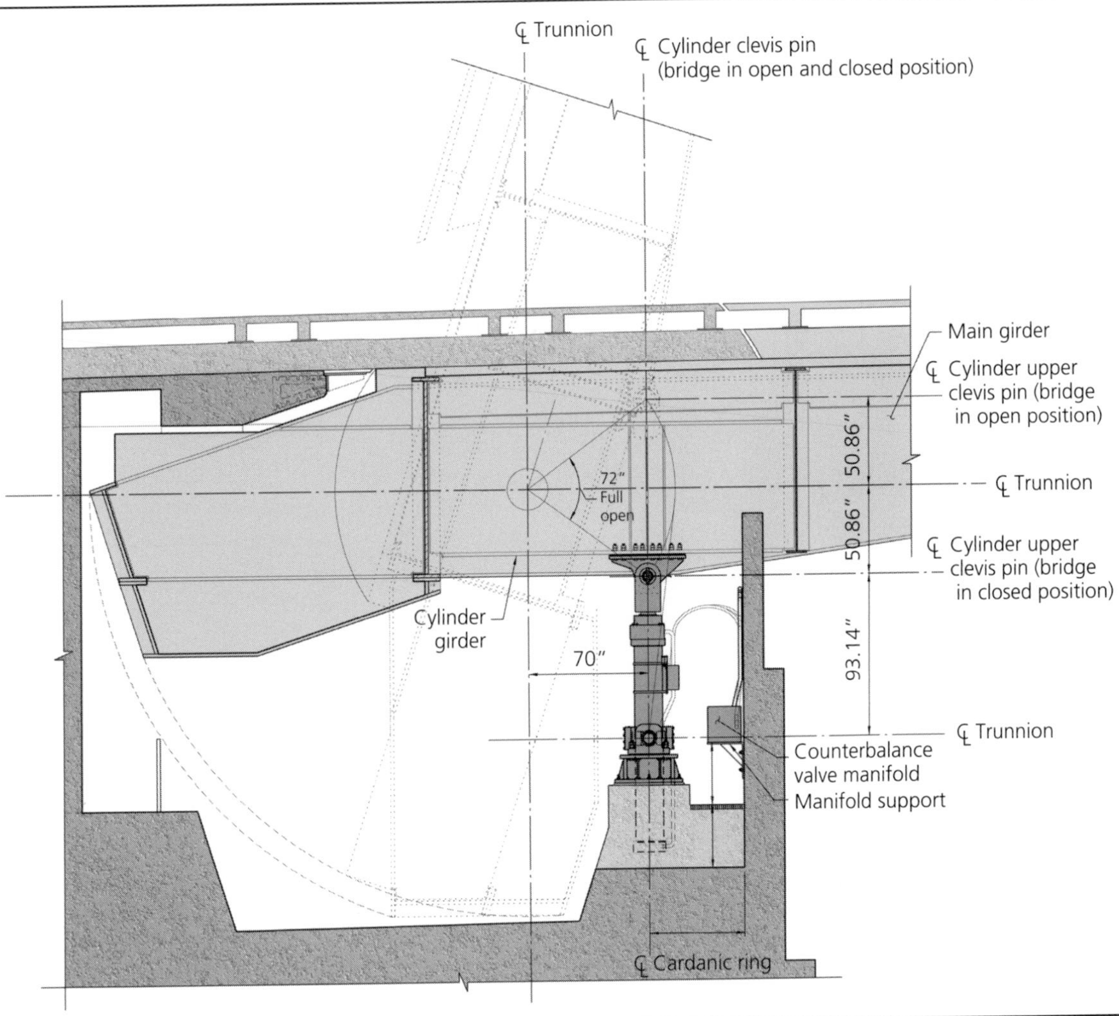

rod buckling dominates the design, and/or in inefficient systems (high-flow–low-pressure). Many of the configurations also leave the cylinder rods exposed and subject to falling roadway dirt and debris while the bridge is not in operation. The cylinders are also too often oriented towards the horizontal rather than the vertical without due consideration of the loads imparted on the bearings and seals due to the mass of the cylinder. The orientation of hydraulic cylinders also affects the resulting loads on trunnions or other machinery supporting the bridge.

Unlike gear drives, which require precision field alignment to exacting tolerances (e.g. rack and pinion gear tooth alignment tolerances), properly designed hydraulic cylinders offer installation tolerances that are much more forgiving. For typical bridges, the performance will not be compromised if the cylinder geometry is obtained to within plus or minus an eighth of an inch or even greater. This is not to suggest that substandard construction should be accepted, but this wider range of tolerances is more forgiving of typical installation challenges. It is noteworthy that the fabrication tolerances for

Figure 14.8 Cylinders arranged in push–pull tandem pairs

high-quality hydraulic components, including custom-fabricated cylinders, are similar to those applied to manufacturing of precision enclosed gearing.

The key to developing a hydraulic cylinder drive that has forgiving installation and alignment tolerances is founded in selection of the cylinder mounting and kinematics of the cylinder motion. These parameters are also critical in designing a system that can accommodate deviations in the prescribed movable span motion and structural deflections that occur due to loading, whether caused by wind, vehicles or extreme events such as earthquakes or vessel collision.

Cylinders should be mounted so that their connections to the fixed and moving structure are free to rotate in all directions, not just the directions of primary motion. This is necessary to assure that the cylinder is only loaded axially as a two-force member. Otherwise, even small variations in alignment could impart significant moments at the attachment point, which in turn could damage the seals, accelerate the wear of piston and/or rod bearings, or even cause buckling of the cylinder rod.

The simplest and most common mounting configuration is to design both ends of the cylinder with spherical bearing attachments. A typical cylinder of this design is shown in Figure 14.9. The blind end of the cylinder is fabricated with a male clevis equipped with a spherical bearing (i.e. self-aligning clevis) that will be connected by a clevis pin to a female clevis weldment, anchored to the fixed pier. Similarly, the rod end features a female clevis (i.e. forked clevis) that will be connected by a clevis pin to a male clevis equipped with a spherical bearing that is attached to the movable span. This configuration of cylinder and cylinder supports allows free motion in the plane of expected movement

Figure 14.9 Cylinder with self-aligning clevis on the blind end and forked clevis on the rod end

of the cylinder in which the cylinder will pivot about the blind-end clevis pin as the bridge rotates. The spherical bearings provide almost unrestrained movement at each attachment to accommodate small out-of-plane movements of the movable span as might occur due to steel fabrication and installation tolerances or wind loads acting on the movable span during operation.

Another method for mounting cylinders without restraining motion at the attachment is the use of a cardanic ring. A cardanic ring mount, such as is shown schematically in Figure 14.7 and depicted in Figure 14.10, provides two axes of rotation, similar to a gimbal, at an intermediate point on the cylinder tube, thereby reducing the length between support points. Cardanic rings may be appropriate where space is limited or where buckling length must be reduced as a design consideration.

The ideal orientation for a hydraulic cylinder is vertical, with freedom of rotation at the support and attachment points. In this state the piston and rod bearings and associated seals are subjected primarily to hydraulic fluid pressure. If a cylinder is oriented horizontally the weight of the cylinder causes bending in the cylinder assembly. Internally this bending is resolved into force couples, with the reaction points located at the rod and piston bearings. It also produces bending moments in the rod and cylinder body. Bending of a cylinder reduces its buckling strength as it produces eccentricity of the load with regard to the cylinder centre of gravity. For small hydraulic cylinders, such as those used to actuate span locks, the cylinders are short and cylinder weight is not significant compared to the other design considerations. However, for bridge drives the cylinders can be large and the strokes significant. If bridge drive cylinders are located more than 15° from the vertical the cylinder may require special design considerations, such as a stop tube, and/or the weight of the cylinder must be accounted for in determining the buckling resistance. A stop tube is a cylindrical spacer fit over the rod and located between the piston and the cylinder head. This spacer blocks the piston from moving too close to the cylinder head. This increases the minimum distance between the rod bearings and the piston bearings, reducing the loading on the bearings due to cylinder bending, thereby improving bucking resistance.

Figure 14.10 Cardanic ring-mounted cylinder

14.5.1.1 Hydraulic cylinders for movable bridges

Tie-rod type hydraulic cylinders, as produced to requirements established by the National Fluid Power Association (NFPA) and/or ISO, are available from many manufacturers. These cylinders are available in a wide variety of models including various combinations of pressure ratings, rod materials, seal types and mounting styles and are adequate for most movable bridge auxiliary systems. For bridge drives, custom cylinders are more appropriate due to their more robust design and ability to accommodate specific features needed for bridges. The following are some of the more common custom features specified for bridge drive cylinders.

- Rod-end and/or blind-end clevises: custom rod end and blind-end clevises are typically required to accommodate the range of movement desired for bridge drive cylinders. Custom clevis designs may also be necessary to meet bridge design code requirements for structural adequacy.
- Rod-end and/or blind-end spherical bearings: in most cases, custom rod-end or blind-end spherical bearings are required to provide unrestrained rotation of the cylinder ends.
- Cardanic ring mounts: cardanic rings are custom cylinder mounts that must be carefully designed and detailed in the cylinder fabrication drawings.
- Custom piston rods: since many movable bridges are located in aggressive coastal environments, custom rod materials or coatings are often specified. For small cylinders stainless steel rods may be a viable option. Stainless steel rods are often used for auxiliary bridge equipment. For larger

cylinders stainless steel is not economical so nickel–chrome plating of cylinder rods is recommended. Other coatings, such as ceramic, plasma spray, high-velocity oxygen fuel (HVOF) spray coated and plasma transfer arc (PTA) welding are also available. These technologies are largely proprietary and should be specified only following careful consultation with an experienced cylinder manufacturer.

- Custom rod seals: movable bridge applications are somewhat unique in that the normal operating pressures are relatively low compared to the maximum holding pressures. Therefore, the cylinder seals should be selected to function under a wide range of pressures.
- Cylinder accessory mounts: cylinders for bridge operation should be equipped with load holding valves mounted directly or hard-piped (connected with rigid pipe or tube rather than flexible hose) to the cylinder. Accessory mounting brackets are often required to support cylinder manifolds. If these mounts are to be welded to the cylinder, they should be installed by the cylinder manufacturer prior to final honing of the cylinder rather than added on after cylinder manufacturing.
- Custom cushions: most hydraulic cylinders are available with cushions at both ends of travel. However, manufacturers standard cushions are relatively ineffective in cushioning movable bridges. Custom cylinder cushions are recommended for all bridge drive applications and discussed in more detail later in this chapter.

Hydraulic cylinders for bridge drives must perform several functions and therefore must be equipped with appropriate valves and piping. The recommended configuration is to mount a custom cylinder manifold on each cylinder, pipe the manifold directly to the cylinder ports, and equip the manifold as shown in Figure 14.11. This configuration features the following

Item 4.0 Hydraulic cylinder
Item 4.1 Cylinder manifold

Figure 14.11 Hydraulic cylinder: manifold schematic for bridge drive

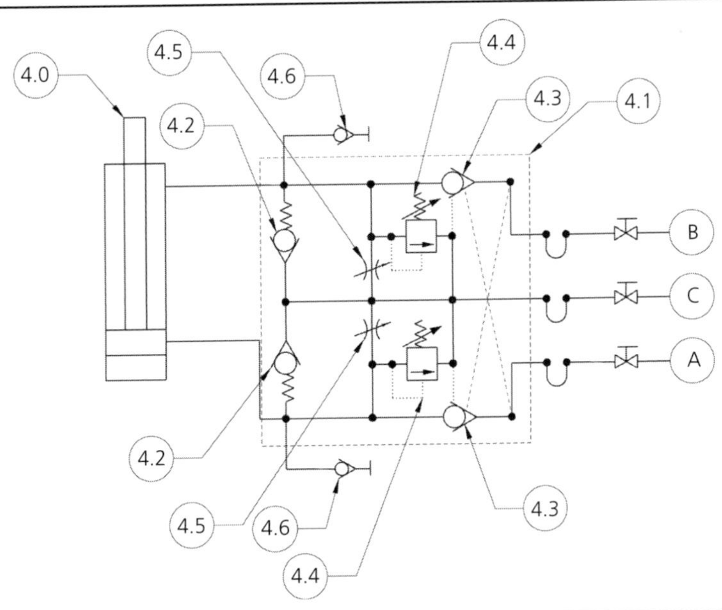

Item 4.2 Anti-cavitation check valve (spring loaded)
Item 4.3 Pilot-operated check valve
Item 4.4 Cylinder relief valve
Item 4.5 Manual release needle valve
Item 4.6 Test port.

Line A connects the system to the blind port of the cylinder. Flow from the pump through Line A acts to extend the cylinder. When the cylinder is retracted, return flow passes through Line A as it goes back to the tank. Similarly, Line B connects the system to the rod port and flow through Line B acts to retract the cylinder. Line C connects the cylinder manifold to the reservoir and functions as a drain line and an anti-cavitation line.

PO check valves function to hold the cylinder in position when it is not being actuated. For movable bridges, the PO check valves mounted in the main drive cylinder manifold should be designed and detailed to hold the movable span in a fixed position, even if the span is subjected to loads up to and including the holding load. PO checks are generally leak free for short periods, for example during the time a movable span is held open for vessels to pass. However, over time, PO checks may leak. Therefore, if a movable span is to be held open for an extended period of time, as is the case for a railroad bridge that is normally open, an independent holding device is recommended. Each PO check has a pilot line that is cross-ported to the opposing port. When neither Line A or B is pressurised, the PO checks are closed, trapping the fluid in both the blind and rod ends and preventing the cylinder from moving. When fluid pressure is applied to Line A, pressure conducted through the pilot line releases the PO check connected to Line B and allows flow to exit the rod end of the cylinder. The PO check valves in the cylinder manifold must be externally vented (vented to the drain line and dumped to the tank as shown) so that they are not affected by any pressure that could be trapped in the system between the cylinder and the hydraulic power unit (HPU). Similarly, pilot pressure to the PO check valves should be ported from the supply side of the circuit, not the cylinder side. To prevent inadvertent pressure from being trapped on the supply side, the directional valve should be spring centred in the steady state and the centre spool position drained to tank. Otherwise, trapped pressure of even a small amount can cause the PO check to open and allow the cylinder to inadvertently drift.

In most all cases, the pressure in the cylinder required to hold the bridge against the specified holding load will be greater than the pressure required to operate the bridge. For this reason, the cylinder manifold is equipped with a pair of relief valves, one connected to the blind-end port and one connected to the rod-end port. The settings of these relief valves is critical as it will govern the design pressure for the cylinder assemblies (cylinder, cylinder piping, and cylinder manifold) and any machinery or structures subject to cylinder loads. These valves must be either non-adjustable or factory set to the required holding load and equipped with tamper proof locks. Setting the cylinder relief valves such that the cylinders can hold the bridge against the holding load allows the system relief valve to be set lower (i.e. to meet the operating requirements). Consequently, the rest of the hydraulic system, other than the cylinder assemblies, can be designed for the lower pressure.

If the cylinder is subjected to a load that exceeds the holding load, for example an extreme wind gust, the cylinder, cylinder piping and manifold will be protected from overload by the cylinder relief valves, as will be any connecting machinery or structures. Fluid relieved by a cylinder relief valve will be made up in the opposite end of the cylinder by drawing fluid from the reservoir through Line C and the anti-cavitation check valves. These valves limit the potential for air to be drawn into the system.

The manual release needle valves provide a means of manually releasing fluid from either end of the cylinder. These can be used to manually lower a bridge that has sufficient imbalance to overcome friction. Manual release valves are also useful in construction to manipulate the cylinder length during connection to the structure. In addition, on bascule and lift bridges, manual release needle valves provide a means for performing simple bridge balance checks by way of drift testing.

It is recommended that each line to the cylinder be equipped with a shut-off valve (ball valve) and a hose. In addition to allowing cylinder movement, such as is required for bascule or swing bridge cylinders that pivot with span rotation, the hoses provide isolation between the dynamic vibrations of the movable span and the field piping rigidly mounted to the pier. Hoses also provide flexibility in locating field piping, thereby improving constructability. The ball valves provide an easy means of isolating a hose or cylinder for future maintenance or replacement. Ball valves also provide a quick and safe means of isolating the HPU from the cylinders so that HPU testing can be conducted without risk of inadvertent cylinder movement.

14.5.1.2 Custom hydraulic cylinder cushions

A unique advantage of hydraulic cylinder bridge drives is the ability to equip the cylinders with custom cushions capable of significantly reducing the speed of the movable span at either end of travel. Cylinder cushions are built into the cylinder as shown in Figure 14.4. They are failsafe and require no maintenance. Properly designed and constructed, cylinder cushions can slow a movable span from full speed to 10% speed or less in just a few inches of stroke. Cylinder cushions provide added protection against slamming the bridge at the end of travel regardless of failure of other systems, even in the event that the bridge is driven at full speed into the cylinder cushion(s).

The basic components of a cylinder cushion are the cushion spear, cushion spear chamber and cushion ports. Referring to the blind end of the cylinder in Figure 14.4, normal flow of fluid from the cylinder passes through the cushion spear chamber and out through the cylinder port. When the cushion spear enters the cushion spear chamber, fluid flow through the cushion spear chamber to the port is blocked as the spear fills the area of the chamber. Fluid is forced to take an alternate path to the cylinder port through the cushion ports which are restricted, either by a fixed orifice or needle valve. The pressure drop created by cushion port restriction increases the pressure in the blind end of the cylinder acting upon the piston. When the cushion ports are properly sized, the cushioning pressure will be sufficient to slow the bridge. To slow the bridge, the cushion pressure, reflected at the pump, must be sufficient to cause the pump to compensate. Once the pump compensates, the span will slow in a controlled manner, metered by the cushions.

As with counterbalance valves, cushioning is dynamic and self-modulating. Referring to Equation 14.8, recall that the pressure drop through a fixed orifice varies in proportion to the square of the flow rate. Therefore, even if the cushion port has a fixed orifice, the response to motion is non-linear. If the cylinder is moving at full speed as the cushion engages, the cushion pressure will be high, as is needed to slow a bridge approaching the end of travel position at full speed. If the cylinder slows, the cushion pressure will reduce exponentially. This inherent non-linear response is beneficial. If the cushion is to be effective, it must engage before the bridge reaches the limits of travel. However, it is not desirable for the cushion to impede normal seating operation. Fortunately, experience has shown that cushions can be sized to significantly slow a bridge at full speed and have little or no impact on the bridge at seating speed.

Cushion performance is also affected by the details of the cushion. In general, the spear is machined with a slight taper at the leading edge. This results in gradual restriction of flow, buffering the initial effect of the cushion. The spear also has an annular clearance within the cushion spear chamber that acts like a cylindrical orifice which is progressively more restrictive as the spear engages further. More than one cushion port can be provided. For example, two cushion ports bypassing the cushion spear chamber can be fabricated, each with a different orifice or needle valve. By porting an internal port to connect to a point within the cushion spear chamber the cushion can be staged or sequenced. As the cushion spear passes a port in the wall of the cushion spear chamber, it blocks it, and that port is removed from the available paths for fluid. The best cushion performance is achieved with some degree of progressive cushioning, designed to provide increased resistance as the cylinder moves closer to the end of travel position.

Because of the number of factors affecting cushioning, including machining tolerances, it is recommended to provide at least two staged orifices. It is also recommended that the orifices either be adjustable needle valves or replaceable orifice plugs so that the cushions can be shop adjusted to provide the desired cushioning pressure/cylinder speed performance ratio. If needle valves are used, they should be made tamper proof after being set. Cylinder cushion designers must keep in mind that the reserve stroke of the cylinder will not normally be effective for cushioning. For example, if a cushion is 4 in. in length and 0.5 in. of reserve stroke is provided, only 3.5 in. of cushioning will be effective in application.

14.6. Open-loop hydraulic cylinder bridge drives with proportional directional control

Open-loop hydraulic cylinder drives are the most common movable bridge hydraulic drive. Drives of this type are capable of a wide range of applications and can be used to operate movable bridges small and large. Their common use stems from the adaptability of the design to various types and sizes of movable bridges as well as their relative simplicity. The primary disadvantages of open-loop hydraulic cylinder drives as compared to other drive systems, including electro-mechanical drives and closed-loop hydraulic motor drives, are that they are less efficient. The advantages of open-loop hydraulic cylinder drives can be many, depending upon the specific bridge design. As previously discussed, cylinder drives are more forgiving of fabrication and installation tolerances than other drive systems. This advantage in constructability often results in lower initial cost.

For small to mid-sized movable bridges, such as two-lane double-leaf trunnion bascule bridges with a span length of 150 ft or less between centreline of trunnions, a simple hydraulic cylinder drive with proportional valve control performs well. This system, shown in Figure 14.12, consists of the following key components

Item 1.0 Pressure compensated axial piston pump with horsepower limiting
Item 1.1 Squirrel cage induction motor
Item 2.0 Main manifold
Item 2.1 Proportional pressure relief valve
Item 2.2 Proportional directional control valve
Item 2.3 Counterbalance valves
Item 2.4 Check valve
Item 2.5 Pressure filter
Item 3.0 Reservoir assembly
Item 3.1 Suction ball valve
Item 3.2 Flow meter
Item 3.3 Return filter

Figure 14.12 Open-loop cylinder drive with proportional valve flow control

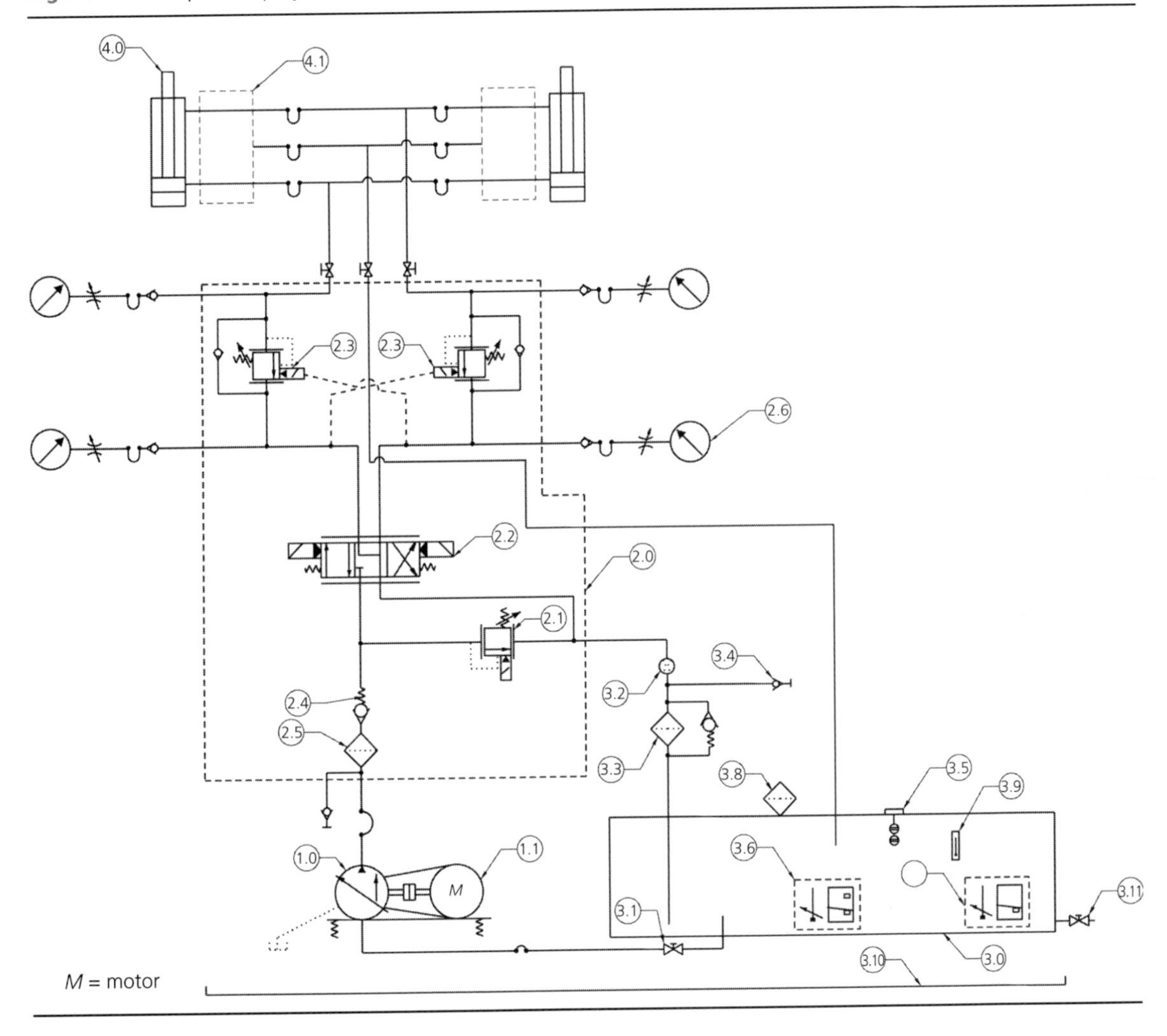

Item 3.4 Checked fill valve
Item 3.5 Dual float switch
Item 3.6 Temperature switch
Item 3.7 Temperature switch
Item 3.8 Reservoir breather
Item 3.9 Fluid level site gauge
Item 3.10 Drip pan
Item 3.11 Reservoir drain ball valve
Item 4.0 Hydraulic cylinders
Item 4.1 Cylinder manifold (see Figure 14.10).

Unlike the previous circuit shown in Figure 14.2, the system shown in Figure 14.12 provides a means to control flow rates. It also provides a means (i.e. a counterbalance valve) to dampen oscillations and counter overhauling loads. The pump features pressure compensation, horsepower limiting and a set stroke limit. As long as the combination of pressure and flow does not exceed the pre-set horsepower limit, the pump will run at the set stroke limit and produce the corresponding flow. When the pump starts, the relief valve will be unloaded (open) and all flow will recirculate back to the reservoir.

To initiate movement of the cylinders a control command will be sent simultaneously to the proportional directional valve and the system proportional relief valve. The proportional directional valve will shift to provide metered flow to the blind end of the cylinders. Simultaneously, the proportional relief valve will begin to close allowing pressure to build in the circuit. As the pressure builds in the blind end of the cylinder, pilot pressure will open the opposite PO check valve to allow fluid to exit the rod end of the cylinder. Pressure in the main manifold on the cylinder's blind end side of the circuit will build and pilot open the counterbalance valve that allows fluid to return through the directional control valve and back to the reservoir. As soon as the pressure in the cylinder is sufficient to move the load, the movable span will begin to accelerate. By throttling flow at the proportional directional valve, bridge acceleration will be controlled. Throughout the above sequence, the pump will be fully stroked and produce full flow, unless the system pressure exceeds the pump compensator or power limiting settings.

If the bridge accelerates too quickly and tends to run faster than the flow through the proportional directional valve will permit, pressure in the blind end of the cylinder will drop because the flow cannot keep pace with the cylinder motion. This drop in pressure will be sensed almost instantaneously by the counterbalance valve by way of the pilot line. The valve will close partially, restricting flow out of the rod and slowing the cylinder until the pressure in the blind end side of the circuit catches up. Once the bridge reaches full speed, a steady state condition will ensue. The blind-end pressure and counterbalance valve position will seek a balanced state unless wind load or other external loads cause the system to readjust.

To decelerate the movable span, the proportional directional control valve will be given a command to shift towards the centre position. This will increase the pressure drop across the valve in both directions as the valve meters flow in and out. When the pressure drop across the directional valve increases sufficiently, pump flow will be diverted over the relief valve, reducing flow to the blind end of the cylinder. Simultaneously, the pressure piloting the counterbalance valve connected to the rod end of the cylinder will be reduced as the flow from the pump is insufficient to maintain pressure. As a result, this counterbalance valve will close partially, restricting flow and providing back pressure for controlled deceleration of the movable span.

An alternative procedure for the example given above would be to set the system pressure relief valve above the pump compensator setting and start the pumps with the relief valve energised (set to maximum system pressure). In this configuration, the pump will fully compensate upon starting, such that no flow is produced. As the proportional directional valve is shifted to allow flow to the cylinders, the pump will stroke up (stroke will increase from the fully compensated position) and supply flow. The pump stroke will modulate to produce flow as per the hp-limiting settings. When decelerating, the directional control valve will be shifted to create a pressure drop sufficient to cause the pump to partially compensate, again modulating as per the hp-limiting settings. When the proportional directional control valve is fully closed, the pumps will de-stroke to the fully compensated position. This alternative with pumps starting at full compensation provides a system that produces less heat than when the pumps start at full stroke. This is because a fully compensated pump produces only a fraction of full power. Conversely, a pump producing full flow, through even a full open relief valve, can generate significant heat. Another advantage is that in this alternative configuration, the proportional relief valve can be replaced with a fixed relief valve unless there are other reasons for using a proportional valve, such as establishing a seating pressure. Seating pressure is a reduced pressure implemented to avoid imparting large forces in the structure when parking the movable span against the fixed structure. A disadvantage of the alternative configuration is that when in idle (pumps started, awaiting a motion command) the pumps are not producing flow which could be processed through filters, heaters or heat exchangers to perform fluid conditioning.

14.6.1 Accumulator circuit for emergency stop or power failure

One drawback to use of PO check valves to hold cylinder position, as shown in Figure 14.11, is that a loss of pilot pressure while the movable span is in motion can result in sudden closing of the PO check valves and a corresponding sudden, forceful stopping of the span. Loss of pilot pressure could result from a loss of system power or an emergency stop (E-stop) that removes power from the drive system. To prevent abrupt stopping of the span, an accumulator circuit and uninterruptible power supply (UPS) is recommended. The UPS is required to provide power to maintain the command that shifts the directional control valve. This valve must be kept in its shifted position to provide a conduit for fluid to the cylinders. Fluid power to pilot the PO check valves (as well as the directional valve if pilot operated) will be supplied by the accumulator. The configuration of the accumulator circuit is shown in Figure 14.13 and includes the following components of the design example given at the end of Chapter 14

Item 2.11 Fill/discharge valve (needle valve with reverse flow check valve)
Item 2.12 Manual shutoff valve (ball valve)
Item 2.13 Accumulator
Item 2.14 Automatic discharge valve (solenoid operated)
Item 2.15 Manual discharge valve (needle valve).

While the pump is running, fluid will pass through the check valve, item 2.11, and fill the accumulator. If power is lost, the fluid in the accumulator will supply the pressure line by passing back through the needle valve in item 2.11. Item 2.14 is an automatic discharge valve that discharges the accumulator back to the reservoir when the power unit is not in use. Needle valve, item 2.15, provides a manual means to discharge the accumulator and shutoff valve, item 2.12, provides a means to isolate the accumulator as may be desirable during testing and start-up operations.

Experience has shown that the accumulator should be sized to provide approximately two seconds of full speed flow to the system. This is sufficient to bring the span to a smooth controlled stop. The pre-charge pressure in the accumulator should be set above the pressure required to shift the directional valve (if pilot operated). The accumulator should be sized so that the required flow is provided as the pressure in the accumulator drops from normal system pressure to the pre-charge pressure. The accumulator should not be oversized as this will increase the time for the bridge to stop once an E-stop has been executed.

14.7. Open-loop hydraulic cylinder drives with pump control

Proportional directional control valves are a relatively inefficient means of motion control. Flow is controlled by regulating the pressure drop across the directional valve. For small bridges open-loop hydraulic cylinder drives with proportional directional control are often selected because their simplicity outweighs their inefficiency. However, for large bridges operated with hydraulic cylinders, higher flows are common and system inefficiencies can become onerous. An alternative, more economical drive is an open-loop hydraulic cylinder drive with pump control.

The circuit for a cylinder drive with pump control is similar to that shown in Figure 14.12 except that the directional valve is not proportional. The valve position is not regulated, but is simply shifted to

Figure 14.13 Open-loop cylinder drive schematic with proportional valve control

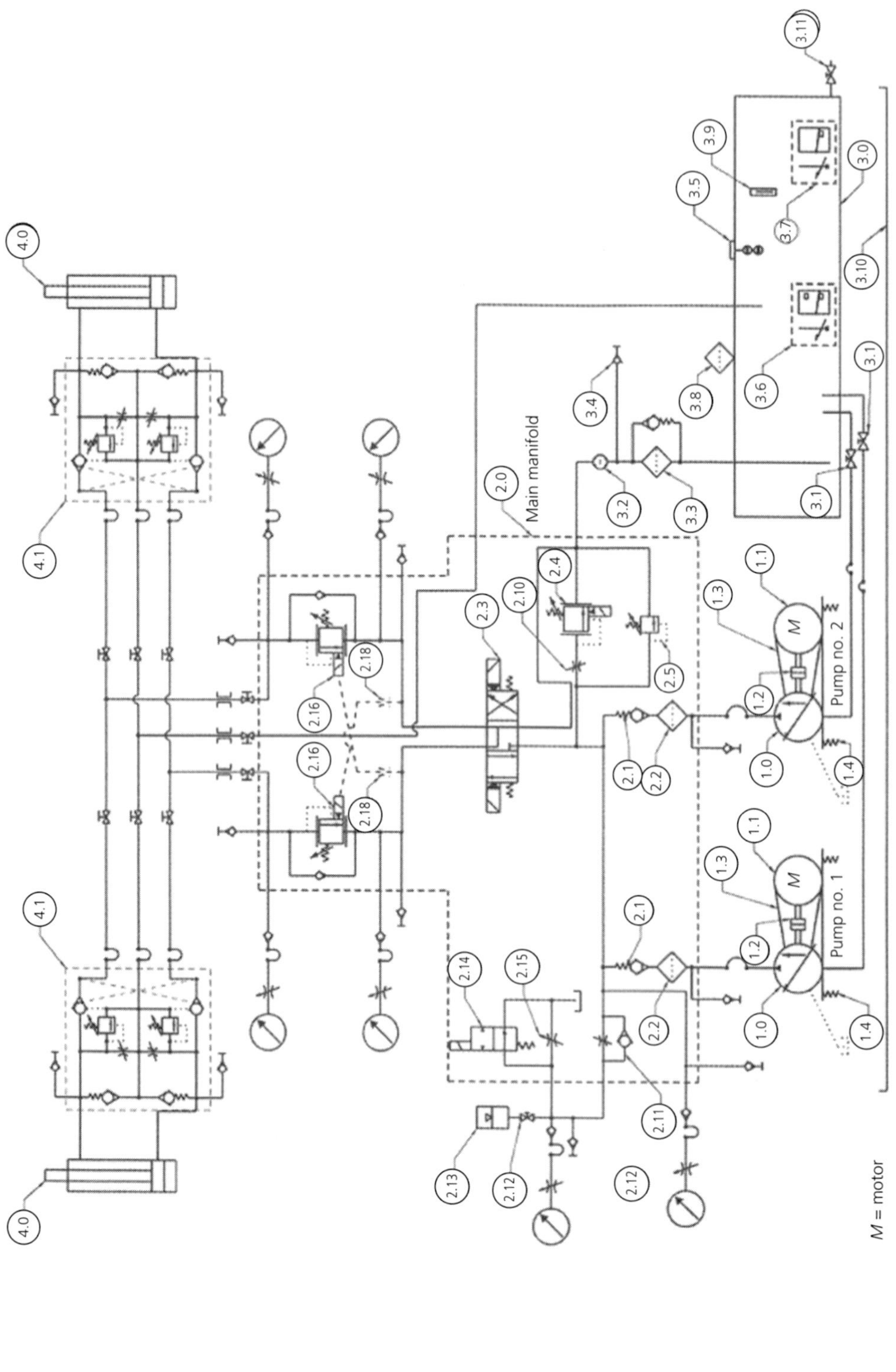

M = motor

divert fluid to one direction or the other when not centred. Since this valve is not used to regulate flow, it can be sized to minimise the pressure drop through the valve.

Flow in an open-loop hydraulic cylinder drive with pump control is controlled by regulating the pump output. The stoke of the pump is gradually varied from zero to full-speed stroke and back to zero to provide motion control of the movable span. The advantages of this configuration over that of proportional directional control include

- The pump starts unloaded, with the stroke set at zero, so the power to start the pump is minimised.
- Power drawn by the pump is minimal until a command signal is issued to stroke the pump (as compared to the typical proportional directional control circuit where the pump is either fully compensated or produces full flow while the system is idling).
- The pressure drop through the directional valve can be substantially reduced.

Pump control does have some disadvantages compared to proportional directional control, primarily with regard to system complexity. Power is required to control the pump. This typically is provided by a small charge pump, piggybacked to the main pump and an independent relief valve to set the control pressure. These are addition components not required for a system with proportional directional control.

With pump control it is important to set the pump limits such that the pump does not inadvertently de-stroke during normal operation, as could be the case if the pressure at the pump were to exceed the pressure compensation or hp-limiting settings. In these scenarios pump-controlled systems are much more likely to exhibit dynamic oscillations than are systems with proportional directional control. It is recommended to set compensation or hp-limiting pressure settings a minimum of 300 psi above the calculated pump pressure at maximum constant velocity load.

14.8. Hydraulic motor drives

Hydraulic motor drives must be designed so that the motor is not a holding device and the holding function is provided by brakes. Bridge drive motors should be carefully selected and configured to avoid use of motors or motor circuits that are inherently load-holding. If the hydraulic motor holds the load without being pressurised, then it will restrain, up to its physical limits, any loads applied to the bridge, even loads that exceed the capacity of the drive machinery (similar to worm gear drives which are not permitted for use in span drive machinery by the AASHTO (2007) specifications for this reason). In movable span mechanical drives, the brakes function to limit the maximum loads that the machinery will be subjected to in restraint of high wind loads. A load-holding motor would violate this design principle as the resistance provided by a hydraulic motor, locked against rotation, would exceed the maximum torque produced by the motor in operation and the brake torque.

14.8.1 Open-loop hydraulic motor drives

Schematics for open-loop hydraulic motor drives are similar to those for open-loop hydraulic cylinder drives. However, there are two significant differences between hydraulic motor and cylinder drives, as follows.

- Unlike a cylinder, motor drives are hydraulically balanced in that the actuator ratio is 1 : 1, so the flows and pressures raising and lowering a bridge are similar.
- Motor drives have brakes to hold the bridge so the drive motor is not subject to holding loads. These differences allow the controlling design pressure to be much closer to the rated pressure of the hydraulic system.

As previously discussed, working at higher pressures and lower flows increases efficiency. For these reasons, hydraulic motor drives are generally much more efficient than hydraulic cylinder drives. The drawback to hydraulic motor drives is that they generally are connected to drive machinery that is similar to that used in electric motor drives, such as enclosed gearing, couplings, bearings, and rack and pinion sets. In general, a hydraulic motor system is more complex than an electric motor with variable speed control, so the hydraulic motor drive must offer an offsetting advantage over the electric motor drive if it is to be recommended.

A common advantage available in hydraulic motor drives is found in the use of LSHT motors. These large displacement motors can produce high torque at speeds up to 160 rpm. When coupled with a small displacement pump operating at high speed, the ratio of the motor to the pump provides a mechanical advantage similar to that achieved through gear reduction in an electro-mechanical drive.

14.8.2 Closed-loop hydraulic motor drives

For most bridge drive applications, open-loop control provides adequate control of the load and satisfactory efficiency. However, in some cases, specific project constraints or design considerations may demand a drive that exceeds the control capability of the open-loop system or requires a more efficient power delivery than can be achieved with an open-loop circuit. Where a bridge requires compact drive machinery and/or precise motion control, a closed-loop hydraulic motor drive is a viable alternative to an electro-mechanical drive. A closed-loop hydraulic motor drive is shown in Figure 14.15. Closed-loop drives provide exceptional power in a relatively small, compact design and are capable of motion control approaching that of DC motors and vector-controlled AC (alternating current) motors (see Chapter 15). They are also much more efficient than open-loop drives, which could be important where power comes at a premium, such as when retrofitting an existing bridge that has limited electrical service available, or where heat generation is of concern.

A typical, simplified, closed-loop motor circuit is shown in Figure 14.14. The basic circuit consists of a variable displacement pump coupled to a fixed displacement motor. In this configuration the pump is commonly referred to as a hydrostatic pump and the accompanied circuit a hydrostatic transmission. Key elements identified in the circuit are

1 Electric motor (typically a 1800 rpm AC squirrel cage motor)
2 Hydrostatic pump, including variable displacement pump, pump stroke piston, charge pump and control pump
3A Hydrostatic transmission, including flushing valve
3B Stroke control unit (proportional with position feedback)
4 Heat exchanger on flushing drain line
5 Hydraulic motor (fixed displacement)
6 Brake release control circuit
7 Hydraulic released brake
8 Reservoir.

Unlike the open-loop motor circuit, this circuit does not have a directional control valve or a counter-balance valve. In the open-loop circuit these valves provide control of the hydraulic motor's (or motors') direction of rotation, dynamic braking and damping to prevent dynamic oscillation. In the closed-loop drive all fluid control functions are provided by the pump and hydrostatic drive, and by the hydraulic motor.

Figure 14.14 Closed-loop hydrostatic bridge drive with hydraulic motor actuator

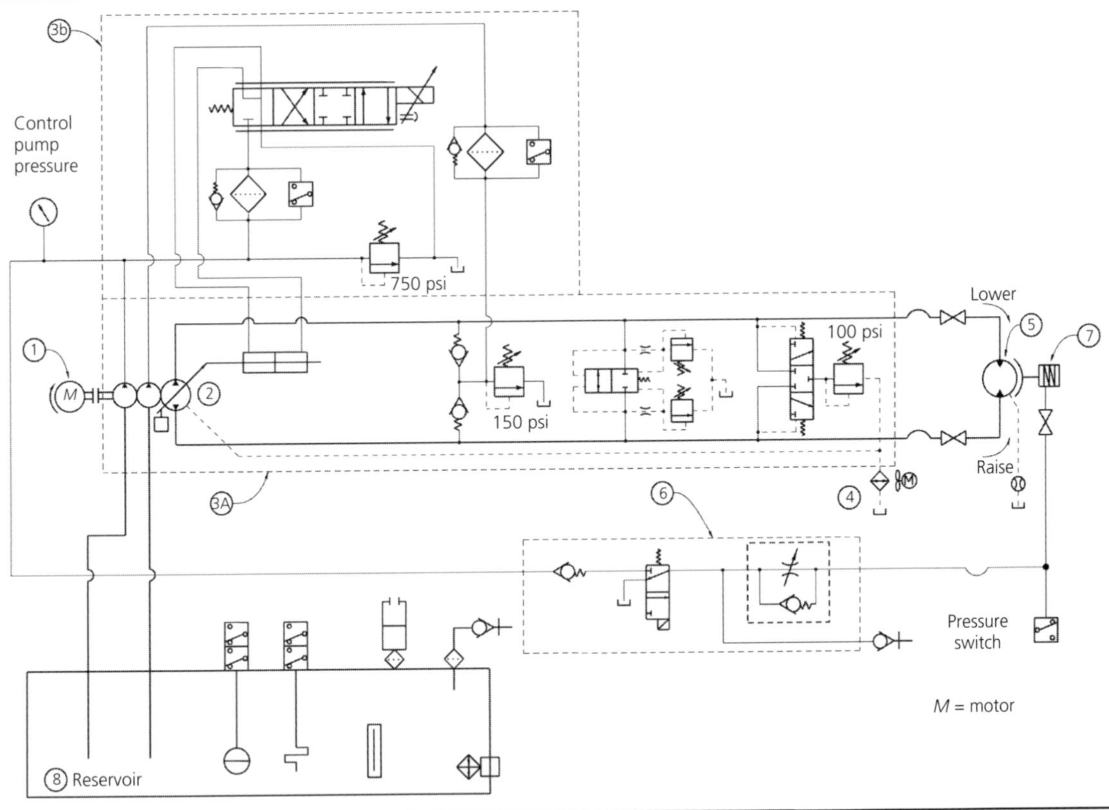

Figure 14.15 Closed-loop hydraulic motor drive; motor coupled to planetary gear reducer

Like open-loop axial piston pumps, closed-loop pumps are of the bent axis or swashplate design. With open-loop pumps the axis or swashplate can be pivoted to vary outflow from zero to full in one direction. Open-loop pumps feature an inlet port sized for suction flow and a smaller outlet port sized for high-pressure flow. Conversely, the stroke of closed-loop pumps can be varied from full flow in one direction to full flow in the other direction. When the pump is centred, the stroke is zero and fluid is not pumped to either port. The ports on a closed-loop pump are the same size to accommodate bi-directional flow.

In Figure 14.14, if the pump and motor are of the same displacement, *d*, then the motor will turn at the same speed as the pump, less speed lost due to volumetric efficiency of both pump and motor. However, if the motor has a larger displacement than the pump then the motor will turn at a slower speed than the pump, but produce proportionally more torque than is imparted to the pump by the electric motor. The ratio of the motor displacement to the pump displacement creates a mechanical advantage similar to that conveyed by a speed reducer. This demonstrates the key benefit of a closed-loop hydraulic motor drive over an electro-mechanical drive. Some or all of the gearing required to develop a mechanical advantage in an electro-mechanical drive can be eliminated.

As an example, consider a vertical lift bridge for which each pinion is driven by a 75 hp, 900 rpm AC motor (870 rpm at full load), through a drive train with a ratio of 72.5 : 1 and a drive train efficiency of 92%. Full load torque (FLT) of the motor is 453 ft-lb. Pinion speed is 12 rpm and torque at the pinion shaft due to FLT is 30 100 ft-lbs. At 125% FLT, the design torque applied to size an electric motor to operate against maximum starting torque, torque at the pinion shaft is 37 700 ft-lbs.

This electro-mechanical drive could be replaced with a closed-loop hydraulic motor drive mounted directly to the pinion shaft. Using a maximum operating pressure drop across the motor of 3000 psi as design criteria, Equation 14.2 can be rearranged to solve for the required displacement of the LSHT motor as follows

$$d = \frac{T \cdot 2 \cdot \pi}{\Delta P} = \frac{37\,700\ \text{ft} \cdot \text{lb} \cdot \frac{12\ \text{in}}{\text{ft}} \cdot 2 \cdot \pi}{3000\ \text{psi}} = 950\,\frac{\text{in}^3}{\text{rev}} \tag{14.12}$$

A LSHT hydraulic motor is selected with the next largest displacement available. The motor specifications, obtained from the manufacturer for this example, are as follows

Motor displacement = 1070 in3/rev
Motor volumetric efficiency = 98%
Motor overall efficiency = 94%

Flow required to operate the motor at 12 rpm is calculated using Equation 14.5 as follows

$$Q = \frac{12\ \text{rpm} \cdot 1070\ \text{in}^3}{231 \cdot 0.98} = 56.7\ \text{gpm} \tag{14.13}$$

Rearranging Equation 14.2, the pressure drop across the motor required to produce the full design torque is

$$\Delta P = \frac{37\,700\ \text{ft} \cdot \text{lb} \cdot \frac{12\ \text{in}}{\text{ft}} \cdot 2 \cdot \pi}{1070\,\frac{\text{in}^3}{\text{rev}}} = 2660\ \text{psi} \tag{14.14}$$

The minimum required pump displacement is calculated using Equation 14.4, assuming a pump motor speed of 1750 rpm and a pump volumetric efficiency of 94%

$$d = \frac{Q \cdot 231}{n \cdot \eta_v} = \frac{56.7\ \text{gpm} \cdot 231}{1750\ \text{rpm} \cdot 0.94} = 8.0\frac{\text{in}^3}{\text{rev}} \tag{14.15}$$

A pump with a maximum displacement larger than 8.0 in^3 will be selected and the stroke limiter will be set to 8.0 in^3 so the pump will produce 56.7 gpm.

In this example, the mechanical advantage of the electro-mechanical drive, developed using an 870 rpm motor and a 72.5 : 1 gear ratio, is replaced by the pump/motor ratio so that the same torque and speed can be delivered at the pinion shaft without the use of gear reduction. The power required to drive this system at full speed and at the equivalent of 125% full load torque of the original electric motor is calculated as follows, assuming the losses in the circuit are 50 psi and overall pump efficiency is 90%

$$P_{\text{hp}} = \frac{56.7\ \text{gpm} \cdot (2660 + 50)\ \text{psi}}{1714 \cdot 0.90} = 99.6\ \text{hp} \tag{14.16}$$

In addition to main pump power, power for auxiliary pumps is also a consideration in determining the size of the electric motor required to power the closed-loop hydrostatic drive. Figure 14.14 shows a control pressure pump and a charge pump as is typical. Both pumps are typically gear pumps, piggy-backed onto the main pump. The control pressure pump must provide sufficient pressure to operate the pump stroke piston and enough flow to overcome leakage. The control pressure pump is commonly required to produce approximately 3 gpm at 750 psi. The charge pump must produce sufficient flow and pressure to continually replenish fluid lost from the circuit through leakage and fluid flushed from the circuit for cooling and filtration purposes. Typical charge pumps provide 10–15% of full pump flow at a pressure of 150 psi. In the above example the power required to operate the auxiliary pumps, assuming an 80% total efficiency is calculated as follows

$$P_{\text{hp}} = \frac{(3 \cdot 750) + (0.15 \cdot 56 \cdot 150)}{1714 \cdot 0.80} = 2.6\ \text{hp} \tag{14.17}$$

This power must be added to the power calculated for the main circuit in determining the size of the electric motor. The result is 102.2 hp to power the hydrostatic drive. Considering that electric motors can be sized for starting torque loads at 125% of full load torque and applying a 5% reserve power factor to the design, the electric motor needed to drive the above hydrostatic drive could be sized at $((102.2 \cdot 1.05)/1.25 = 85.9)$, say 100 hp. Therefore, in this example, the equivalent to a 75 hp electro-mechanical drive is a 100 hp hydrostatic drive (with some conservatism due to rounding up the next available electric motor size). In this example, the designer should consider the increase in required drive power of the hydrostatic drive as compared to the advantages of eliminating most, or all, of the mechanical drive train.

14.9. Design loads for machinery

Hydraulic systems apply loads to machinery that they are connected to. The loads for which these elements must be designed are determined by the physical size of the hydraulic actuators and the controlling design pressures. For machinery, the applicable design limit states, as defined by AASHTO (2007), are service, overload and fatigue.

At the service limit state, machinery connected, directly or indirectly, to hydraulic motors or hydraulic cylinders should be designed for the torque or force produced by the actuator at the normal working pressure available to actuate the motor or cylinder.

At the overload limit state, machinery connected, directly or indirectly, to hydraulic motors or hydraulic cylinders should be designed for the torque or force produced by the actuator at the maximum working pressure available to actuate the motor or cylinder.

At the fatigue limit state, machinery connected, directly or indirectly, to hydraulic motors or hydraulic cylinders should be designed for the torque or force produced by the actuator when resisting the maximum constant velocity load. For this loading the resistance should be based on the endurance limit of the material of the machinery component being designed.

In determining the above design loads, the efficiency of motors or cylinders should be neglected.

In the above design load determinations, the setting of a relief valve that limits the holding pressure in the system was not applied. Only the pressure available to motivate the actuator was considered. This criterion is valid for auxiliary drives or bridge drives that do not have a holding function. However, for actuators that have a load holding function, such as bridge drive hydraulic cylinders equipped with PO check valves, the maximum setting of the pressure relief valves in the cylinder manifolds must be used to determine the forces acting upon connected machinery components, such as clevis pins and clevis bearings. Where the actuator has a load-holding function, the design load for connected machinery, to be applied at the overload limit state, should be the torque or force produced by the actuator at the maximum working pressure at the motor or cylinder as established by the load holding relief valve. As before, in determining the design loads, the efficiency of motors or cylinders should be neglected.

14.10. Design loads for structure

Hydraulic systems apply loads to connected structural elements. Structural elements of bridges subjected to loads imparted by a hydraulic actuator should be designed for strength, service and fatigue conditions. The design loads below are calibrated for use with load factors and structural resistances calculated as per AASHTO (2012).

At the strength limit state, structural elements subject to direct loading from hydraulic motors or cylinders should be designed for the larger of the following

- Force produced by the cylinder at 150% of the maximum working pressure as determined by the drive circuit or the holding circuit, whichever is greater. The 150% factor is applied to account for pressure spikes which may occur before a relief valve can fully respond to a sudden load such as a large wind gust acting upon a bascule or swing bridge held in the open position.
- Calculated maximum operating load plus 100% impact. The 100% impact factor is applied to account for dynamic pressure spikes which may occur with sudden stopping, starting or wind gusts during operation.

At the service limit state, structural elements subject to direct loading from hydraulic motors or cylinders should be designed for the force produced by the actuator at the normal working pressure available to actuate the motor or cylinder.

At the fatigue limit state, structural elements subject to direct loading from hydraulic motors or cylinders should be designed for the torque or force produced by the actuator when resisting the maximum constant velocity loading. For this loading the resistance should be based on the endurance limit of the material.

14.11. Synchronising or load sharing in hydraulic systems

For movable bridge drives with more than one actuator, the actuators are either synchronised (i.e. the relative positioning or displacement of the actuators is controlled) or the system is designed to share loads between actuators. The choice of synchronisation or load sharing is a function of the type of bridge and the desired level of control. In general, bridges that are not subject to skewing during operation, such as swing and trunnion bascule bridges, are configured with load-sharing hydraulic actuators. Conversely, bridges that are subject to skewing during operation, such as vertical lift and certain rolling-lift bridges, are generally configured with synchronised hydraulic actuators.

Load sharing is an inherent advantage of hydraulic systems. Load sharing in mechanical drives is typically implemented through a differential gear assembly within the drive train (see Chapter 13). For hydraulic motor or hydraulic cylinder drives, equalisation of the load among actuators is typically as simple as feeding the actuators from a common supply. If the actuators are identical and the piping runs to the actuators are of the same size and length, then the forces or torques produced by the actuators will be essentially equal. This arrangement is commonly achieved by porting multiple actuators from a common manifold and configuring piping so that the common manifold is located midway between actuators. It can also be accomplished by running the piping to a location midway between actuators before splitting off to each actuator.

Synchronising of actuators in a mechanical system may be achieved by physically connecting the actuators, as is the case for two pinions driven off the same shaft or connected shafts. This same approach can be used in hydraulic motor drives. For bridge drives that do not require 'precise' synchronisation, flow dividers can be used to synchronise actuators. The definition of 'precise' is subject to determining the operational parameters of the specific movable bridge under consideration, particularly those related to acceptable skew offset between actuators. Commercially available flow dividers are typically limited to smaller flows than are required on a movable bridges and provide synchronisation to a tolerance of no better than $\pm 3\%$, which limits their use on movable bridges. Custom flow dividers, composed of paired axial piston pumps mounted on a common shaft, have the capacity to handle typical movable bridge fluid flow rates. However, their accuracy in equally dividing the flow may still not be sufficient for many applications. If a flow divider is used, an easy means to re-synchronise the actuators should be provided. This could be accomplished by providing a means to release pressure from the actuators each time the movable span is seated after an operating cycle.

For hydraulic cylinder bridge drives where flow dividers are not a viable solution, or where active skew control is desired, a system to monitor the position of the movable span and adjust actuator movements independently can be employed. Synchronisation of physically independent hydraulic actuators is similar in nature to synchronisation of physically independent mechanical drives in that it involves the following

- provision of independent control of actuator speed
- implementation of a control system with position sensing
- inclusion of a motion controller of some type in the control system

- establishment of a master actuator
- synchronisation of the other actuators (slaves) to the lead actuator.

With open-loop circuits, independent control can be applied within the branch of the circuit that feeds each actuator. Consider the following example.

Problem: A rolling-lift bascule is designed to be operated by two hydraulic cylinders. Each cylinder is oriented horizontally at the elevation of the centre of roll and is attached to a pintle that is located in the web of each of two main girders. The pintles are also located at the centre of roll and support the cylinders just outboard of the main girders. The cylinders retract to roll the leaf open and extend to roll the leaf closed. Synchronisation of the cylinders is a design criterion to mitigate the effects the drive system may have in skewing the movable span on the tracks.

Solution: A common HPU is employed for economy and simplicity. The HPU is of open-loop design with a pair of pressure-compensated pumps that feed a common manifold. The pumps are set to deliver a constant flow and pressure to the manifold during operation. No flow control is provided in the pumps or main manifold other than a system relief valve and the pressure compensation feature of the pumps. For independent direction and speed control, each cylinder is equipped with a proportional directional valve mounted to the cylinder manifold. Linear variable differential transformers (LVDTs) are mounted on each cylinder to provide feedback on the position of the cylinder rod end. The bridge control system includes a programmable logic controller (PLC) and proportional–integral–derivative (PID) controller. The control system provides commands to the proportional valve on both cylinders to accelerate the leaf to full speed, move at constant velocity, and decelerate to a stop. Through position feedback from the LVDTs the control system adjusts the control signal provided to the slave cylinder to maintain a specified tolerance in the position of the rod end of the slave cylinder relative to the master cylinder.

In the above example it is important to note that the hydraulic system must have adequate capacity to drive the slave cylinder fast enough to maintain pace with the master cylinder. The master cylinder cannot be given a command that uses half the power capacity of the system or there will not be sufficient reserve capacity to compensate should the slave cylinder fall behind the master.

14.12. Design example

Design a hydraulic cylinder drive system for a single-leaf trunnion bascule highway bridge based on the following parameters

Design specification	AASHTO LRFD movable highway bridge design specifications (AASHTO, 2007)
	No ice load due to warm climate location
Operational requirements	Raise or lower in 67 seconds
	5-second acceleration and deceleration ramps
	2-second creep speed travel upon seating
	Full angle of rotation = 72°
	Bridge is normally left in closed position
Trunnion bearings	24.8 in. dia. plain bearing (bronze bushing)
Hydraulic cylinder	2 cylinders per leaf
	Mill duty, 3600 psi pressure rating
Hydraulic power unit	Open-loop with proportional directional control
	3000 psi minimum components rating

Bridge geometry and loading	Bascule leaf length trunnion to tip = 87.5 ft Bascule leaf deck width = 47.0 ft Bascule leaf weight = 1920 kips Unbalanced moment = 2 kips at the tip of the span

The bascule pier and bascule leaf arrangement is shown in Figure 14.7. Selection of the geometric positioning of the cylinder relative to the trunnion and the resulting kinematics is often a matter of trial and error. The farther the cylinder connection to the leaf is located from the trunnion, the smaller the forces will be in the cylinder. The drawback is that smaller cylinder forces come at the price of a longer cylinder stroke, which brings into play cylinder bucking as a limiting element of design. Minimising cylinder stroke by moving the cylinder connection close to the trunnion also has limits. When this dimension gets too small, the cylinder force acts to lift the leaf off the trunnions and/or the applied force is so close to the centre of gravity of the leaf that oscillations occur during operation. As a general rule of thumb, the cylinder attachment point should be located on a radius from the centre of rotation of the leaf of between one fifteenth and one tenth of the trunnion to tip span length. Experience has shown that an attachment point located on a radius one-twelfth of the span length from the trunnion is a good starting point.

For the design example, with a trunnion to tip length of 87.5 ft, a cylinder attachment point located on a radius of about 7.25 ft is selected. Based on detailing, the actual dimension is 7.21 ft. The connection to the leaf is positioned at half the opening angle, or 36° below the trunnion. The cylinder is mounted to the bascule pier using a cardanic ring. This positions the cylinder pivot point up out of the counter-weight pit and reduces the buckling length of the cylinder.

Based on the leaf geometry and mass, the AASHTO operating and holding loads are calculated with the following results

Maximum constant velocity torque (dynamic friction, unbalance, 2.5 psf wind normal to the deck)
820 kip-ft (raising) (span nearly full closed)
610 kip-ft (lowering) (span nearly full open)

Maximum starting torque (static friction, unbalance, 10 psf wind on any vertical projection)
1800 kip-ft (raising) (span nearly full open)
1840 kip-ft (lowering) (span nearly full open)
Holding torque (20 psf wind on vertical projection)
2920 kip-ft (wind pushing leaf closed) (span nearly full open)
2840 kip-ft (wind pushing leaf open) (span nearly full open)

Note that for hydraulic cylinder drive design it is important to examine cylinder loads acting in both compression (pushing the leaf open, resisting wind pushing the leaf closed) and tension (pulling the leaf closed, resisting wind pushing the leaf open) as the pressures and flows are different. With the selected cylinder geometry and kinematics, only the full open and full closed positions are significant for the purpose of sizing the hydraulic cylinders. From the selected geometry the effective moment arm of the cylinder in both positions is calculated by multiplying the radius of attachment about the trunnion by the cosine of 36°. The result is 5.83 ft. If the cylinder pivot point were not located directly below the attachment point or the attachment point were not half the angle of opening below the trunnion, then different effective moment arms would need to be computed for the full open and full closed positions. The cylinder buckling length is calculated as the length between the cylinder pivot point (axis of the cardanic ring) and the cylinder to leaf attachment point with the leaf in the full open (72°) position. The buckling length, L, is 195 in. The working stroke is the length the cylinder must extend as the bridge rotates from full closed to full open. In this case the working stroke of the cylinders is 101.72 in.

Selecting cylinder size is often an iterative process. In this case, an initial size is selected by examining two cases

(a) the pressure required under the maximum starting torque for lowering the span and
(b) the holding load, wind pushing the leaf down.

A cylinder with 3600 psi rating has been selected and this full pressure can be used for holding, subject to buckling capacity. However, 3000 psi is selected as the design pressure for the system other than the cylinders. This is a typical value used in bridge drives because there are a wide variety of components available rated for use at up to 3000 psi and it is also a good practical limit for sizing of piping and hoses. Some reserve capacity is needed for the operating system to account for inefficiencies, so under maximum starting torque a value of 10% below maximum working pressure, or 2700 psi, is selected as an initial limiting value.

For case (a), the minimum required rod-end area necessary to maintain working pressure within the established maximum is found by setting the pressure equal to the maximum and solving for the resulting rod end area as follows

$$A_r \min = \frac{1840 \text{ kip-ft} \cdot 1000 \text{ lb/kip}}{5.83 \text{ ft} \cdot 2700 \text{ psi} \cdot 2 \text{ cylinders}} = 58.4 \text{ in}^2 \text{ minimum rod-end area/cylinder} \quad (14.18)$$

For case (b), the maximum compressive load in a cylinder is calculated by dividing the holding-load-resisting wind pushing the leaf closed by the effective moment arm to the cylinder line of action and the number of cylinders as follows

$$F_b = \frac{2920 \text{ kip} \cdot \text{ft}}{5.83 \text{ ft} \cdot 2 \text{ cylinders}} = 250.3 \text{ kips cylinder buckling load} \quad (14.19)$$

Cylinders are commonly provided with the blind area approximately twice that of the rod-end area. In other words, the cylinder ratio is roughly 2 : 1. Doubling the minimum rod-end area and solving for the diameter nets a cylinder with a minimum bore diameter, $D_{b,min}$ as follows

$$A_b = 2 \cdot 58.4 \text{ in.}^2 = \frac{\pi \cdot D_{b,min}{}^2}{4} \therefore D_{b,min} = 12.2 \text{ in} \quad (14.20)$$

A common cylinder size just slightly larger than this is a 12.60 in. bore (320 mm) with an 8.66 in. dia. rod. For this cylinder the blind area is 124.7 in^2, the rod area is 65.8 in^2 and the ratio is 1.9 to 1. Dividing the torque values above by the effective area of the cylinder results in the cylinder forces and pressures listed in Table 14.1.

Notice that the holding torque that results from wind acting to push the leaf open beyond full open produces a pressure in excess of the design value established for the cylinder. However, this condition will be ignored in design as the bridge is configured with a bumper block, a physical restraint that would

Table 14.1 Summary of hydraulic cylinder forces and pressures

Loading condition	Torque: kip-ft	Cylinder force: kips	Effective cylinder area: in^2	Location of pressure	Cylinder pressure: psi
Constant velocity torque	820	70.3	124.7	Blind end	560
Constant velocity torque	610	52.3	65.8	Rod end	790
Maximum starting torque	1800	154.3	124.7	Blind end	1240
Maximum starting torque	1840	157.7	65.8	Rod end	2400
Holding torque	2920	250.3	124.7	Blind end	2010
Holding torque	2840	243.4	65.8	Rod end	3700

prevent the leaf from rotating open beyond the full open position. All other pressure values in the table are within the design parameters.

The next step is to check the cylinder for buckling. A good means of performing an initial buckling check is to calculate the Euler buckling load, F_E, as if the rod spanned the full buckling length between attachment points, and using a factor of safety of 3.0 to determine an allowable buckling load.

$$F_E = \frac{\pi^3 \cdot E \cdot D_r^4}{64 \cdot L^2 \cdot K} \tag{14.21}$$

Because the cylinder is unrestrained from rotation at the attachment points, K is 1.0. Therefore

$$\frac{F_E}{FS} = \frac{\pi^3 \cdot E \cdot (8.66 \text{ in})^4}{64 \cdot (195 \text{ in.})^2 \cdot 1.0 \cdot 3.0} = 690 \text{ kips} > 250 \text{ kips} \tag{14.22}$$

Because the cylinder is mounted to pivot at an intermediate point along the length of the tube, buckling is found not to control for an 8.66 in. diameter rod. If the cylinder were mounted with a blind-end clevis attached to the pier, the cylinder would need to be about 166 in. long between attachment points (clevis pin to clevis pin) in its retracted position in order to provide the required working stroke. This minimum retracted cylinder length has been determined after a review of published dimensions for mill-type cylinders and discussion with cylinder manufacturers. The minimum cylinder length must be adequate to account for the internal works of the cylinder, including piston, head, cap and custom cylinder cushions, as well as the dimensions for the rod-end clevis. The buckling length for that configuration would be the stroke plus the collapsed length or 268 in. Checking buckling for this condition yields the following

$$\frac{F_E}{FS} = \frac{\pi^3 \cdot E \cdot (8.66 \text{ in})^4}{64 \cdot (269 \text{ in.})^2 \cdot 1.0 \cdot 3.0} = 360 \text{ kips} > 250 \text{ kips} \tag{14.23}$$

In this case the blind-end clevis mounting could have been used which would have eliminated the need for the cardanic ring. However, this would have placed the blind-end clevis assembly in the bottom of the counterweight pit which is not desirable from a long-term maintenance standpoint.

Some owners have required a bridge to have a minimum of four hydraulic cylinders to provide redundancy. This approach is generally not necessary as bridges with two cylinders are normally capable of being operated by a single cylinder under maintenance conditions. Increasing the number of cylinders also has other drawbacks. Not only is the complexity of the system increased, but efficiency often suffers as well. This is because as the cylinders become smaller, buckling becomes the controlling element of design. For example, if the bridge in the design example had four cylinders, the minimum cylinder bore size would drop to 8.66 in. A typical rod size for this bore would be 5.51 in. in diameter. A rod of this diameter would not have sufficient buckling resistance, even using a cardanic ring mounting, if the previously calculated Euler buckling load were used. AASHTO (2007) presents a more comprehensive approach to calculating buckling resistance as compared to the Euler method. This alternative method accounts for the stiffness of the cylinder body, not just the rod. Using this less conservative method would yield adequate bucking resistance for this smaller rod size for the cardanic ring mounting, but not for the blind-end clevis mounting. As discussed previously, system efficiency is improved by working with lower flows and higher pressures. If bucking controls the cylinder design and the rod size is increased to satisfy buckling criteria, then the cylinder bore will need to be increased correspondingly to maintain pressure within the design criteria. Consequently, with an increase in cylinder bore, the fluid flow required to operate the bridge will increase, thereby reducing the efficiency and increasing the cost for construction, operation and maintenance.

Focusing back on the design example, the next step is to determine the size of the pump(s) needed to operate the bridge. For redundancy the hydraulic power unit will be equipped with two pumps that feed a common manifold. This will allow the bridge to operate at approximately half speed in the event one pump is removed for servicing.

The hydraulic schematic for the design example is shown in Figure 14.13. Bridge control will be 'open-loop', without speed control. The speed of the bridge will be calibrated to a given set of control commands, but will not be continuously monitored and adjusted during operation. Because of this, the pump only needs to provide sufficient flow to extend the cylinder by its full stroke in a given time interval. The volume of the cylinder, V_c in gallons, that must be filled to extend the full stroke is calculated as follows

$$V_c = A_b \cdot 2 \text{ cylinders} \cdot \text{Stoke (in)} \div 231 \frac{\text{in.}^3}{\text{gal}} = 124.7 \text{ in.}^2 \cdot 2 \cdot 101.72 \text{in.}^2 \div 231 \frac{\text{in.}^3}{\text{gal}} = 110 \text{ gal} \quad (14.24)$$

To calculate the flow, Q, in gpm the cylinder fluid volume is divided by the time of operation less half the ramp time (derived by integrating flow with respect to time over the full stroke of the cylinder). The creep speed time is also deducted and the small travel at creep speed, which is generally 10% of full speed, is neglected.

$$Q = \frac{110 \text{ gal}}{(67 \text{ sec} - \frac{(5 \text{ sec} + 5 \text{ sec})}{2} - 2 \text{ sec})} \cdot \frac{60 \text{ sec}}{\text{min}} = 110 \text{ gpm} \quad (14.25)$$

Therefore, each pump must be capable of delivering 55 gpm. Each pump will be powered by a standard AC squirrel cage induction motor with a synchronous speed of 1800 rpm. With estimated motor slip, the speed of the motor when loaded is taken as 1750 rpm. The minimum pump displacement, dmin, is determined from the following basic relationship, where η_v is the volumetric efficiency of the pump.

$$d_{\min} = \frac{Q \frac{\text{gal}}{\min} \cdot 231 \frac{\text{in}^3}{\text{gal}}}{n \,\text{rpm} \cdot \eta_v} \tag{14.26}$$

Solving for displacement, using 55 gpm per pump and a volumetric pump efficiency of 0.95, yields a minimum pump displacement of 7.64 in.^{3} or 125 cm^3. Recommended practice is to provide a minimum of 5% reserve displacement capacity to account for contingencies in design and construction. Therefore, although 125 cc pumps are available from several pump manufacturers, a slightly larger displacement pump of the next larger commonly available size, 135 cc, is selected for design. The selected pump is a variable displacement axial piston pump with pressure compensation.

To determine the parameters and select components for the remainder of the system the design flows and pressures must first be established for each branch of the circuit. First the design flows are determined. As previously established, the flow produced by each pump will be 55 gpm. This is therefore also the design flow for the suction lines between the reservoir and pump, the hose connecting the pump to the main manifold, the pressure filter, and the pump outlet check valve. Within the manifold, the pump flows are joined. As a result, the design flow for the system relief valves, proportional directional control valve and counterbalance valves is 110 gpm. Piping and hoses connecting the manifold to the blind end of the cylinder will see half the full system flow in the raising cycle (extending the cylinder) or 55 gpm. Design flow for the rod end is equal to the blind end flow divided by the cylinder ratio or 30 gpm.

At this point in design, the pressure at the pump required to operate the bridge has not been determined. The inefficiencies in the system must be determined first. Using 3000 psi as a predetermined minimum rating, each of the components is selected by reviewing available product data. Once the components are selected the pressure drop in each component is derived from the manufacturer's data at the appropriate flow rates. For the counterbalance valves, the pressure drop at full stable flow is assumed to equal the pressure drop in the circuit between the proportional valve and the cylinder port, adjusted for the cylinder ratio. Examining the raise cycle for example, the pressure drop for the flow out through the counterbalance valve (through the bypass), ball valves, lines and PO check valves (piloted open) is estimated at 140 psi. The assumed pressure drop for the return flow through the counterbalance valve is 140 psi, multiplied by the cylinder ratio of 1.9, or 265 psi. This pressure drop is more than the pressure drop found on the counterbalance valve data sheet for 60 gpm flow with the valve piloted full open, so the 265 psi will be used as the controlling pressure drop. Examining the lowering cycle flow, the pressure drop for the flow out from the proportional valve to the cylinder is estimated at 100 psi. The assumed pressure drop for the return flow through the counterbalance valve is 100 divided by 1.9 or 53 psi. In this case however, the pressure drop found on the counterbalance valve data sheet for 110 gpm flow and the valve piloted full open is 200 psi. Therefore, the larger value of 200 psi will be used in design.

The design flows having been established, system pressure losses can be estimated based on representative components (determined by reviewing product literature for viable commercial components) and piping (by sizing piping for the design flow and calculating pressure losses through the estimated piping lengths). These values, although not presented in detail in this example for simplicity, are summarised in Table 14.2. A total pump efficiency η_o of 0.85 was used in calculating these values. For the cases involving operation at maximum constant velocity load, the pressure at the pump is the sum of the load induced pressure at the cylinder (blind end for raising the bridge or rod end for lowering the bridge) and the system pressure losses. For design purposes, a 20% reserve power factor is applied. For the cases involving maximum starting load the pressure losses are neglected, as the system must only provide

Table 14.2 Summary of power requirements

Load case	Load-induced pressure at cylinder	Flow at pump	System pressure losses	Required system power without/ with reserve power factor
Raising maximum constant velocity load	560	110	520	82/98 hp
Lowering maximum constant velocity load	790	110	980	133/160 hp
Raising maximum starting load	1240	110	0	78/94 hp
Lowering maximum starting load	2400	110	0	151/181 hp

sufficient pressure to initiate movement. Although the pumps are of the variable displacement type, because flow is controlled at the proportional valve, the pump will always produce 110 gpm, unless the pressure raises enough to compensate the pump. When the proportional valve is restricting flow to the cylinders, the excess flow will be dumped over the relief valve. Therefore, power must be available to run the pump at full flow up until that point. The power required of the electric motors is calculated as follows for each load case

$$P_{\mathrm{hp}} = \frac{Q \cdot P}{1714 \cdot \eta_{\mathrm{o}}} \cdot \text{reserve power factor} \tag{14.27}$$

For the system as configured to meet the above power requirements, each pump must be coupled to a 100 hp motor (the next standard size available above 90 hp). Even if the 20% reserve power factor is backed out, the overall efficiency of the system as configured is poor and can be roughly calculated as follows. The power to operate the bridge at full speed is determined from the basic flow and pressure

$$\frac{110\ \text{gpm} \cdot 560\ \text{psi}}{1714} = 36\ \text{hp} \tag{14.28}$$

The power provided without the reserve power factor would be two 75 hp motors, or 150 hp. The resulting efficiency is 36 hp/150 hp or 24%.

The efficiency of the hydraulic drive can be improved by implementing horsepower limiting. With use of horsepower limiting the pumps are set to produce the flow and power required for maximum constant velocity load operation and the pressure required to produce the maximum starting load. The power required is now only 50 hp per pump, derived from the summary of power requirements, Table 14.2. The flow of the pumps is limited to 55 gpm by the maximum stroke setting. The pressure compensator is set at 2400 psi. These two limits will override the horsepower limiter. However as long as flow and pressure are below these limits, the pump will perform along the constant power curve shown in the pump performance curve of Figure 14.16. Flow will automatically vary with pressure at the outlet of the pump to maintain the power limit. From the pump performance curve the flow for the various load cases can be calculated as is summarised in Table 14.3. With use of horsepower limiting the motor size was reduced from 75 to 50 hp and the efficiency improved from 24% to 36%. Further

Figure 14.16 Horsepower-limited pump performance graph

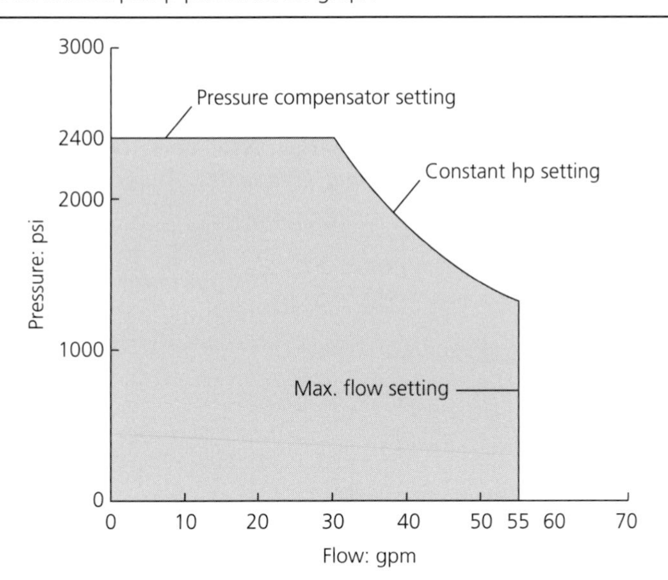

Table 14.3 Summary of operating performance for two 50-hp pressure-limited pumps

Load case	Load-induced pressure at cylinder	System pressure losses	Flow at pumps	Function controlling flow
Raising maximum constant velocity load	560	520	110	Stroke limiter
Lowering maximum constant velocity load	790	980	82	Horsepower limiter
Raising maximum starting load	1240	0	110	Stroke limiter
Lowering maximum starting load	2400	0	60	Horsepower limiter

improvements in efficiency could be realised by use of an open-loop drive with pump control as discussed above in Section 14.7.

REFERENCES

AASHTO (American Association of State Highway and Transportation Officials) (2007) LRFD movable highway bridge design specifications, 2nd ed. AASHTO, Washington, DC, USA.

AASHTO (2012) LRFD bridge design specifications. AASHTO, Washington, DC, USA.

Birnstiel C (2008) Movable bridges. In *ICE Manual of Bridge Engineering* (Parke G and Hewson N (eds)), 2nd edn. ICE, London, http://dx.doi.org/10.1680/mobe.34525.

Frankenfield TC (1984) *Using Industrial Hydraulics,* 2nd edn. Hydraulics & Pneumatics Magazine, Cleveland, OH, USA.

ISO (International Organization for Standardization) (1999) ISO 4406: 1999 Hydraulic fluid power – fluids – method for coding the level of contamination by solid particles. ISO, Geneva, Switzerland.
Parke G and Hewson N (eds) (2008) *ICE Manual of Bridge Engineering*, 2nd edn. ICE, London, http://dx.doi.org/10.1680/mobe.34525.
Rabie M (2009) *Fluid Power Engineering*. McGraw-Hill, New York, NY, USA.
Stewart HL and Philbin T (1984) *Pneumatics and Hydraulics*, 4th edn. Bobbs-Merrill, New York, NY, USA.

Movable Bridge Design
ISBN 978-0-7277-5804-0

http://dx.doi.org/10.1680/mbd.58040.371

Chapter 15
Electrical system design

William Bowden

15.0. Introduction

Most movable bridges of more than a few metres' span are electrically powered and controlled. Usually, the high-speed–low-torque output of the electric motor, or motors, is converted to the low-speed–high-torque output necessary to move the span by mechanical gearing, as described in Chapter 13. Such a span drive is termed electro-mechanical.

Sometimes, the electric motor drives a hydraulic pump to pressurise a fluid (usually oil) and the fluid is then used to drive low-speed–high-torque, hydraulic motors or hydraulic cylinders to create the force necessary to move the span. Such span drives are called hydraulic drives and are described in Chapter 14.

For electro-mechanical span drives, electrical, and more recently electronic, controls are used to control motor speed and other duty-cycle events such as operation of auxiliary equipment. For the hydraulic drives, electrical and/or electronic controls are necessary to control pressures in hydraulic actuators and also the other duty-cycle events of auxiliary devices.

The assemblage of electric and electronic devices that forms a control package has become known as a 'drive'. Hence, for example, the industry speaks of an 'SCR drive' for a wound rotor motor. The motor and the gearing are the mechanical span drive, the motor of which is controlled by the 'SCR drive'.

Control systems have evolved since the first movable bridge was built. Not only has the equipment grown more complicated but concerns about safety, reliability and operation have impacted the designer's choices as to the manner of span operation. Devices such as solid state drives, programmable logic controllers, and touchscreens have taken bridge electrical design into a new century. Many of the devices are adoptions from other industries, such as crane and motion control. However, it is wise to look at the predecessors of modern control systems before looking at current-day devices.

As explained in earlier chapters, there are three types of movable bridges in common use today. They are the bascule, swing and vertical lift bridge. The type of span was selected early in the design phase and was influenced by many factors such as traffic, required vertical and horizontal clearances, and cost. Depending on the type of span, the selection of the prime mover was limited due to the power available, space constraints, torque required to move the span, and machinery design.

The prime mover was an internal combustion engine, a DC motor, or a wound rotor motor (due to its ability to control torque). The control system consisted of relays and drum switches which controlled the torque and speed of the motor. While the designs were somewhat crude by today's standards, much can be said for the simplicity of design and the fact that many of these bridges are still operating today, a century after they were commissioned.

Designs were left to the individual designer with no national recommended practice until the American Railway Engineering Association (AREA) and the American Association of State Highway Officials (AASHO) published design standards in the early part of the 20th century (see Chapter 5). The standards have evolved over the years so that they are now are a reference on all bridge projects in the USA and internationally (AREMA (American Railway Engineering and Maintenance-of-Way Association), 2013 and AASHTO (American Association of State Highway and Transportation Officials), 2007). They recommend minimum design requirements, with the designer given the opportunity to provide a more sophisticated design.

When designing the electrical part of a movable bridge, there are four areas of primary importance that should be of concern to the designer. They are the prime mover, interlocking, safeties and auxiliary devices. The most important is the selection of the prime mover.

15.1. Prime mover

Many systems have been used to move a span throughout the history of movable bridge engineering. From a pure manual capstan and chain to today's complex systems the industry has evolved based on the technology available. Early designs utilised the equipment available at the time such as steam engines, internal combustion engines and DC electric motors. Today, the industry has many different choices in how to move a span. One common way is the electric motor drive system using an electric motor as the prime mover, with primary and secondary gear reduction. A second is the direct drive diesel with gear reduction. Another is a hydraulic drive system using cylinders or low-speed–high-torque motors. Choice of the prime mover was often dictated by the expertise of the designer, the owner, and the bridge location. Each of these applications has its advantages and disadvantages. New to the scene, flux vector drives used with vector-rated squirrel cage AC motors have replaced both direct current (DC) motors and AC wound rotor motors in most applications.

15.1.1 Direct current (DC) motors

Direct current shunt wound motors have been in use on movable bridges for the better part of two centuries. See (Oberg, 2004, p. 2469) for description of DC shunt wound motors. Even with today's advances in technology, a DC motor is still a good application for a movable structure. One of the advantages of the DC shunt wound motor is that speed varies in proportion to the armature voltage, when full voltage is applied to the field windings. This allows for constant torque with horsepower varying as a function of the speed. They are capable of supplying a constant torque at any speed between zero and the rated rpm.

Before the advent of solid state devices, DC motors were controlled by a motor generator set that converted incoming AC voltage to DC or by a DC utility feed, which was not uncommon in the last century. The rotation of the AC motor drove a DC generator. Rectification was accomplished by the commutator of the generator and the generator output was controlled by adjusting the generator field strength (Millermaster, 1970).

A DC electronic drive provides adjustable speed control. Speed is directly related to armature voltage and torque is proportional to armature current. The shunt field is normally supplied by a fixed voltage power supply internal to the drive. The DC drive converts incoming three-phase voltage to a DC voltage output through the use of power SCR convertors. These convertors are controlled by a regulator that controls the speed based on the desired input through speed feedback by comparing the desired speed at any instant to the feedback from a tachometer generator.

The standard DC drive is comprised of a disconnect switch, incoming branch circuit protection, an isolating contactor, DC power module, regulator, and control inputs for speed and torque selection. On movable bridge applications, a regenerative reversing drive is supplied whereby the DC drive acts as a four-quadrant controller. These drives are capable of not only controlling the speed and direction of motor rotation (two quadrants) but also the direction of motor torque (two quadrants). When the load is directed normally, both the rotation and the torque are in the same direction. However, in an overhauling condition, the motor torque opposes the direction of rotation, thereby providing a controlled braking force to the load. The ability to switch between modes is a function of the regulator. Most drives have a built-in current limiter to prevent excess torque being applied to the machinery. This current limiter setting is nominally set to about 150% of rated motor torque.

15.1.2 AC wound rotor motors

Wound rotor motors, as shown in Figure 15.1, have been used on movable bridges for almost 100 years. See Oberg (2004, p. 2470) for a description of the AC wound rotor motor. During this time there has been little change in the technology, with the exception of the introduction of thyristor drives in the 1960s. The thyristor drive enables the designer to shape the speed–torque curve based on the load of the bridge. Most bridges built in the past 70 years have wound rotor motors as the prime movers.

One of the prime benefits of using a wound rotor motor on a movable bridge is the ability to select the required torque for span operation (Borden, 1996). Selecting the correct secondary resistance gives the designer flexibility in choosing the speed torque characteristics under which the bridge will operate. Each bridge presents unique problems in application. A swing bridge will have different operating characteristics than a bascule bridge. How a span is balanced and external environmental factors, such as wind and or snow, will also impact torque requirements. The ability to control torque is an important design feature because machinery stresses may be minimised and the total power required to operate the span kept within normal parameters. As can be seen from Figure 15.2, little or no resistance in the rotor circuit produces a speed torque curve that closely approximates that of an AC squirrel cage motor. As resistance is added to the rotor circuit, the slip increases depending on the load and the point at which maximum torque is produced is reduced relative to speed.

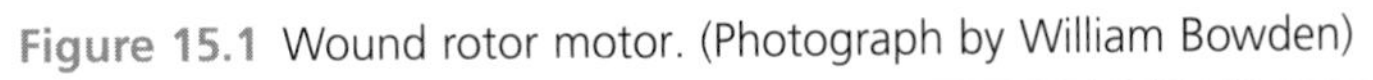

Figure 15.1 Wound rotor motor. (Photograph by William Bowden)

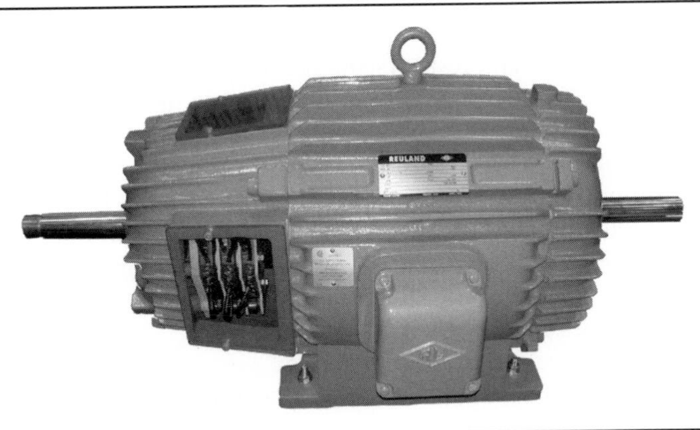

Figure 15.2 Wound rotor motor speed – torque characteristics

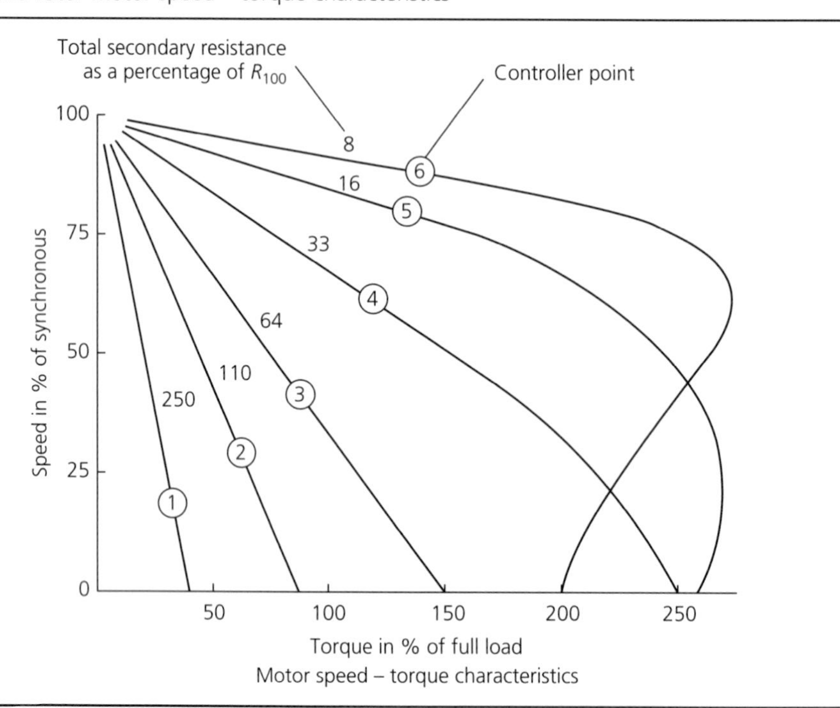

Total resistance presented to the rotor slip rings determines the speed torque curve of the motor. This resistance comprises the resistance of the field wires between the rotor and the secondary resistors (which are located external to the motor, sometimes remotely) and the secondary resistors themselves. Base resistance is determined by using the formula R_{100} equals open circuit rotor volts divided by $\sqrt{3}$ times the rotor full load current. As can be seen from Figure 15.2, the more resistance introduced into the rotor secondary, the more it impacts the starting torque and the motor slip. This added resistance also reduces starting current as the torque curve has been shifted lower. It is not unusual to start the motor with a resistance of 200% of base resistance which limits the starting torque to about 50% of rated torque. As resistance is shorted out, the percentage of full load torque increases and the motor slip decreases.

Proper selection of the motor secondary resistors and the shorting points will result is a series of torque limits as shown in Figure 15.2. These torque limits correspond to the controller positions (points) on a drum master control switch. It should be remembered that the control of the bridge span through the wound rotor motor and secondary resistors is in reality a torque control and not a speed control. Speed follows the maximum torque allowed for the particular point selected. The resistors are chosen based on the starting torque needed as well as the period of time the resistor will be in the circuit. The accelerating steps are designed so that available torque is above the nominal 100% requirements of the load. Torque peaks are normally limited to 180–200% of the load, thereby requiring more power points and accelerating steps in the design.

In a conventional design, the secondary control comprises a drum switch, shorting contactors, resistors and stepping timers. The timers served as a lower limit in allowing a minimum of time to elapse before the next shorting contactor is energised and shorts out more resistance. The resistors are selected based on the torque or current required for the various torque points. The rating of the resistor is based on the design so that available torque is above the design load accelerating torque. The ampacity of the resistor is normally based on a continuous duty cycle even though design criteria would allow a duty cycle of 'off' and 'on' periods.

The operator moves the controller through the power points until the maximum torque is available to the span motor. Timers are normally added to the design so as to stage each controller point to prevent the operator from moving the controller too quickly. The resistors can be shorted by use of power-rated contacts on the drum switch or by use of shorting contactors. Common usage is by using shorting contactors. Another advantage in using a wound rotor motor is its ability to limit the current inrush that is common with an AC squirrel cage motor. By starting off in a reduced torque mode, the motor draws less current than if started across the line. An application issue that the designer must be aware of is the difference in maximum ratings of the electro-mechanical devices on a bridge. For example, a wound rotor motor per the requirements of the National Electrical Manufacturers' Association (NEMA) should be capable of a breakdown torque of not less than 275% of full load torque (FLT). This motor which has the capability of a minimum of 275% full load torque is being connected to mechanical shafting and gear reducers that in many cases have a design rating of 200% of FLT, maximum. Thus, unless the designer is aware of the conflict, the motor can produce torque in excess of the design ratings of the mechanical components. Overpowering mechanical systems is not uncommon, but is to be avoided.

15.2. AC thyristor drives

The middle of the last century saw the development of solid state devices, among them the introduction of the thyristor. This electronic switch was applied to wound rotor motors for control of voltage applied to the motor terminals. Solid state thyristor drives give the designer the ability to design superior performance over an existing switched contactor design using drum controllers. The thyristor drive provides step-less acceleration, the ability to control torque loading, and built-in control for overhauling loads. Speed of an AC motor can be controlled in two ways. Varying the voltage input to the induction motor will alter the motor's torque and thereby its speed (AC thyristor). The second way is to vary not only the voltage to the motor but also the frequency (AC flux vector). This will change not only the torque but also directly the motor speed (Hughes, 1992).

In a thyristor drive, the speed of the motor is controlled through the use of gating amplifiers and the thyristors. A mirror set of thyristors was introduced so that a raise-and-lower function was possible without the use of reversing contactors. The drive motor combination is a closed loop system consisting of the wound rotor motor, a tachometer for speed reference, the AC thyristor drive and secondary resistors to shape the speed torque curve.

Each manufacturer has their own design features but the following basics are common to all designs. The thyristor drive consists of a power section containing 10 thyristors, gating amplifiers for control of the thyristors, a speed control section, a voltage reference section whereby the tachometer feedback voltage is compared to the desired speed reference, and a power supply. Additional components such as an isolation contactor, incoming circuit protection and control transformers are normally supplied.

The thyristors modulate the voltage applied to the motor on a half-cycle basis. The half cycle allows voltage above the gating signal to turn on the thyristor allowing this part of the cycle to pass to the motor. All three phases turn on the same angle producing an adjustable voltage to the motor terminals.

In varying the gating signal and the firing angle of the drive, an adjustable voltage is presented to the motor which is used to develop a speed torque curve. Feedback from the tachometer generator into the reference section of the drive allows the drive to accelerate to the selected speed and maintain a close tolerance based on the feedback voltage of the tachometer.

The drive motor combination on a bridge faces three different torque conditions. The first condition for a bascule or a vertical lift is normal raising whereby the motor is supplying torque to the machinery. The second condition is at seating on a bascule or vertical lift where the drive and motor are holding the span in the seated position producing zero rpm at an adjustable (usually 80–100%) torque. The last torque condition is where the span is in an overhauling load and the drive must produce a counter torque to slow the span to an acceptable speed. This is accomplished by firing the opposite thyristors to slow the span down and bring it to a safe operating speed. A swing bridge presents its own set of conditions. It is slow to accelerate due to the mass and slow to decelerate for the same reason.

Normally, the acceleration ramp is set to seconds. Hunting is kept at a minimum during acceleration and at the selected speed. Deceleration is usually the same ramp as for acceleration. There are applications, such as a swing bridge, where the drive must produce counter torque to slow the span down and this is done by firing the opposite thyristors.

15.3. Flux vector drives and motors

Flux vector drives (FVD) and vector motors are the latest improvement to prime movers. The technology involved includes the ability to operate at full torque at zero speed. This is a major improvement over previous designs using a squirrel cage motor and adjustable frequency drive (AFD) which were difficult to apply to a bridge. In addition, the combination also gives the ability for close speed regulation, torque regulation, and the ability to be operated in a closed loop configuration for precise control. Previous generations of adjustable frequency and adjustable voltage drives locked the force and direction of the magnetic flux in the motor. In today's drive motor combination this lock is broken allowing independent torque and speed settings.

Most applications on a bridge use what is called a closed-loop system. In this system, information as to motor performance is sent from the motor by an encoder to the drive and from the drive to the motor. The information that is sent from the motor to the drive consists of shaft position, shaft speed, and direction of rotation. The FVD takes this information and processes it against internal parameters to derive an output that is, in turn, supplied to the motor terminals.

Motors used in a flux vector application are based on a NEMA 'Design A' or 'Design B' speed torque curve. For design operating characteristics of squirrel cage motors and induction motors see Oberg (2004, p. 2470). While there is a correlation between torque and current, it is not always linear. The current of an AC motor is made up of two parts. The first is the current to magnetise the motor windings. The second part is the current required to produce the output torque. The combination of these produces full load current. Flux vector motors are special squirrel cage motors that have been developed specifically for applications with a flux vector drive. Unique features of vector motors include the special insulation needed for the higher voltages and harmonics of the drive, the ability to supply full torque at zero speed and the ability to maintain a constant current at all speeds. Flux vector systems can be designed either as open-loop or closed-loop systems. A closed-loop system requires encoder feedback to determine shaft position, speed, and direction of rotation. This information is passed to the flux vector drive by an incremental encoder and maintains a closed loop system that provides two-way communication between the flux vector drive and the flux vector motor.

A three-phase flux vector drive consists of a three-phase rectifier, a DC bus, and an inverter to convert the DC back to three phase. The rectifier converts the three incoming phases to a DC voltage which in turn is filtered using a wound coil inductor or capacitor depending on the design. Depending on the supplier, an inductor, a capacitor or combination of these components smoothes the DC signal (reduces voltage ripple) in the DC bus part of the FVD. The inverter takes this DC voltage and converts it back to a three-phase AC voltage where the drive control system can vary the frequency of the output voltage as well as control the voltage at the motor load. The design of the vector control exploits the relationship that exists between torque and slip frequency (Holtz, 2002).

Due to the possibility of an overhauling load on a bridge, the drives are furnished with a dynamic braking resistor. This braking resistor is used during periods of required braking and produces a braking torque within the motor during braking. The drive ramps the frequency to zero. The energy from the rotational force of the motor and the load are driven back through the inverter to the DC bus which can see voltages as high as 800 volts. If the voltage at the DC bus is higher than a predetermined amount, the resistor is pulsed to dissipate this energy until the bus voltage is brought into a normal window.

The drive manufacturer normally determines the power rating (watts) needed to prevent overheating during braking duty. The peak braking current is determined by the specified resistance value. Each drive manufacturer specifies a resistance range with a minimum to prevent over-current and damage to the drive and a maximum value to give adequate lower dissipation capability.

Before replacing a wound rotor motor control system with a flux vector drive system, the torques required to control the span should be known. Because the characteristics of a wound rotor motor and a squirrel cage motor differ, replacement of a wound rotor with a flux vector squirrel cage motor unit should be checked against torque measurements (by means of strain gauging) in order to determine the torque required under all conditions. In many cases the wound rotor is capable of supplying more torque than the squirrel cage. NEMA specifications state that the wound rotor motor should be capable of supplying a minimum of 275% of FLT. A squirrel cage motor operating under flux vector control is usually limited to a maximum of 150% of full load torque. There are conditions such as ice and snow load which add to the torque requirements that may not be apparent under cursory inspection.

The drive motor combination should be tested on a dynamometer at the factory for a number of reasons. Preliminary settings for the drive can be preloaded leaving only final adjustments to be done on site. The drive and motor should be tested for 150% overhauling load to 150% motoring load using various speed points, such as 50%, 75%, and 100%. A written test report should be submitted to the designer for approval.

15.4. Power synchro unit (synchro-tie)

One problem with a vertical lift bridge is keeping the span level as it raises and lowers. Various factors such as lubrication and condition of the guide rails, and rope slippage may cause the span to skew. Normally, some skew of the span is permissible. The designer calculates the amount of normal allowable running skew and the excessive skew which could cause damage to the span or the auxiliary equipment.

A power synchro unit (synchro-tie) system was the only automatic way to control span skew on a tower drive vertical lift bridge. The synchro-tie was first installed on a vertical lift bridge in the USA in 1931, for the Katy Railroad at Boonville, MO. The immediate response to the slightest skew movement results in a torque transmission to the opposite side of the span so as to maintain the span within a normal range of permissible skew. Other systems using resolvers, inclinometers, and the like normally wait until the span has entered a skew window before applying a corrective signal in order to minimise hunting and over-correction. A PDI (process, derivative and integral) loop is created to correct span position but, to prevent the span drive from oscillating, output is delayed. In contrast, the power synchro unit (synchro-tie) gives an immediate active response to a skew condition rather than a delayed operation waiting for an operational window.

A power synchro unit is similar in application to a line shaft that connects two pieces of machinery. In each case torque is transmitted from the delivering end to the receiving end with the rotation being the same. With a power synchro unit, if the rotation of each shaft does not match, the electrical characteristics of the motors will apply torque to the receiving end to bring the rotational shafts back to electrical zero.

A power synchro unit consists of two wound rotor motors, typically with the same electrical characteristics as the prime mover. The stators of these wound rotor motors are connected to the same power source with the rotors connected to each other. Once the rotors have been phased out and power is applied, no shaft rotation will occur because the rotor voltages are equal. As one shaft starts to rotate, the voltages become unequal and current flows which produces torque to bring the rotor voltages back into an equal state. An angular turn on the driving shaft will produce and equal turn on the receiving shaft. The more the relationship between the driving shaft and the receiving shaft differs as to rotation, the more torque is transmitted in the attempt to bring both shafts back to the identical rotation Nowacki (1933).

The starting method for a power synchro unit is unique in that the shafts must be synchronised when both ends of the span are seated. Two-phase stator voltage is applied to both motors to phase out both motors. Voltage is applied to the third phase on the first motor and after a brief time delay to the second motor. This locks the shafts in synchronisation.

A benefit of the power synchro unit system is that the span can be moved with one of the prime movers out of service. The active lifting motor transfers torque to the opposite side through the synchro unit to raise the end with the non-functional motor. This will impose an increased load on the active motor but because the time of operation is usually less than two minutes, the active motor should be able to bear the additional load.

There are design issues with a power synchro unit system that are not encountered with other methods. These include available space for the additional motors, costs of motors, wiring and control equipment. It is up to the designer to make the appropriate decision. Other systems using positional devices are less expensive and will work if applied correctly.

15.5. Auxiliary motors

There are applications on a bridge that require smaller squirrel cage motors such as those shown in Figure 15.3. These include roadway and pedestrian gates, span locks, wedges and end lifts. Normally these motors range in horsepower from fractional to fifteen. The motor is usually provided with a small disc brake and some means for hand-cranking the motor so that the device may be operated in case of a power failure. Safety limit switches are provided for the hand crank to remove electrical power when the hand crank is inserted.

Figure 15.3 Span lock motor. (Photograph by William Bowden)

Span lock devices lock the span in either a closed or open position. On a bascule bridge they can be located at the heel or the toe of the bascule span or both places. The span lock motor drives a lock bar into a receiver on the span which locks the span to the structure, or vice versa. An end lift device is used on a swing span to raise the tips of the movable span to make them level with the approach roadway. In operation, the end lifts are lowered so that the span can open or close without hindrance. A centring wedge is also used on a swing span to support live load at the pivot pier. These devices are operated after all traffic control devices are in the proper position to prevent roadway traffic.

Some designers select a two-speed squirrel cage motor as an auxiliary prime mover. This gives the user the ability to move the draw in case of failure of the main prime mover or its drive. The fast speed is used between the nearly open and nearly closed positions, while the slow speed is used between the nearly open and fully open and nearly closed and fully closed positions. Care should be taken when designing the seating sequence as maximum torque is available at motor stall speed.

15.6. Brakes

A structure that moves needs some means of stopping or holding that structure. This is accomplished on a bridge by the use of brakes. Brakes are rated in dynamic and static braking force that is expressed in either foot pounds or Newton metres.

Brakes operate under two different conditions. The first condition is when the brake is called to stop the moving span, usually under an E-stop condition. Under the second condition, the brake acts as a holding brake to prevent span movement. Most design applications in use today do not use the brake to slow down or stop the span during normal bridge operation. The controls on the prime mover normally perform this function. However the brake still must have the capacity to stop the span under an E-Stop condition.

The most common types of brakes used on a movable span are a thrustor drum, disc brakes or band brakes. The brake drum or disc is acted upon by a set of brake shoes which in turn are controlled by a thrustor mechanism.

Some old installations have band brakes where the braking force is applied by hand, solenoid, compressed air or fluid power. Disc brakes have been introduced lately into the industry because of their small size relative to braking torque capacity. Drum brakes are the most common. Disc and drum

Figure 15.4 Disc brake. (Photograph by William Bowden)

brakes are spring-set and hydraulically released (see Figures 15.4 and 15.5). The drum for the drum brake is not shown. The braking action takes place by the force generated by the brake shoes when they are pressed against the revolving drum or disc by springs. The drum brake can be further subdivided into two types: thrustor- or solenoid-released (or actuated).

The thrustor mechanism is a small hydraulic pump connected to a push rod which moves when the hydraulic pump is running. An orifice valve controls fluid movement which is important especially when the pump motor is shut off and the brake begins to set. This valve controls the amount of time it takes from removal of thrustor motor power to the time the brake shoes engage the drum or disc.

Figure 15.5 Drum brake. (Photograph by William Bowden)

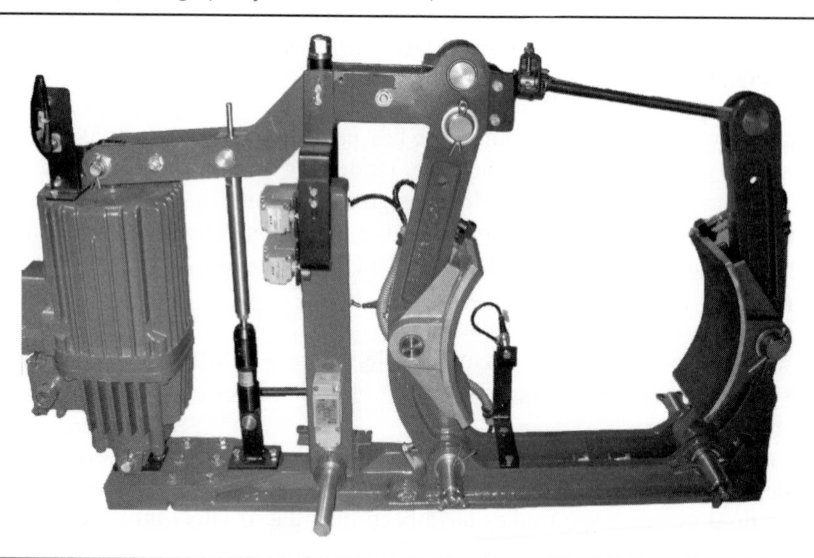

The nature of the bridge machinery and the shafting involved requires a gradual stopping. If a solenoid released brake is utilised, the action of the brake on the rotating surface is almost instantaneous, thereby stressing the shafts and gearing. A thrustor brake applies pressure to the rotating surface over an adjustable period of time (about 3–5 seconds). This time delay prevents the instant grabbing of the rotating drum or disc. Increasing pressure is applied over a small period of time to the brake drum to bring the spinning drum to a stop. This lessens the impact of bringing the structure to a sudden halt. Solenoid brakes are normally found on small-hp motors used for locks, wedges and centring devices where a sudden stopping of the shaft will not cause damage to the connected load.

Most movable spans have brakes installed in two locations in the gear train. The first is usually at the output of the prime mover and is of a lesser torque rating. The other brake is usually located after the secondary gear reducer and acts on the low speed shafting. This brake will have a larger torque rating. Industry nomenclature calls the first brake a motor brake and the second a machinery or service brake. If a thrustor actuator is used, the motor brake will have a short time delay while the machinery brake will be set about two seconds later.

A hand release is normally supplied with the brake mechanism. This feature is useful if operation of the span is required and no electric power is available to operate the brake or if there is a problem with the electrical part of the brake. The hand release is specified as either right hand or left hand. This determines the location of the hand release in relation to the shaft being controlled. The drum or disc size of the brake wheel is determined by the torque rating required and the diameter of the shaft that passes through the wheel. New standards require that the diameter of the brakewheel hub be 2.5 times the diameter of the shaft passing through it. There are many rating organisations such as the Association of Iron and Steel Engineers (AISE) which set standards for brakes such as mounting dimensions, centreline of the brake wheel to the base plate and minimum ratings.

Brakes used on movable spans are normally supplied with three limit switches. These limit switches are normally connected to the bridge control system to advise when the brake is set, when the brake is power released, and when the brake is hand released.

Brakes are supplied with a corrosion-resistant weatherproof enclosure, for two reasons. The first is to protect it against the elements and the second is to protect maintenance personnel from the rotating shaft and wheel. The enclosure should be designed so that the maintenance personnel have access to the brake mechanism, especially the shoes, for inspection and replacement without removing the whole cover.

15.7. Limit switches and resolvers

Many types of limit switches are used on movable bridges. They range from a rotary cam, plunger and lever to proximity limit switch. Each one has a unique reason for being used in that application. Rotary cam limit switches such as depicted in Figure 15.6 are primarily used for span position indication (nearly closed, nearly open and fully open). The input to the cam is through a shaft that is coupled to another shaft. As this shaft turns, the cams inside the rotary switch will open or close depending on the settings of the cam. Precision is usually within one degree of input rotation. A vertical lift span will normally require a set of rotary cam limit switches in each tower which monitor the turns of the sheaves by means of six contacts. The contacts are wired in a series-parallel combination to a master permissive relay which, in turn, activates the raise or lower commands if the span is skewed. The amount of permissible skew is determined by the input gear ratio to the rotary limit switches and the

Figure 15.6 Rotary limit switch. (Photograph by William Bowden)

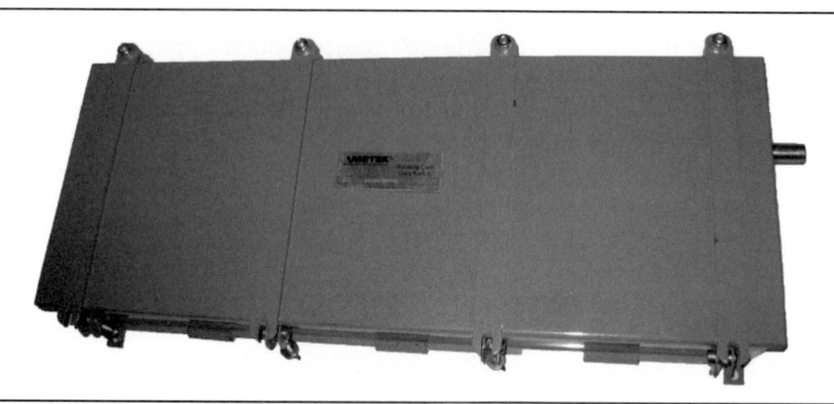

switch is usually designed to trip at between 18 and 24 inches (457–610 mm) of lift span differential movement between towers.

Lever limit switches are available in many types and are used for a variety of applications. They are found, for example, on brakes to indicate position of the brake (set or released) and its hand release. They are also used to sense position of a fully open or over-travelled span, the driven or pulled position of a span lock, and the position of other devices such as end lifts and centring devices on a swing span. These switches are available with potted cables that are environmentally sealed to prevent the entrance of water and other contaminants. Figure 15.7 shows a lever-operated limit switch.

Figure 15.7 Lever limit switch. (Photograph by William Bowden)

Figure 15.8 Plunger limit switch. (Photograph by William Bowden)

Proximity switches are a different type of limit switch as there is no external means of actuation as with a lever or rotary limit switch. They are tripped by a close proximity to ferrous metal, usually 25 mm or less, or by a magnet. The advantage of this limit switch is that there is no physical connection or external means of actuation. They are used primarily to sense a span-seated position.

Plunger limit switches are actuated by a shaft that presses against a plunger assembly that, in turn, trips a spring-loaded armature to transfer the contact state (see Figure 15.8). There is about a 2 in. adjustable over-travel available so that the limit switch can be adjusted for proper seating. The switch is mounted with the shaft pointed downwards so that water and other contaminants do not enter the enclosure. The enclosure is normally of cast aluminium, rated waterproof. The switch is available with 2–6 contacts. The plunger limit switch is used for detecting the span-seated condition.

The environment should be considered when selecting limit switches. If used outdoors, they should be of a construction that will not deteriorate in the harsh marine environment. Usually, a stainless steel enclosure or other suitable enclosure is specified. In addition, care should be taken to ensure that the actuator will not be fouled by foreign matter, such as, road salt, dirt, debris or other solid material which would impede operation.

Resolvers are rotational devices that convert rotational turns into an analog electrical output usually 1–10 volt DC or 4–20 ma. This electrical output is then inputted into a programmable logic controller (PLC) or analogue convertor which is then used to provide control points for system operation. The resolver can be single-turn or multi-turn. Resolvers are used for positional information. An inclinometer performs a similar function as a resolver. When attached to the span it transmits an analogue signal indicating the angle off-level of the span.

An encoder is a digital device that also converts rotational turns. The output is based on counts, usually 1024 per revolution, that is transmitted back to a drive if motor mounted or to a convertor or PLC input if measuring span position.

15.8. Traffic control devices

Highway bridges present unique design problems for vehicular and pedestrian traffic, depending on the type of span and the space available. Common to all types are the need for traffic control devices such as traffic lights, warning gates and barrier gates. Various agencies such as AASHTO, recommend needed devices, depending on the type of bridge under consideration. Some bridges by their design, such as double-leaf rolling bascules, allow the designer to omit barrier gates. Safety of pedestrian and vehicular traffic makes the proper design and implementation of gates imperative. Normally, the sequence for bridge operation is to switch the traffic signals to red. The green traffic light extinguishes and the amber or yellow lamp comes on for a period of about six seconds. After this time the amber light goes out and the red traffic light comes on. At this point the warning gates can be operated. Normal design does not allow operation of the warning gates or barrier gates unless the traffic signals are red. The bridge operator should visually inspect the bridge to make sure that all vehicles and pedestrians have exited the span.

Many bridge designs have both oncoming and off-going warning gates. The reason is not only to stop the roadway traffic in the oncoming lanes but to also prevent a vehicle from approaching the span from the wrong direction. These warning gates are normally lowered with the 'oncoming' lowered first to stop traffic with the 'off-going' lowered after all traffic has left the span. When the span is being opened to roadway traffic the 'off-going' gates are raised first and after they are fully open, the 'oncoming' gates are raised. The most common type of warning gate is the semaphore type that pivots from vertical to horizontal on closing. The internals of the gate normally have an electric motor, gear reduction, lever arms, bearings, limit switches for position and the housing. Figure 15.9 depicts a warning gate at the right side of the view and a barrier gate to the left, both in the open position.

Barrier gates are typically provided to prevent vehicles from entering the open draw. They are not just warning gates but are designed to prevent a vehicle from travelling past a particular point. They are rated for an impact load travelling at a designated maximum speed. The design of the gate allows the mechanism to bring the vehicle to a stop without the vehicle passing through the barrier gate. There are different types of barrier gates in use, among them the semaphore type, an assembly that spans the roadway and that drops across it and locks, a vertical barrier that lowers across the roadway (to the left in Figure 15.9) and bollards which rise from the approach roadway. The semaphore type is the most common. In most cases the arm or assembly is locked into the opposite gate or anchored into a cut-out on the abutment. Operation for a semaphore barrier gate is the same as a warning gate. Normally the oncoming sides are lowered first with the off-going sides following once the oncoming gates are lowered. Another type of barrier gate drops horizontally across the roadway from two fixed vertical stanchions. These stanchions anchor the movable netting, or wire, stretched across the roadway in the raised and lowered positions.

Both warning and barrier gates are normally supplied with steady and flashing lights mounted on the gate arms. These lights will flash when the gates are not in the fully open position in order to alert the motorist that there may be an obstruction in the roadway. Most designs include the gate arm lights that

Figure 15.9 Warning gate and vertical barrier gate. (Photograph by William Bowden)

also come on when the traffic signals have been turned red to alert the motorist. A warning bell is also supplied to give a pedestrian audible warning that the gates are about to move.

Wind loading presents a design problem, especially in areas where hurricanes or high winds prevail. The designer should check local and national standards to ensure that the gates supplied meet these requirements.

15.9. Electrical power distribution

Power distribution is the means of allocating required electrical power to various locations on a bridge. These requirements include lighting, span movement, heating, and information circuits such as fire alarm and security. By far, span movement is the largest electrical load. A bridge that is open to highway traffic consumes little load, mainly lighting and heating. However, once the span starts to move the power requirements increase substantially. The power distribution should be designed to include the power required under the most severe loading for proper bridge operation.

Power is distributed from the public electric utility feed through a main disconnect with circuit protection and a revenue meter. After the service passes into the operator house, it is normally split through a distribution panel or motor control centre to the various electrical loads on the bridge. These loads take the form of incoming line voltage which will supply power to span drive motors, brakes, span locks, and warning and barrier gates. Control voltage is normally derived from another transformer which converts the voltage to a lower voltage such as 120 VAC. This voltage is used for the traffic lights, lighting, control system and other miscellaneous devices such as alarms and CCTV equipment.

Many matters should be taken into account when designing the power distribution system. Among these are an alternate power supply, means of transferring this backup power, surge suppression, economical routing of the cables and conduit, and safety. Normally, the backup power supply is by means of an emergency internal combustion engine-generator or secondary utility power source. This backup power is connected to an automatic transfer switch which will transfer power upon loss of the regular utility feed. When normal power is restored, the automatic transfer switch senses the restored source and transfers the load back to the primary source. Care must be taken in the design of the system to prevent a return to the normal source if the bridge is in operation as this could create a sudden power loss which would impact bridge operation. Normally there is a lockout circuit to prevent return operation to the normal position of the automatic transfer switch when the bridge control system is energised.

With the advent of today's solid state devices such as programmable logic controllers (PLCs), motor drives and resolvers, any line disturbance or spike could cause damage to these devices. Certain areas of the country are more prone to lightning storms and line disturbances than others. In these cases, a transient voltage surge suppression system installed in the incoming power feed may protect these devices. In addition, all wiring originating in field devices, such as limit switches, should also be protected where these wires enter the building.

Control circuits at 120 volt or less should not be run in the same raceways as power conductors. Analogue signals should always be run in separate conduits. Each circuit or wire should be identified with a permanent identification tag at each end. Twisted shielded wire and low voltage conductors should not be field spliced. When any cable used for instrumentation crosses a waterway, cable length should be continuous without splices. All terminations should land on corrosion resistant terminal blocks with permanent wire markings.

All power and control cables rated less than 2000 volts should be tested with a megger meter to measure insulation resistance. This test should be made between each conductor and ground and between each conductor and all other conductors. Minimum acceptable resistance shall be in excess of 100 megohm.

15.10. Control system hardware

Next to selection of the prime mover, design of the electrical control system hardware is the most important task facing the designer. The owner must have input at this very important stage of engineering. The control system hardware can range from a simple system consisting of relays and drum switches to the opposite of a complex computer controlled system utilising a PLC with human machine interface (HMI). With this complexity the designer is able to take advantage of the inherent abilities of the PLC to not only replace relays in the control scheme but to add elements that would have been impossible to replicate in a relay-based system. There is a price to be paid for this complexity in that a higher level of engineering and maintenance is required to not only programme the PLC but also to maintain it over the life of the bridge. The life span of a relay-based system is 40–50 years with proper maintenance. The new PLC systems and HMI-based systems have a built-in product obsolescence that makes getting replacement parts a challenge even a short while after installation.

Early designs of bridge control systems were rudimentary in that fewer than 40 relays and three or four timers were needed for a two-leaf bascule bridge. Operator controls were manual with manual interlocking and safeties at a minimum. The control system consisted of control desk, relay cabinet, auxiliary panel for the auxiliary devices, starters, and a power distribution panel that contained the incoming protective devices, circuit breakers, and span motor control elements. Wound rotor motors were controlled by either switched secondary resistors or by thyristor drives.

Today's systems are much more complex. With the addition of a PLC, the designer can take the level of sophistication to any level desired. Analogue and digital devices which were unheard of 50 years ago are now commonplace on bridges. They are used for span position, motor feedback, and current and voltage recording. The output of these devices feed into an analogue input on the PLC so that the PLC can use them not only in the control logic but also in recording and data logging what is happening on the bridge. The PLC allows for a greater level of interlocking and safeties in the design which, due to the complexity of the circuits, were unheard of 50 years ago. The PLC can also output analogue signals to various devices such as meters, drives and data logging systems to control and record decisions made by the control system.

The control desk or console is the command centre for bridge operation. The console contains the operating controls as well as the necessary indicators, meters and bypass devices. The desk is usually constructed of stainless steel sheet and has a sloping pinnacle that contains the required meters, position pilot lights, and position indicators. The meters would normally be a system voltmeter with selector switch, system ammeter with phase selection, and power meter. The desk meters that measure voltage, current, watts and rpm are being replaced with a combined meter that displays volts, KW, amps, power factor and frequency. These meters are capable of storing the information and downloading it at a later date for review. This ability is invaluable since a baseline can be established and compared to performance sometime in the future. There is also a position indicator that gives the position of the span(s) in degrees of feet (m). For many years this was accomplished by the use of Selsyn transmitters and receivers. These devices are no longer made and the function has been replaced with digital meters that may also contain a bar graph that indicates degrees or feet (m).

The base of the desk contains the operating switches and pushbuttons as well as status pilot lights for the various devices being monitored. The indicating lamps normally have a means of testing them for burned out bulbs such as a press-to-test feature or a master test function. LED bulbs have replaced the standard incandescent variety due to long life and more resistance to vibration. Even the standard control desk layout is being replaced by a touchscreen (HMI) which duplicates all the functions of a normal control desk. The touchscreen has a number of screens that the operator uses to control the bridge. Figure 15.10 shows a touchscreen on a control desk for a rolling bascule and Figure 15.11 depicts a HMI for a vertical lift bridge. Devices such as pilot lights, pushbuttons, selector switches, and metering that would normally be found on the typical control desk are duplicated on the screen of the HMI. The graphic representation gives the operator a real-time sequence as the pilots change colour as the sequence of operation progresses from step to step. Analogue information is constantly updated in real time. Even the span moves on the screen as the bridge goes through the opening and closing sequence.

These screens have not only the usual graphic pushbuttons and indicators but also alarm information that can be scrolled on the screen if an unusual event occurs.

The touchscreens can also data-log the running parameters of the drive system and store them for later recall as an aid to maintenance. Typical drive parameters are shown in Figure 15.12.

Screens can be added for maintenance use to aid in the location of faults or problems. The screen shown in Figure 15.13 illustrates the functioning parts of the PLC system as well as the communications

Figure 15.10 Touchscreen for rolling bascule bridge. (Canadian National Railway Company)

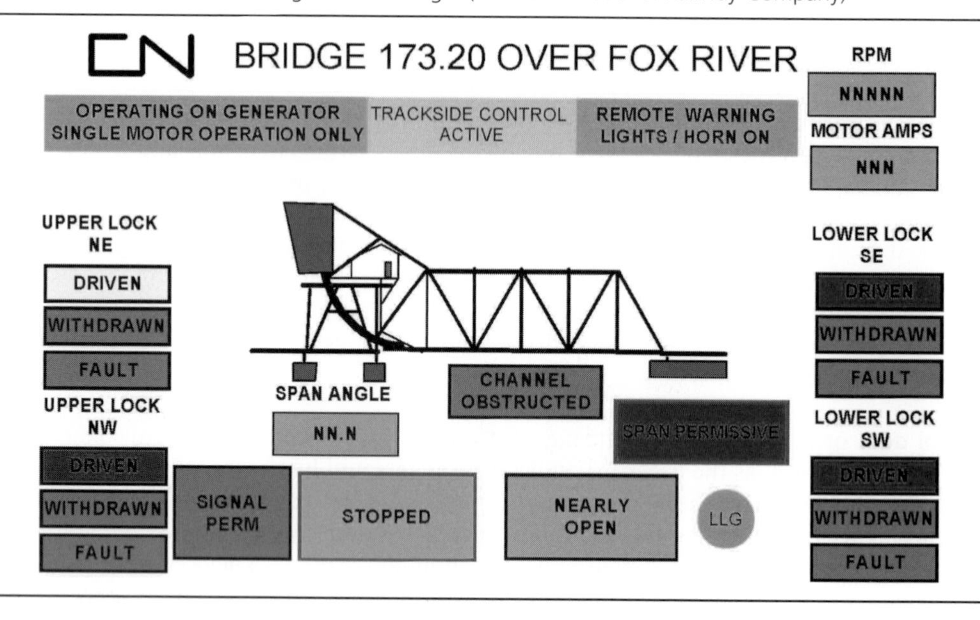

Figure 15.11 Touchscreen for vertical lift bridge. (Norfolk Southern Railway)

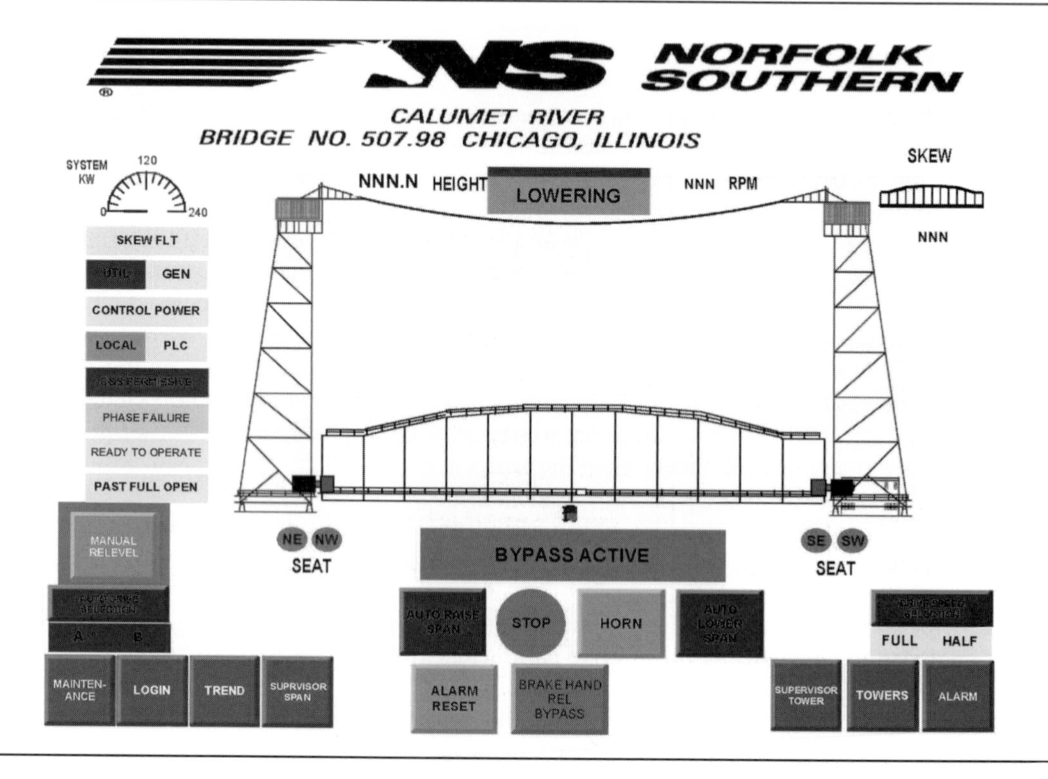

Figure 15.12 Typical drive parameters

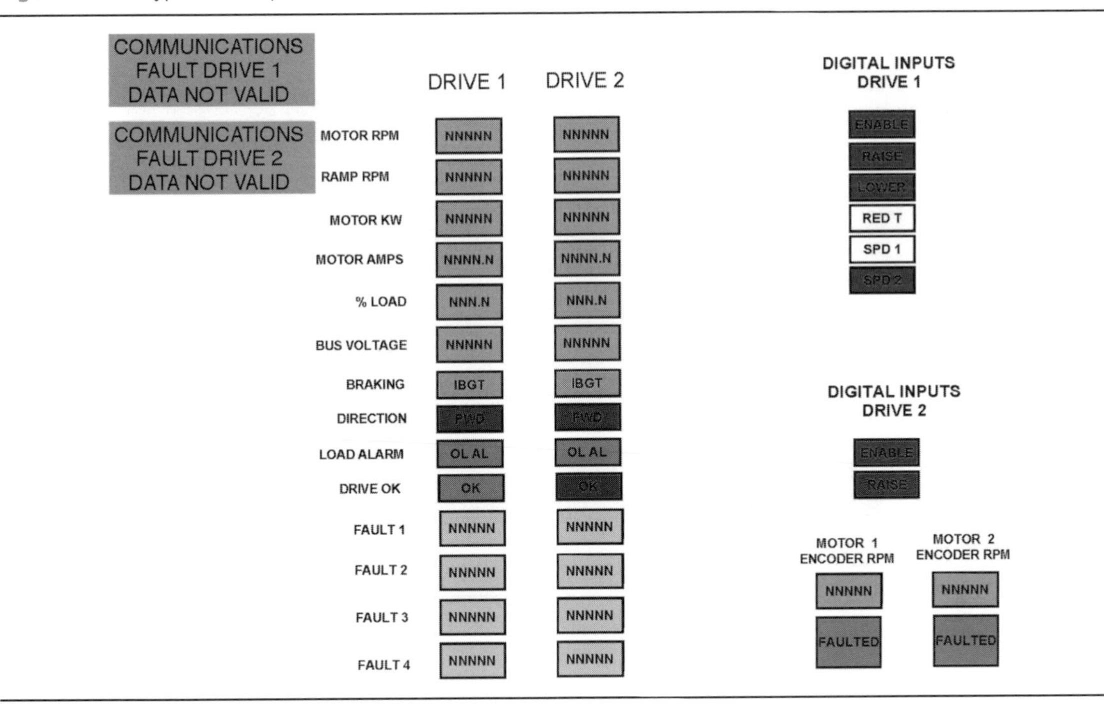

Figure 15.13 PLC component layout

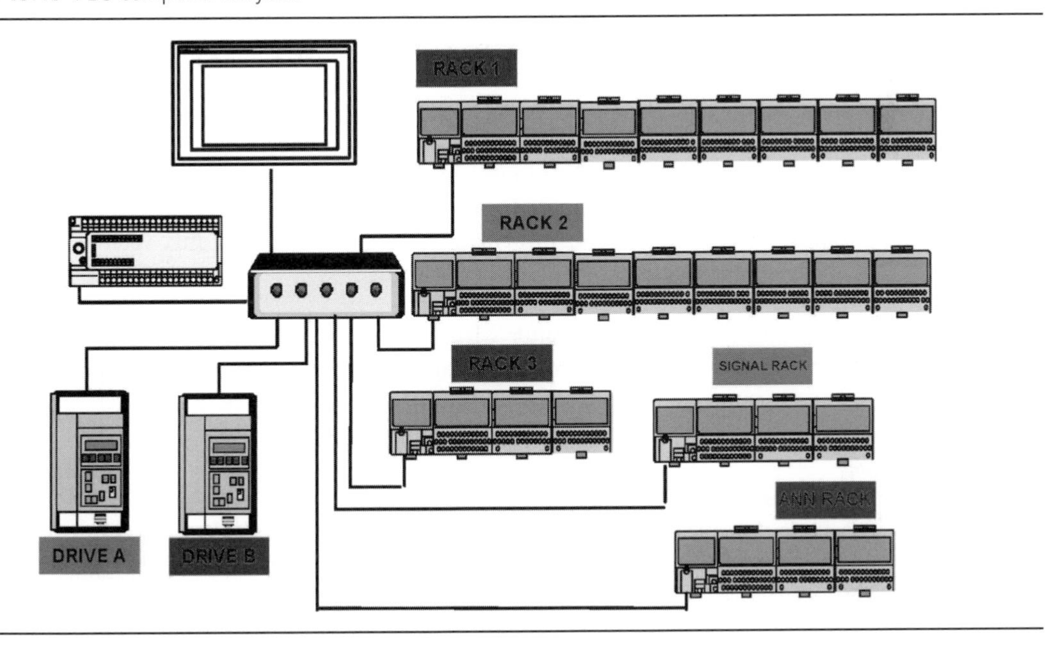

Figure 15.14 PLC rung

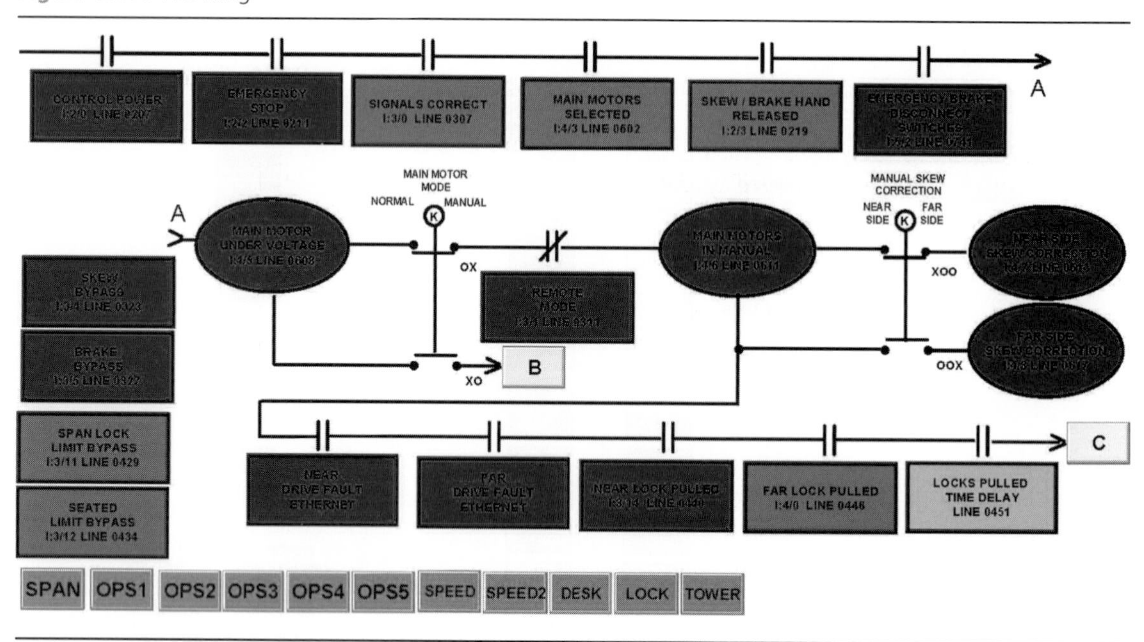

between the various components. Troubleshooting of the PLC program is possible by illustrating a typical rung in the program (see Figure 15.14) and highlighting problem areas.

15.11. Remote controls

A technology borrowed from the crane industry that is appearing in the bridge industry is the use of remote control systems for various applications. This encompasses the use of fibre optic cables, radio controls, and microwave communications. Certain cities (such as Milwaukee, WI) control many bridges from a central point, usually a 'controlling bridge house', on the same river. A duplicate control desk, as found on the remote controlled bridge, is installed on the controlling bridge. The connection between the two is fibre optic cable. A PLC system is installed at both locations for the control of the remote bridge.

In another instance, radio control systems are used to avoid the replacement of existing submarine cables by sending and receiving control signals from the shore to the opposite shore or to the centre pivot pier. Radio controls are also used for information transfer between a flux vector drive and the motor-mounted encoder that is located on the opposite shore.

Ethernet systems eliminate the majority of inter-wiring between the various sub systems within the overall control system. In the past there would be a great deal of inter-wiring between the control desk, the logic control panel, the drive systems and the motor control centres. This inter-wiring can be replaced with Ethernet cable that is wired between the major components.

15.12. Interlocking

Sequence interlocking is important for proper operation of the bridge control system. The goal is to remove as much operator error as possible with attention paid to safety. On a vehicular bridge the various auxiliary devices must be operated in a proper sequence. On a bascule or lift bridge, operational steps are as follows, depending upon the design and devices installed:

To raise the span

- Turn traffic lights to red.
- Lower oncoming warning gates.
- Lower off-going warning gates.
- Lower oncoming barrier gates (if used).
- Lower off-going barrier gates (if used).
- Pull heel and/or toe span locks (if used).
- Release brakes.
- Energise span drive prime mover.

To lower the span, the sequence is in reverse

- Lower span to seats.
- Drive span locks.
- Raise off-going barrier gates (if used).
- Raise on-coming barrier gates (if used).
- Raise off-going warning gates.
- Raise on-coming warning gates.
- Turn traffic light to green.

A swing bridge would also have interlocking for centring devices and end lifts.

Design of a control circuit with proper interlocking would make it virtually impossible for the operator to lower any gate before the traffic signals are Red or to lower the off-going warning gates before the oncoming gates are lowered. This is important so as not to trap a vehicle on the span with the off-going gates lowered and the oncoming gates in the process of lowering. In like manner, the span locks should not be pulled until all vehicular traffic has cleared the span and pedestrians and vehicular traffic are prevented from entering the area of span movement. A suggested sequence of interlocking appears in AASHTO (1998).

There are several additional interlocks that are part of the normal design of the control system. For example, if any of the warning or barrier gates are not in the fully raised position, the traffic lights go to red signal immediately. This warning is to prevent an automobile or truck from coming in contact with a gate that is not in the full raised position and has presumably a member obstructing the traffic lane. The waterway navigation lights are controlled by a limit switch that turns the waterway lights from red to green when the span(s) are at the maximum angle of opening and the waterway is clear for traffic.

Redundancy of the control system must also be considered by the designer. Depending on the owner's requirements, there may be no redundancy or complete redundancy or any amount in between. Consideration must be for redundant PLC processors (hot backup or switchable), duplication of input/output cards, or a relay backup. The question should be asked, 'At what point is a single-point failure acceptable for bridge operation?'

15.13. Bypass switches

Bypass switches are a necessity in bridge design (Bowden, 1996). Because most device failures on a bridge are limit switch-related, a means must be provided for the continued operation of the bridge with a failed limit switch. In many cases the device is in its correct location but the limit switch does not indicate that the position has been reached. Therefore, a bypass switch is necessary to allow continued operation. Most bypass switch applications short out the limit switch interlock to give the control system the ability to continue its operation. Placement of this bypass is critical in system design.

Nameplate wording for the particular bypass switch is important. Is the bypass switch labelled the device being bypassed or the function that is being allowed? For example, a span lock-driven limit switch has failed. The bypass switch could be labelled 'span lock-driven' bypass. Or it could be labelled 'barrier gate' bypass. The problem comes up when the same limit switch interlock is used in upstream and downstream circuits. This interlock should not only be in the raise barrier gates circuits but also in the raise-warning gate circuits. The intent is to give the operator the clearest means of understanding the function of the switch.

Cascading interlocks present problems. A cascading interlock is one normally used in a circuit with the understanding that its normal function is working in that circuit and therefore must be working in other circuits. When this contact is bypassed, this assumption disappears. For example, a span-seated limit switch. Normally, this limit switch will close when the span is seated giving a permissive to drive the span locks. It is also assumed that in order to raise the barrier and warning gates that the span is seated. If the span-seated limit switch fails and the bypass switch is used to drive the span locks, what about the barrier and warning gates? Do the gate interlocks look at just the span lock-driven limit switch or do they also look at the span seated limit switch? This is a question every designer should ask.

There have been cases where, with bridge operation stopped for no apparent reason, the operator will activate all the bypass switches in order to let the operating sequence continue. This is a dangerous condition. The designer must ask the question of the owner what should be the policy for multiple bypasses and design their preference into the system. Current operating philosophy is not to allow more than two bypasses to be in operation at any one time. Some owners have gone even further in preventing operator error by introducing a single-shot sequence whereby the bypass switch in use is electrically reset to an off state at the end of a particular operation. The operator must then physically turn the bypass switch to the off position and then back on if the bypass is still needed. Not only does this make the operator think as to the function but also prevents a new operator coming on site from operating the bridge without knowing that there is a problem.

There are other matters besides the design of the bridge, that the designer should consider. Among these are issues that concern documentation, spare parts, and testing. Documentation is not just the project plans and specifications but also the final documents required such as-built drawings, spare parts list, troubleshooting guides, manufacturers' addresses and service literature. The goal is to assemble a set of documents that will be of assistance to the service personnel years after the completion of the bridge.

15.14. Documentation

No bridge construction project is complete without the manuals that finish the project. These manuals are a resource for the continued maintenance and operation of the span for many years to come. It is suggested that the requirement be divided into two parts for the electrical system. The first manual is for the operators. The manual should include a description of the control devices and indicators located on the control desk and other devices used by the operator, a step-by-step sequence of operation, fault indications and their significance, sequence interlocking, use of bypass switches and auxiliary systems information.

The second maintenance manual should contain a device schedule, a list of the spare parts supplied, troubleshooting flow charts, manufacturers' addresses and literature, preventative maintenance procedures, an inspection schedule, a printout of drive settings, PLC program and as-built drawings.

15.15. Spare parts

With the complexity of the control system, replacement parts become a consideration for the designer. Standardised parts such as pushbuttons, pilot lights and relays are available from a number of different manufacturers. These parts are often a direct replacement between manufacturers, for example, a 30 mm pushbutton from different sources will fit in the same hole.

The designer should concentrate on replacement parts that have an extremely long lead time or that become technically obsolete. This would include PLC cards, HMI screens, a complete motor drive, resolvers, encoders and limit switches. These devices should be stocked in a known location that is accessible to maintenance personnel.

15.16. Testing

Testing of the span control system begins at a shop test. This test seeks to replicate the actual installation on site using the major parts of the control system factory interwired so as to simulate the sequence of span operation. It is at this test that the logic of bridge operation is examined and tested. The designer tests not only the designed method of operating the bridge but also tests for unintended consequences. Controls are operated out of sequence in order to examine what the results of such action would be. Interlocking and safety circuits are tested to prove the validity of the logic. It is at the shop test that the adherence to contract specifications is verified.

Onsite testing checks that the field wiring has been installed correctly. It is during the onsite test that the final settings for limit switches are implemented. Adjustments to drive parameters to reflect the dynamics of the span are implemented. A formal test procedure that tests every aspect of the bridge operation should be prepared and become part of the official documentation for the project.

REFERENCES

AASHTO (American Association of State Highway and Transportation Officials) (1998) Movable highway bridge design specifications. AASHTO, Washington, DC, USA.

AASHTO (2007) LRFD Movable highway bridge design specifications, 2nd edn. AASHTO, Washington, DC, USA.

AREMA (American Railway Engineering and Maintenance-of-Way Association) (2013) Movable bridges – steel structures Ch. 15 in *Steel Structures, Manual for Railway Engineering*. American Railway Engineering and Maintenance of Way Association, Lanham, MY, USA.

Borden L (1996) Torque characteristics of wound rotor motors, revisited. *Proceedings of the 6th Biennial Symposium of Heavy Movable Structure,* Clearwater Beach, FL, USA.

Bowden W (1996) Use and abuse of bypass switches. *Proceedings of the 6th Biennial Symposium of Heavy Movable Structures*. Clearwater Beach, FL, USA.

Holtz J (2002) Sensorless control of induction motor drives. *Proceedings, American Institute of Electrical Engineers*, **90**: 1359–1394, http://dx.doi.org/10.1109/JPROC.2002.800726.

Hughes R (1992) AC adjustable voltage vs adjustable frequency control for movable bridge applications. *Proceedings of the 4th Biennial Symposium of Heavy Movable Structures,* Ft Lauderdale, FL, USA.

Millermaster RA (1970) *Harwood's Control of Electric Motors*, 4th edn. Wiley, New York, NY, USA.

Nowacki LM (1933) Induction motors as Selsyn drives. *Transactions, American Institute of Electrical Engineers*, **33**: 1721–1726, http://dx.doi.org/10.1109/EE.1933.6430514.
Oberg E (2004) *Machinery's Handbook*, 27th edn. Industrial Press, New York, NY, USA.

Movable Bridge Design
ISBN 978-0-7277-5804-0

http://dx.doi.org/10.1680/mbd.58040.395

Index

Page locators in *italics* refer to figures separate from the corresponding text.